RISKS CHALLENGING PUBLICS, SCIENTISTS AND GOVERNMENTS

SELECTED AND REVISED PAPERS FROM THE ANNUAL MEETING OF THE SOCIETY FOR RISK ANALYSIS–EUROPE, COMO, ITALY, 12–14 SEPTEMBER 2005

Risks Challenging Publics, Scientists and Governments

Editor

Scira Menoni

Politecnico di Milano, Milan, Italy

CRC Press
Taylor & Francis Group
Boca Raton London New York Leiden

CRC Press is an imprint of the
Taylor & Francis Group, an **informa** business

A BALKEMA BOOK

Published by:
CRCPress/Balkema
P.O. Box 447, 2300 AK Leiden, The Netherlands
e-mail: Pub.NL@taylorandfrancis.com
www.crcpress.com – www.taylorandfrancis.com

CRC Press/Balkema is an imprint of the Taylor & Francis Group, an informa business

ISBN 13: 978-0-415-58072-4 (hbk)
ISBN 13: 978-0-203-84783-1 (ebk)

Visit the Taylor & Francis Web site at
http://www.taylorandfrancis.com

and the CRC Press Web site at
http://www.crcpress.com

Typeset by Vikatan Publishing Solutions (P) Ltd., Chennai, India

Cover photo: A damaged monument after an earthquake. Courtesy of Claudio Margottini.

Risks Challenging Publics, Scientists and Governments – Menoni (ed)
© 2010 Taylor & Francis Group, London, ISBN 978-0-415-58072-4

Table of contents

Risks challenging publics, scientists and governments: An introduction ix
S. Menoni

Organization xvii

Emergency preparedness: From contingency plans to crisis management

Organisation and structure of Chemical, Biological, Radiological and Nuclear (CBRN) special firemen units in Italy 3
A. Pugliano

NEDIES: A European portal for the generation and dissemination of lessons learned from disasters 9
E. Krausmann, M. Christou, S. Scheer, S. Mara, J. Hervas & M. Delaval

Multi-criteria decision aid for nuclear emergency management: Problem structuring based on a stakeholders survey 17
C.O. Turcanu, B. Carlé & F. Hardeman

The Potenza Province emergency plan: The seismic vulnerability map of buildings for the construction of earthquake damage scenarios 27
A. Attolico, A. Bixio & S. Pacifico

Seismic risk assessment: A software procedure for the simulation of damage scenarios in emergency management 33
A. Attolico & A. Bixio

Pilot project for a global safety plan on economic areas 41
C. Fiévez, D. Dupuis, X. Dugnoille & C. Delvosalle

Environment and public health: Looking for new risk assessment tools

A framework for decision support on HSE regulations 49
T. Aven, F. Asche, P. Lindoe, A. Toft & H.S. Wiencke

Nutritional balance, the key to reducing the risks to the health of school children 57
R. Nelson, D. McCleery, N. Windrum, J. Phelan, A. Markey & R. Sweetnam

Release of toxic substances into the atmosphere: Forecast, prevention and protection of the human health Organizational model for the management of Civil Protection interventions 63
G. Evangelista, G. Francioni & E. Tonelli

Rapid evaluation of ecological risk in soil with DNA biosensor 69
G. Bagni, M. Mascini, E. Sturchio, P. Boccia & S. Marconi

Bioindicators to detect environmental damage after "fall out" of pollutants 77
E. Sturchio, P. Boccia, S. Marconi, P. Ferrazza, C. Beni, R. Aromolo, B. Felici, B. Ficociello, L. Casorri, A. Gatteschi & C. Conti

The soil contamination with cadmium after applications of sewage sludge 85
K. Hyblerova, F. Bozek, J. Hlusek & T. Losak

Embedding social and economic perspectives into risk assessment and management

Perceptions of risk 93
D. Zaruk

Perceived ways to adapt to the potential effects of the climate change in the northernmost Europe 101
M. Rusanen & T. Rusanen

Integrated risk analysis: How to consider coping capacity? 105
C. Frischknecht, J.-J. Wagner & F. Romerio

Use of a national population database to aid major hazard modelling and risk estimation 119
H. Balmforth, H. McManus & A. Fowler

Interdisciplinary quality-of-life parameters as a universal risk measure 123
U. Köhler, D. Proske & M. Curbach

Improving information of major technological risks by research in social sciences 131
A. Van Zanten

Whistleblowing, risk communication as core feature of sustainable management 137
B. Rohde-Liebenau

The contribution of economic and public-private cooperation instruments to the prevention and mitigation of natural hazards 145
G. Pesaro

Estimation of economic losses according to the earthquake scenarios for Istanbul 155
S. Kundak

Evaluation of the possible urban planning and insurance policies to create a risk reduction culture in Turkey 167
A. Taylan

Risk mitigation criteria in land use planning and critical infrastructures siting and design

The "seismic behaviour" of urban complex systems 179
A. Galderisi & A. Ceudech

Spatial planning processes, territorial planning law and flood risk in the region of Valencia (Spain) 191
J. Olcina Cantos

From structural to systemic: Proposals to make French flood management more consistent and robust 205
P. Pigeon

Environmental liability as a prevention tool: A GIS application 211
C. Simeoni & S. Bellagamba

Nation's critical infrastructures: Inherent vulnerabilities and economic fall backs of their unavailability 215
M. Martelli & P. Spagnesi

Safety information systems related to transport of dangerous goods in Norway 223
O. Njå & J. Vatn

Methodology based on indicators for comparison of risks results from diverse energy systems 233
A. Colli, C. Kirchsteiger, A.L. Vetere Arellano & B. Ale

Implementation of safety management systems in nuclear power plants in Germany and results of a German research project 247
B. Ernst, H.P. Berg & H.P. Balfanz

Determinants of investment risk in processes of transformation and modernization of polish heat energy sector 255
E. Sitek

Environmental conflicts in Hungary—the case of the used battery reprocessing plants 259
R. Szanto

Knowledge based tools for stability verification during pressure equipment life cycle 267
P.A. Bragatto, P. Pittiglio & S. Ansaldi

Author index 275

Risks Challenging Publics, Scientists and Governments – Menoni (ed)
© 2010 Taylor & Francis Group, London, ISBN 978-0-415-58072-4

Risks challenging publics, scientists and governments: An introduction

S. Menoni
Dipartimento di Architettura e Pianificazione, Politecnico di Milano, Milan, Italy

Editing a book is rarely considered hard work; it certainly requires less effort than organising, structuring and writing every single word of an entire text. Nevertheless, editing may be challenging as well: it is not just about providing a short foreword to an assembly of articles; it means instead a lot of thinking to frame the index, find the most suitable location for each paper, making clear the rationale to the reader. Introductions to proceedings and/or collection of papers generally take for granted that the text is useful, meaningful and adding value to the already existing scientific production in a given domain. It may seem rather odd to start by questioning this fundamental assumption. Nevertheless it must be recognized that there is a certain level of dissatisfaction with the way current scientific literature is produced, peer reviewed and finally selected for publication (see Lee, 2006; Young et al. 2008). One of the criticism derives also from the rather long time passing between the production of papers and their final publication. In this respect the present volume is no exception, as it carries the title of a conference held already four years ago in the context of the annual meeting of Society for Risk Analysis-Europe. Certainly, the perspective to have a published paper constitutes nowadays an important incentive for attending a conference, particularly in times of limited budgets and strong pressures to be more productive set by academic bodies in general.

Despite of the long delay, it can be said that the process conducive to this final product cannot be considered valueless, particularly as an independent peer review process was set up, with readers in different fields of expertise asked to comment on the submitted articles. Most reviewers did not attend nor presented papers to the conference. Most papers were accepted with revision, which was undertaken by authors, a few were accepted as they were, a limited number refused, some were deemed by their authors to have become old by the time passing from the conference to the actual publication. This process can be hold to have a certain value per se: while a brief presentation of fifteen/twenty minutes may offer just hints about a project that has just started or the short summary of a completed one, writing and revising a paper requires to re-think the entire subject, to develop the topic to the point it becomes clear and understandable to readers. Furthermore, in the context of the Society for Risk Analysis, articles must be accessible to people of different fields, sharing the interest for risk assessment and management, but in areas as different as chemistry, urban planning, sociology, engineering. What justifies therefore the effort and the commitment to get to the final publication is the belief that the conference brought some ideas that deserve to be circulated and offers a way of thinking about theoretical and applied research that may be interesting for the scientific community working in risk and disaster related fields.

The notes that follow try to justify the latter statement. In particular, the title chosen for the conference, though admittedly a bit pompous, was meant to attract abstracts and articles looking at risks in a more comprehensive fashion, considering them not just as a technical or social or political or economic matters, but rather as originating and challenging the various disciplines. The title alluded to the importance of context aspects, of what engineers use to define as "margin conditions", which are generally taken as given and rarely discussed, but which instead would deserve much more care, particularly when uncertainties are large.

1 RISKS CHALLENGING SCIENTISTS

As mentioned above, one important feature of the Society for Risk Analysis is its being multidisciplinary, gathering experts in a variety of fields, who share a common interest in hazards, risks, eventually disasters studies. In many branches of human knowledge the concept of risk has been recognised as worthy to be studied since mathematical modelling became available to estimate probabilities on the one side, and since the scientific culture has accepted the need to tackle the non-deterministic facets of physical and biological systems on the other (Bernstein, 1996).

In the meantime, the massive development and spread of technologies after the Industrial Revolution has made manifest the dark side of unexpected counter and side-effects to be dealt with, adopting caution but also being able to identify the range of variation of potential errors and surprises.

At its start, SRA and its conferences attracted mainly scientists and professionals working on risks associated to technology and modern life, either in forms of feared accidents or long term impacts on human health and on the environment.

In the last decade or so, experts dealing with other types of risks, including natural hazards scholars, approached the Society. The reason is easy to explain; tools used to analyse natural and technological hazards are often rather similar, particularly when the perspective of decision makers is taken into account. For example, land use planners cannot avoid considering all hazards, both natural and technological, threatening the areas subject to their plans and programs.

Furthermore, the scientific community is increasingly challenging the notion of difference between the two, with some authors denying there is such a thing like "natural risks" (O'Keefe et al. 1976; Pigeon, 2003). In addition, the distinction is blurred by the potentiality of so called na-tech, accidents in plants or lifelines triggered by earthquakes, floods and other phenomena, which have been recorded in recent events, since the introduction of the term in literature (Showalter & Fran Myers, 1992).

The enlargement of the field of interest brought fresh and new insight on the "risk society" (Beck, 1992), and significantly contributed to the articulation and the large spectrum of meanings associated with terms like vulnerability, coping capacity, risk.

«The outcomes of efforts to define risk analysis as a coherent field and to promote education in the field remained a challenge» have observed Thompson et al. (2005, p. 1381), illustrating the history of SRA. The statement still holds true in contemporary efforts, carried out inside and outside the Society. One may even question the need for demarcating neatly the field of risk analysis, as it would be perhaps better to move the objective from the creation of a clear cut field of studies to a more open arena for truly interdisciplinary work.

Event though SRA has been traditionally open to a variety of disciplines, this does not mean automatically that what is presented at conferences or in the Journal of Risk Analysis is interdisciplinary work. Instead, many times, risk is analysed and represented in individual fields, as in some papers of this book, focussing on risk in energy production (see Colli et al.) or in the transportation of hazmat (see Nja et al.).

Interdisciplinary work requires instead a variety of expertises bringing their knowledge and proficiency to tackle a loosely defined complex problem, which would be better addressed by efforts carried out at the boundaries of consolidated science, rather than at its core (Pulver & VanDeveer, 2009). Many times, such work at the margin, though necessary to tackle the complexity of issues at stake, is not so welcomed by scientists, for a variety of reasons.

First, because many times cross-boundaries problems are posed by decision makers who must set standards, define rules, distribute resources in the face of risks and sometimes in the aftermath of a crisis or a disaster.

In such circumstances, scientific work becomes inevitably involved in "political" aspects, something scientists are rather reluctant or incapable to do.

Second, because genuinely interdisciplinary research is not rewarding in academia and hardly accepted in traditional disciplinary journals. Difficulties in getting interdisciplinary research published have been denounced even in authoritative arenas (Lee, 2006) and is the common experience of many of us. Reviewing and having reviewed an interdisciplinary paper is challenging for various reasons, including the vast knowledge one single person cannot cover to address all the issues, and the need to focus more on the how models, expertises and data are integrated than on the specific results.

The interaction of different aspects and elements make some environmental and particularly risk problems complex and impossible to solve for an individual expertise: partial or limited solutions may often move the risk in space and time, potentially augmented. The challenge becomes greater as such interdisciplinary efforts are developed for regulatory purposes, as shown by Jasanoff (1990). The latter author also suggests that specific rules should be followed while reviewing papers describing such research, different from those generally used for purely scientific publications.

The challenges scientists face in developing regulatory research, though, are not limited to its acceptability in traditional, academic, arenas. There are several other obstacles and constraints which have been addressed by Salter (1988), Funtowicz and Ravetz (1990), Stirling (2007). With different accents, they all point at the divergent interests of scientists and decision makers, at the different interpretation given to the same words and sentences by a variety of stakeholders.

In the last 15 years or so, since the articles and books quoted above appeared, changes have occurred, and some relevant initiatives have been carried out to narrow the gap between scientists and publics with respect to fundamental issues like risks menacing communities.

The conference originating the present book aimed at addressing, among he others, exactly this topic, in the belief that a conference could host more easily experiences in this particular, sometimes uncomfortable, arena. Contributions such as those provided by Zaruk and Van Zanten address the question of how social perspectives may be better embedded in risk assessment and management, taking into consideration the position of different stakeholders.

2 RISKS CHALLENGING GOVERNMENTS

While looking at how good scientists have become in addressing complex issues with large social and political stakes, it may be relevant to ask also the question regarding the flip side of the coin, that is how good governments, and particularly the European Union as an entity per se, are in making use of available knowledge and scientific expertise. In this regard, the legitimate question is whether or not and to what extent science is really influencing decision making in the European Union. In this respect, according to Allio et al. (2006) there is room for improvement. After having reviewed four Extended Impact Assessments, 10 Explanatory Memoranda and the 2002 Communication on the Collection and Use of Expertise, the Authors concluded that there are fallacies in the way the EU Commission makes use of scientific advice. The most relevant are: expertise that is used by decision makers does not undergo any compulsory peer review process; there is no mandatory requirement to make use of scientific advancement after a given period, to re-assess regulations and laws.

On the other hand, scientific advice is not free from "structural limitations", including the inevitability of subjective judgement regarding several risk issues at stake, the fact that «scientific evidence is not able to answer "value based questions"» (Allio et al. 2005, p. 7) in the presence of uncertainties in scientific findings. At the same time it is certainly frustrating for scientists to realize that new knowledge is not necessarily conducive to better risk mitigation strategies (White et al. 2001; Weichselgartner & Obersteiner, 2002). In the meantime, one has to recognise the inherent difficulties implied by risk research, which are common to practically all environmental sciences. Whilst other disciplines, even theoretical, may sometimes measure their success through technological applications, the main practical output of environmental research is often given in the form of recommended behaviours and policies. Technologies may be part of the solution of individual problems, but in the more general picture, the overall diagnosis which is provided by environmental sciences, including risk assessments, require societies to change behaviours, strategies and to a certain extent also revise priorities and preferences. This leads to two main consequences. On the one hand, successful prognosis are such only if societies decide to adopt the required steps, which inevitably will show their efficacy only in the long run, and only after being "tested" by a natural extreme or an accident.

On the other, the success of scientific understanding of hazards, vulnerabilities and risks must be connected to an equally profound understanding of how communities and individuals may respond in the short (in the case of an impact) and in the long terms (as far as non structural measures are concerned).

Sheila Jasanoff (1993) called for bridging between "hard" and social sciences as the object of analysis, risk, develops at the interface between "natural" and human systems.

As for European policies, the question is not only how well scientific expertise and research results are taken into consideration in the development of new policies, but also to what extent there is a European vision of science, aiming at producing results that may be actually used to support European policies, for example in the environmental field (see McCormick, 2001). In this regard, the situation is rather mixed, with lights and shadows. On the one hand, the European Commission has committed to some important initiatives related to risk governance in general and to the governance of science, particularly when risks are at stake.

The juridical evolution of the precautionary principle is an important example of the European specificity in tackling risks and hazards. Whilst the principle has encountered criticism, particularly from US scientists, it triggered the production of papers, articles, documents tackling crucial and critical aspects that cannot be neglected in risk assessment and management, such as the various types of uncertainties associated to old, new and emerging risks (Godard, 1997). Whilst the precautionary principle affects specifically decision making, the recognition of uncertainties implied by risk assessments and by models used to forecast extreme events and their impact holds important consequences also for the way science is carried out (Funtowicz et al. 2000). In this respect, the report issued by the EU Commission "Taking the European knowledge society seriously" can be quoted as an interesting example of how the complex interface between science and society has been tackled in the EU.

Although the document seems of great significance in potentially paving a European way to deal with risks, the situation "on the ground" is inevitably more articulated. We are referring here to the difficulties in achieving a genuine European research area, going beyond the collection of individual contributions of scientists of different member states. De Elera (2006), in his article briefly describes the ups and downs of the European research area concept, the obstacles in conceiving research as more than a tool for economic development or to promote exchanges of scientists among academic institutions across Europe. The Framework Programme mechanism, though fruitful in many perspectives, does only rarely produce results of European relevance. The results of a recently concluded project in the field of natural hazards, Scenario (*Support on Common European Strategy for sustainable natural and induced technological hazards mitigation,* CN 036979), funded as a specific support action, illustrate the challenges in achieving a European view on current and emerging risks, particularly when aspects like vulnerability are considered. But even in the case of hazards, few projects have provided maps and assessments at the European scale, as until now the preferred solution has been working on test areas in the countries involved in each project.

Following what is stated in the introduction to the Espon project report (Schmidt Thomé, 2006), such difficulty is partially due to the fragmentation of administrations and institutions holding data and information, which are therefore difficult or even impossible to obtain.

Since the VI FP, the EU Commission is requiring to explicit the European added value of projects it finances. Defining the latter may be very challenging. Societies and associations such as SRA-Europe could certainly provide some contributions to the definition of such added value, while recognising the variety of languages and cultures that characterize the Old Continent with respect to other areas where risk studies are advanced, as for example the US or Australia (Poumadore, 1998).

In looking for answers on what does define a "European" way to risk and disaster studies, also the question of how innovative European research is should be posed (see Dosi et al. 2005). Research activities have been inserted in the Lisbon agreement as key factors to foster European economic development. Because of the variety and articulation of research carried out in EU countries, answering the question of how innovative and well positioned in the international arena is European science is perhaps impossible. Inevitably a varied picture appears.

On the one hand the apparent technological gap with the US, which has fostered the EU Commission intervention in the policy of science and research, is questionable as far as given countries and regions are concerned; as for quality assessment, considering some established criteria like the citation index, the picture is rather diverse as well (see King, 2004).

Looking at expenditure, both private and public, certainly Europe as a whole ranks significantly lower than the US, even though the situation is different when individual countries, like France, the UK or Germany are considered.

Again, though, the problem lies more in the fragmentation of funding sources and in the general difficulties to reach cross-border, trans-national research activities, aiming at producing a different result than the sum of individual contributions.

In the last few years a change can be detected at least at the EU Commission level: for example, workshops have been organised to share experiences among project teams; information about activities related to the different bodies dealing with natural hazards and risk in general are much more circulating than before. A way in which SRA-Europe may contribute to this process, is for example dedicating conference sessions to illustrate cooperative research, funded under the Fps and other EU programs. This is not to say that SRA-Europe should become a close club, on the contrary it should provide positive and open forum to enhance understanding and visions about issues that are intrinsically cross border and would require or benefit from a cooperative strategy.

3 RISKS CHALLENGING PUBLICS

The latter statement brings us to the last "actor" mentioned in the title of the conference and the book, the more general "publics" facing risks. In this respect, what Mitchell et al. (2006) suggested for global assessments, can be easily transferred to research in general, when the audience is not restricted to scientists only. In this case, research, besides being credible, must seek also salience and legitimacy. As for Mitchell et al., salient means relevant to potential end-users, while the term legitimate indicates to what extent crucial aspects for communities safety and well being are tackled and reported. A couple of articles in the present volume actually deal with this question (see for example articles by Rusanen or Simeoni).

Among the many publics that could be considered, two will be addressed here: people working on the ground (like civil protection officials or fireman) and the private sector involved in one way or another in risk issues. Although the first are often neglected in conferences and gatherings dealing with risks, the experience of personnel in charge of search and rescue, emergency and crisis management can be particularly valuable in providing crucial information for setting models, calibrating simulations and scenarios. Overall, their direct experience in disaster may also constitute a resource for understanding what is really needed and for what specific purposes. In this respect, articles such as those of Pugliano, illustrating the activity of Italian Firemen Department in dealing with NBCR and that of Krausmann et al. summarizing the activity of the Nedies project, that comprised lessons learnt from a large number of disasters that occurred in European countries in the last decades, are particularly enlightening.

Finally, a couple of words are worth spending regarding the relationship between risk research and private companies in general. Differently from what occurs in other fields, like medicine or mechanical engineering, conferences on risk analysis do not attract a vast audience from the private sector. This is due in part to the lack of a specific target addressed by conferences on risks, and to the fact that few presentations concern practical results of immediate use, like prototypes. Nevertheless, the situation may be changed and perhaps at the advantage of both professionals and experts. For the latter, the larger involvement of the private sector, as for the previously discussed official in civil protection, may provide knowledge and professional expertise from the ground. On the other hand, several branches of the private sector may be interested in joining such conferences: companies dealing with the rather wide range of technologies involved in structural measures (consolidation and hydraulic works, new construction materials and techniques) as well as in communication and monitoring systems, aimed at providing support especially to emergency and crisis management activities. In addition, representatives of industries that are risk prone may want to confront with the latest tools for risk assessment. In the present volume several articles address from a variety of perspectives the question of how to improve current risk analysis and assessment. Such improvement relates both to better analytical tools and to the inclusion of factors (social and/or economic) that have been neglected and poorly developed so far.

According to Nicolais (2007, translation from Italian, p. 122), «innovation is a continuous process, comprising several factors, collective and individual, scientific and technical, cultural and social. Whilst the process cannot be imposed in a top-down fashion, it can be only guided towards the wished goals, as innovation does not disseminate automatically, but propagates by proximity and emulation». This statement is relevant for the reflections conducted here, in particular with respect to the role of scientific research for innovation. Analysing the results of the rather extensive Canargie Mellon survey on the role of science for industries' R&D activities, Cohen et al. (2002) have rather convincingly shown that industries look much more for fresh knowledge, new techniques and instruments rather than specific products in their relationship with universities. Basically, they seem to benefit more from ideas and knowledge resulting from research rather than from direct joint ventures or patenting. The first two sources of information regarding new scientific achievements are publications, conferences and informal exchanges. This is clearly relevant for us, as both publications, as the present one, and conferences, such as those organised annually by the Society for Risk Analysis–Europe among the others, seem to provide industries with opportunities for becoming informed of the last advancements in given fields and for paving the floor for informal contacts.

4 ORGANISATION OF THE BOOK

Reasoning around interdisciplinarity in the context of risk studies has led us to decide the final structure and organisation of the present book. We realised that beyond the intrinsic content of articles, the order in which they are presented gives way to rather different meanings, to rather different interpretations of their same content. Therefore as in filmmaking, where alternative ways of editing scenes provide different stories and pave the floor for sometimes diverging interpretations, so here organising papers according to different paths may be conducive for a variety of readings of the same material. The more traditional way would have been to group them according to the type of threat at stake: natural versus technological hazards, health versus environmental stress factors.

Another structure would be obtained by simply moving the focus to the exposed/vulnerable object or system: cities versus ecosystems, individuals versus communities, individual objects like constructions versus territorial systems, like lifelines and infrastructures.

Otherwise, papers could have been organised so as to respond the suggested topics for the just born Journal of Risk Analysis (Thompson et al. 2005, p. 336): «probability of damage to life, health and the environment; the determination of the existence of a risk; quantification of the risk; and the application of assessments to such things as cost-benefit evaluations, administrative actions, political factors».

Finally, the decision was taken by the editor who holds the entire responsibility for the choice, to make emerge fields of concern that are not necessarily new, but which have been often considered as marginal or simply added as appendixes to the main stream of risk studies.

Some of those issues are the core of other societies which operate in what can be loosely defined as the disaster studies field, like emergency management. Some of the papers under the latter heading clearly show how the scientific understanding of threats and vulnerabilities, technological means and understanding of communities' needs, perceptions, knowledge must be blended in order to get the best out of crisis management, alleviating people's suffering, mitigating losses and preventing events from growing in magnitude and severity.

Another section is devoted to land use, spatial planning and siting issues, which, though more present today than in the past, can be still considered a discipline with weak (sometimes inexistent) connections with risk assessment and management. This for a twofold persistent bias, one on the side of risk scientists the other on the side of planners. The first have considered the second for long as end users and not as partners in risk assessment studies and in prominent position to endorse non structural mitigation. The latter, ironically, have neglected the fact that planning and design have substantially contributed to create and shape vulnerability, abdicating their responsibility to engineers, geologists, seismologists (see articles by Galderisi et al.; Olcina Cantos; Pigeon).

A third group of papers comprises contributions related to ecological and health risks, showing how the same concepts have been interpreted in disciplines like ecology, environmental sciences, medicine and public health (see for example Aven et al.; Nelson et al.). Actually this group of papers represents some of the core disciplines that founded SRA, with their main interest in the risk assessment of pollutants contaminating given environments and neighbourhoods (see for example Bagni et al.; Sturchio et al.).

Finally, another branch more at the core of SRA contributed to the last section, considering how social and economic perspectives have been (or should have) embedded into risk assessment and management in the more recent years. The small subgroup of articles related to economic aspects manifest, though, a significant component of risk that has not been sufficiently explored until now. Considerations of cost benefit analysis were almost everything that could be found in past literature, with few attempts to go deeper in order to understand how economic tools (Pesaro; Taylan) could complement others to achieve mitigation and how costs could be correlated to different types of damages (Kundak).

5 CONCLUSIONS

A conference as the one that prompted the production of this book, can be thought of as part of a chain: there have been conferences before, there will be many others in the future. One question worth to ask is whom would we like to attract and whom would we like to speak to, besides professionals and researchers in the filed of risk analysis and management.

Certainly the possibility to exchange ideas and research results is important, nevertheless something more can be done perhaps in order to a achieve the main goal which is ahead of us: risks mitigation and reduction.

The answer can be divided in two parts: first who are the stakeholders from whom, as a community of experts, we can learn about problems and achievements in prevention. Those certainly include civil protection agencies personnel, as already mentioned, but also a large number of decision makers, public officials, whose work has to do with prevention. The latter may have a word to say about difficulties and constraints in applying an implementing research results as well as adhere to safety standards. Some of them may instead provide success stories of mitigation projects that actually achieved the set target.

Meetings like the annual conference of SRA should in principle constitute opportunities for a variety of stakeholders to get informed about recent advancement in the field of risk assessment and management. The point is that those stakeholders rarely join such meetings for a variety of reasons. One is probably the fact that it is not easy to identify who relevant stakeholders are. Also the EU Commission requires in most projects calls to make explicit who are the potential users of the proposed project and if the latter will be involved in some way in the research work. Again, besides the more obvious officials in civil protection and similar emergency management related agencies, it is challenging to identify those in charge of long term mitigation measures: land use planners, engineers, lifelines managing companies, designers...

Several professional figures based both in private businesses and institutions should work together to achieve the common goal of safer built environments. But, as Handmer correctly puts it (1999, p. 175), an almost insurmountable obstacle is due to the «difficulty of shifting the orientation from events and projects to comprehensive and integrated long-term programmes. Here, resistance is not simply driven by the concerns of specific agencies or constituencies; the whole state budgetary process works to reinforce sectoral planning and existing agency boundaries». Due to the fact that such stakeholders cannot be targeted by specific information campaigns and refer to different sectors in government and market, they are not likely to attend risk related conferences, if not occasionally.

The second part of the answer mirrors the first, referring to the improvement that can be sought for overcoming barriers in implementing what is already known about hazards and risk mitigation.

But what does implementation actually mean? It may address the application of scientific findings; in this respect it is easier to find technical applications rather than behaviours and cultural changes. Nevertheless, the hygiene revolution in the XIX century demonstrated that it is possible to achieve also cultural and behavioural changes, though in longer time and through a tougher process.

But implementation refers also to the laws, norms, and standards deriving indirectly and partially from science. In this respect three points can be raised: whether or not norms and rules correctly translate scientific findings into action; second whether or not norms are actually applicable and coordinated among them so as to shape a coherent legislative body. Finally what are the conditions for making a variety of stakeholders and the community at large respect such rules and norms. Those questions deserve to be tackled to avoid the sad feeling of "Knowing more and applying less" (White et al. 2001) arising after almost each new disaster.

REFERENCES

Allio, L., Ballantine, B. & Meads, R. 2006. Enhancing the role of science in the decision-making of the European Union. *Regulatory Toxicology and Pharmacology* 44: 4–13.

Beck, U. 1992. *Risk Society: Towards a New Modernity*. London: Sage.

Bernstein, P. 1996. *Against the gods. The remarkable story of risk*. New York and Canada: John Wiley & Sons, Inc.

Cohen, W., Nelson, R. & Walsh, J. 2002. Links and impacts: the influence of public research on industrial R&D. *Management Science* 48(1): 1–23.

De Elera, A. 2006. The European reserach area: one way towards a European scientific community? *European Law Journal* 12(5): 559–574.

Dosi, G., Llerena, P. & Sylos labini, M. 2005. *Evaluating and comparing the innovation performance of the United States and the European Union*. Expert Report prepared afor the TrendChart Policy Workshop.

European Commission, 2007. *Taking European knowledge society seriously*, EUR 227000, Directorate General for Research Science, Economy and Society.

Funtowicz, S. & Ravetz, J. 1990. *Uncertainty and quality in Science for policy*. The Netherlands: Kluwer Academic Publishers.

Funtowicz, S., Shepherd, I., Wilkinson, D. & Ravetz, J. 2000. Science and governance in the European Union: a contribution to the debate. *Science and Public Policy* 27(5): 327–336.

Godard, O. (ed.) 1997. *Le principe de precaution dans la conduite des affaires humaines*. Paris :Editions de la Maison des Sciences de l'Homme.

Jasanoff, S. 1990. *The fifth branch: science advisers as policymakers*. Cambridge: Harvard University Press.

Jasanoff, S. 1993. Bridging the two cultures of risk analysis. *Risk Analysis* 13(2): 123.

Handmer, J. 1999. Natural and anthropogenic hazards in Sydney. In Mitchell J. (ed.) *Crucibles of hazards: mega-cities and disasters in transition*. United Nations University Press.

King, D. 2004. The scientific impact of nations. What different countries get for their research spending. *Feature* 430.

Lee, C. 2006. Perspective: peer review of interdisciplinary scientific papers. *Nature* 05034.

McCormick, J. 2001. *Environmental policy in the European Union*. Palgrave.

Mitchell, R., Clarck, W., Cash, D. & Dickson, N. 2006. *Global environmental assessments. Information and influence*. MIT Press.

O'Keefe, P., Westgate, K. & Wisner, B. 1976. Taking the naturalness out of natural disasters. *Nature* 260: 566–567.

Nicolais, L. 2007. Ricerca, innovazione, reti di competenze. T. Maccaro (ed.) *La ricerca tradita. Analisi e prospettive di rilancio*. Milan, Italy: Garzanti.

Pigeon, P. 2003. Réflexions sur les notions et les méthodes en géographie des risques dits naturels. *Annales de Géographie* 627–628: 452–470.

Poumadore, M. 1998. From risk analysis in Europe to European risk analysis: the first ten years of SRA-E (1987–97). *Journal of Risk Research* 1(1): 77–86.

Pulver, S. & VanDevver, S. 2009. Thinking About Tomorrows: Scenarios, Global Environmental Politics, and Social Science Scholarship. *Global Environmental Politics* 9(2): 1–13.

Thomson, K., Deisler, P. Jr. & Schwing, R. 2005. Interdisciplinary vision: the first 25 years of the Society for Risk Analysis (SRA), 1980–2005. *Risk Analysis* 25(6): 1381.

Salter, L., Leiss, L. & Levy, E. 1988. *Mandated science. Science and scientists in the making of standards*. Dordrecht-Boston-London: Kluwer.

Schmidt Thomé, P. 2006. *The Spatial Effects and Management of Natural and Technological Hazards in Europe – ESPON 1.3.1 Executive Summary*, http://www.gsf.fi/projects/espon/Summary.pdf

Showalter, P. & Fran Myers, M. 1992. *Natural disasters as the cause of technological emergencies: a review of the decade 1980–1989*. Natural Hazard Research and Applications Center, University of Colorado.

Stirling, A. 2007. Risk, precaution and science: towards a more constructive policy debate. Talking point on the precautionary principle. *EMBO Reports* 8(4): 309–315.

Thompson, K., Deisler, P. Jr. & Schwing, R. 2005. Interdisciplinary Vision: The First 25 Years of the Society for Risk Analysis (SRA), 1980–2005. *Risk Analysis* 25(6): 1333–1386.

Weichselgartner, J. & Obersteiner, M. 2002. Knowing sufficient and applying more: challenges in hazard management. *Environmental Hazards* 4(2): 73–77.

White, G., Kates, R. & Burton, I. 2001. Knowing better and losing even more: the use of knowledge in hazard management. *Environmental Hazards* 3(3): 81–92.

Young, N., Ioannidis, J. & Al-Ubaydli, O. 2008. Why current practices may distort science. The market of exchange of scientific information: the winner's curse, artificial scarcity, and uncertainty in biomedical publication. *PLoS Medicine*. www.plosmedicine.org.

Risks Challenging Publics, Scientists and Governments – Menoni (ed)

Organization

Editor: Scira Menoni
Editor Assistant: Funda Atun

SCIENTIFIC COMMITTEE

Cesare Alippi
Francesco Ballio
Alberto Guadagnino
Emanuele Lettieri
Diego Magni
Davide Manca
Daniela Molinari
Floriana Pergalani
Renato Rota
Enrico Zio

This book is a collection of articles that were presented on the occasion of the Annual Meeting of the Society for Risk Analysis–Europe, which was held in Como (Italy), 12–14 September 2005. The Society for Risk Analysis–Europe has contributed to funding this publication.

Emergency preparedness:
From contingency plans to crisis management

Risks Challenging Publics, Scientists and Governments – Menoni (ed)
© 2010 Taylor & Francis Group, London, ISBN 978-0-415-58072-4

Organisation and structure of Chemical, Biological, Radiological and Nuclear (CBRN) special firemen units in Italy

A. Pugliano
Firemen, Lombardia Regional Headquarter, Italy

ABSTRACT: Terrorist attacks in the United States and Europe since September 11 and the continuous threats of future attacks using weapons of mass destruction have produced world-wide, and in Italy as well, a substantial change within structures, organisations and bodies constituting the rescue system. This change involves all the system as well as the "philosophy" of approaching rescue. The aim of this paper is to show the effects of this change by discussing the new approach adopted by Italian Fire-fighters.

1 INTRODUCTION

The classical way used by the Italian Fire-fighters in their intervention is based, for all kind of event, on an "automatic" reply mechanism that sends closer team in the hit area on the spot.

This usually is sufficient for the resolution of most emergencies. In few cases, special equipments of support are sent since the beginning or are requested by the team leader, after he was evaluated initial situation and its evolution.

The intervention involving dangerous substances requires that the traditional logic has to be partially modified, emphasising and anticipating the location phase, classification, measuring, risk prevision, protections, abilities and technologies necessary to face and resolve the emergency.

The CBRN emergency assigns more and before that in other cases an essential role to the operating room and requests to the team leader who firstly intervenes a specific ability to evaluate "before" and "safety distance" the type and extension of the scenario, the necessary protections and the intervention possibilities.

2 THE ORGANIZATIONAL MODEL

The organisational model foresees three different levels of response:

- basic team who intervenes firstly with the crew at a Provincial Headquarter level;
- provincial experts, present in each Provincial Headquarter;
- operational CBRN Team, operating at a regional level and present in each capital of the region headquarter.

Obviously the operating levels are appropriate with the overall fire-fighters organisation. It is necessary to remind that in Italy fire-fighters are organized at a national level with central and provincial headquarters, structured in such a way that flexibility and complementary is assured. the 103 provincial headquarters are co-ordinated by 18 regional headquarters and a central department headed by the internal ministry.

There are foreseen 4 level of skills:

- level 0 corresponding to the components of the intervention of the basic brigade;
- level 1 corresponding to the head of the basic brigade;
- level 2 corresponding to the "provincial experts" pertaining to each headquarter;
- level 3 corresponding to a resolute unit specialised who is present in regional ambit.

Level 0 is characterised by:

- the basic know-how of chemical risk, bacteriological and radiological;
- the know how of substances rules and agreements;
- the know-how of the possibilities and technique for the recognition of the substances;
- the know-how of the rules for the delimitation of the intervention areas;
- ability in using the dpi and the surveying and decontaminating equipments issued to the brigade;
- ability to carry out the basic procedures for approaching the intervention, for isolating zones, anti-contamination and decontamination of the same.

Level 1 is characterised by the:

- know-how to analyse the accident in order to evaluate the severity of the initial scenario and

its possible evolution, relevant to the substances, containers and environmental conditions;
- know-how and ability in employing the security schedules relevant to the information on risks and contrast actions;
- know-how and use of the rules and technique for the prompt evaluation of the potential damages and delimitation of the "hot", "tepid" and "cold" zone;
- know-how and use of the rules and defensive technique to control the event – when possible – and to protect people, the environment and goods involved;
- ability to evaluate the qualification of individual protection equipments and preparing the latter for use;
- ability in talking with the operations' room and the superior authority in order to refer on the situation and arrange the most suitable strategy, also in relation to the local emergency plans to the event development of and the potential consequences of undertaken actions;
- ability to evaluate the feasibility in autonomy of simple interventions and to co-ordinate the execution or to plan and practice an initial reply, keeping in mind available staff skills and ability, equipments of personal safety;
- ability in evaluating the progress of adopted actions referring to the expected target;
- ability in managing the post intervention task.

As for the other fields of operating activities into the organisation, also for the CBRN department the presence, in each headquarter, of qualified staff assigned to level 2 is expected. The latter is characterised by the knowledge and abilities necessary for:

- depth analysis and observation of chemical, biological, radiological and nuclear risk, proportionate to control releases or potential releases of substances or dangerous emissions and to support or execute the functions correspondent to the level of complexity required in heavy and complex situations;
- the use of simple equipments and techniques in order to identify or classify unknown substances and to measure the concentration or the value of emission, expressing an assessment of danger with reference to the respective damage threshold;
- collecting and interpreting information drawn from publications, technical sources, computerised databases and remote instruments;
- evaluating the damages suffered by containers in order to appraise the risk level;
- foreseeing the probable behaviour of the released substances and of their containers when several substances are involved;
- evaluating the dimension of the dangerous zone using computerised models, survey equipment or with the aid of an expert in the subject;
- evaluating the feasibility and develop an action plan involving scenarios of modest complexity, coherent with the local emergency plans, the standard procedures, the competence, the ability, equipment and dimension of available resources. Severability it is the ability to place and implement that is plan and the initial response while waiting for more intervention;
- evaluating the progress of the promoted actions;
- selecting the most suitable decontamination procedures;
- concluding an incident drawing up (drafting) a report, preparing evidences (documentation), participating to the de-briefing operations and to the critical review of the developed operations;
- identifying and satisfying the didactic demand and training of the level 0 staff; participating to the training and updating of the level 1 staff;
- co-ordinating the area, including the administration and the maintenance of special equipments;

Complex situations, requiring the handling and control of hazardous materials, the deployment of dangerous operations and special equipments, are managed by increasing both the number of firefighters and the ability level.

The specialised staff intervention, usually under the regional authority, permits to support the incident commander activity and role which belongs to the fire chief of the provincial fire brigade. To this role is also appointed the authority to connect authorities and other figures involved in the accidental scenario.

This represents the maximum operative capacity and permits to perform all the responsibility functions and decision levels relative to the management of the incident and of the area where it occurred.

In light of the above, the 3rd level of ability is requested in order to:

- analyse the complex type of incident, collect and interpret the information drew from instruments, documents and literature, also in collaboration with local responsible authorities or with experts of specific sectors, in order to establish the problem gravity and the potential scenarios developed and their predictable effects;
- identify feasible actions of contrast, evaluating the available resources;
- plan and co-ordinate the execution of the contrast actions (offensive and defensive) considered as suitable, using equipments and/or superior ability to that corresponding to level 1 and 2;
- approve the level of individual protection expected for an action typology;

- put practice the standard intervention procedures as well as the feasible procedures deriving from planning, agreement and conventions suitable for communication and the use of others resources (corporate, volunteers...);
- supervise available resources in accordance to the assigned tasks and the local activities, assuring the technical supervision and in case of logistic support providing, when requested, the indication and the technical elements necessary to give information to the mass media and to the local authorities;
- evaluate the effects of the undertaken actions in relation to the evolution of events and to the purposes, adapting the action plane;
- conclude the intervention, with the adequate transfer of functions control, a de-briefing of the incident, a meeting among the involved forces, giving a feedback of the reaction and the collection of necessary documentation;
- learn, experiment, verify, update material, instruments and procedures;
- participate to the planning and implementation of the most suitable training programs.

In the model described above a basic level to which corresponds a high level of risk awareness is identifiable. This can be translated into a better ability of auto-protection and of a correct activation of procedures, the intervention is connected to the identification of the hazards associated to dangerous substances present on the scenario. This intervention ability at a basic level is requested to all staff operating in the entire national territory. At a provincial level, with an intermediate skill, the ability to analyse the accidental scenario is requested, to determine the problems magnitude and develop action plans. The specialised level activates the use of regional level operative units, able to develop the definitive response also in the case of complex scenarios.

The described model refers to the NFPA standard and represent an adaptation to the organisation of the national fire fighters brigade existing in Italy for sixty years. Under a formal point of view the basic level, unique in the NFPA coding, foresees a double level of ability: the former relates to the team components and the latter to the team leader who is responsible of the basic operational team.

When there are complex emergencies, the resources present the various territorial ambits can be activated through simple procedures of mobilisation, as it already happens for any natural disaster or intervention of complex typology.

It is obvious that in the mobilisation schemes each regional unit CBRN shall be supported from one or more basic teams, from the provincial unit or other regional and provincial units, according to the extension and complexity of the event to be faced.

3 INTERVENTION SCENARIOS

Up to now the general of the organisational criteria have been presented, it is now necessary to establish the classification and the choice of the scenarios and events to take into consideration for drafting the standard procedures. The question is to find a reasonable compromise between the need of re-unify and simplify—that would lead to bring back all the possible case histories to a low number of procedures—and that of specificity and details, that would impose to define a procedure for each type of risk, each category of substances, each typology of scenario and so on.

In relation to the use of substances and/or dangerous agents, it is possible to identify three reference scenarios. They clearly characterise the risk and the various procedures of intervention.

1. recovery without loss;
2. loss in limited area;
3. environmental contamination.

The described scenarios represent the extension and the generalisation of the scenarios classification system, directed towards the management of intervention in presence of materials potentially infected by anthrax spores, adopted by the central direction for the emergency of the fire-fighters department straight afterward the events of September 11.

In particular, the expected scenarios are the following:

1. scenario: presence of suspicious material but devoid in evident traces of dust or liquid;
2. scenario: presence of suspicious material with evident traces of dust or liquid, also dispersed on surface but not affecting the environment;
3. scenario: presence of suspicious material with evident traces of dust or liquid, also dispersed on a surface and in the external environment.

The typologies of event can be quoted in the following three categories:

A. release (or possibility of release) without (or without the risk of)fire or explosion;
B. release (or possibility of release) with (or with the risk of)fire or explosion;
C. recover of dangerous substances (or suspicious) without (or with remote) risk of loss.

The intersection between the above mentioned scenarios and the categories of events, correlated to the typologies of chemical, biological, radiological and nuclear risk, represents the matrix of essential reference matrix on which the response of the national corps to the CBRN emergencies is based.

4 EXPERIENCES NEEDS AND PERSPECTIVES

The typological and dimensional variability of the expected events in large parts of the Country, combined with the predictable complexity and gravity of risks, requires to the entire organisation on the one hand training and ability to guarantee response and on the other hand an exact and in depth qualification.

The requirements of operators' protection and of a timely and resolute intervention represent two fundamental objectives which fire-fighters have always tried to reach.

The necessity to reconcile two different needs—adequate presence in the territory and the adequate specialisation—have lead to define an organizational model based on diverse operative abilities differentiated also across territorial scales and levels.

Large efforts have been required to organise and install in every region of Italy a Regional Operating Unit, exporting experience accumulated in more than ten years of interventions of the piloting fire-fighter centres of Lombardia (Milan), Veneto (Venice), Lazio (Rome) and Puglia (Bari).

Even though the project of development of the described model has not been fulfilled yet, the experience made during the training plans and simulations (conventional and not conventional) and during real accidents (conventional) is enough to draw some useful suggestions in order to identify future needs and to improve a system of response to a terrorist attack.

5 THE TECHNIQUES OF IDENTIFICATION OF THE CONTAMINATING AGENTS

Tools available on the market are sufficiently useful for a field use by non-specialist personnel. Yet the detection and environmental monitoring require knowledge, sensibility, skill and familiarity for the combined use of more detectors. Moreover, ability of evaluation and interpretation in order to avoid misunderstandings and mistakes in the diagnosis of the problem and in the resulting measures is required. Those skills cannot be reached with an occasional use of these instruments. It would be useful to develop simulating software to be connected to other detectors and instruments and to the operational training and exercise centre. These remarks recommend that at least the personnel entrusted with the most serious intervention should be rightly selected and should be assigned only to training and NBCR interventions.

5.1 *Communication*

The possibility and capability to maintain contact with and between operators who enter in the "warm zone" with totally encapsulating chemical-protective clothing is essential. Even though, personnel will never be sent alone in the contaminated area and, in case of more operators, the constant control of the time and of the position of each person must be assured (by an officer or a co-ordinator of the operation), and the reliability of the communication is a very important matter. It is therefore necessary to prepare a suitable local communication system, to be properly used in confined or underground spaces and equipped in order not to engage the personnel hands.

5.2 *Coordination*

The coordination of forces called to manage the emergency, in particular the Emergency Medical Services and the Police is a crucial issue. The Civil Defence National Plan sets the general lines for managing a deliberate terrorist attack entrusting the tasks to every interested Organisation. These lines are therefore taken locally in the Civil Defence Provincial Plans which are promoted and co-ordinated by the Prefects. Yet it is necessary that beside the planning there will be an intensive training program involving all forces. The latter should be aimed one side to the integration of competencies and from the other side to the spread and check of operating procedures concerning the different considered accident scenarios. Such a joined training should also concern the personnel at the emergency calling centres, waiting for the accomplishment of the project of the unified emergency calling centre, where in the same room the personnel work at separate positions.

5.3 *Information to population*

This point, does not concern firemen directly but which is of great importance to mitigate the consequences of a NBCR terrorist attack. Even though the matter is delicate and there are many objective reasons for keeping such information secret, the preventive education of the population with regards to the right behaviour to be taken in case of emergency is absolutely necessary. Since the factor time and the limitation of the induced contamination are basic to mitigate the effects of a NBCR attack, instructions, given by responders, during crises with no prior knowledge on the side of the population, however timely would be absolutely ineffective. And this is even more so in cases of crowded places.

5.4 *Materials*

As far as it concerns the materials it is pointed out the need to develop and to adopt according to the different cases:

- more comfortable protective clothing, (above all, those supplied with internal ventilation);
- communication apparatus for confined and underground areas which do not require the use of operators hands;
- robot systems to detect and monitor contaminated areas;
- software for managing the personnel inside the contaminated area;
- software of simulation to be connect to other detectors and instruments at the emergency calling centre and training centre.

6 CONCLUSIONS

In Italy, since the end of the second world war, public safety has been entrusted to National Firefighters Brigade, through of specific legislation, towards the perils rising from the use of nuclear energy both in ordinary civil use and in the event of radioactive relapse on the national territory in consequence of explosive nuclear devices. This in a more general ambit of the organisation system in the NATO Countries for nuclear, biological and chemical defence.

The operative ability has become consolidated during the years through the technical rescue activity implying the handling of dangerous substances, in occasional scenarios concerning fixed plants, factories, and transportation vectors.

The interventions involving dangerous substances and emissions require without any doubt a constant updating of professional training, equipments and organisation at capabilities.

Recent and still present international developments have reopened new scenarios of response that seemed for a while overcome and have reopened the floor for expertise regarding non-conventional weapons, accentuating problems of competence, preparation and mobilisation of the National Firefighters Brigade, which was pertain to the daily reality.

Last, we are conscious that the improvement of response ability in non conventional type of crises will make the entire fire-fighters' organisation more qualified – nationally organised, as few in the world – to face intervention of technical rescue in the CBRN field, with greater efficiency.

Risks Challenging Publics, Scientists and Governments – Menoni (ed)
© 2010 Taylor & Francis Group, London, ISBN 978-0-415-58072-4

NEDIES: A European portal for the generation and dissemination of lessons learned from disasters

E. Krausmann, M. Christou, S. Scheer, S. Mara, J. Hervas & M. Delaval
Joint Research Centre of the European Commission, Institute for the Protection and Security of the Citizen, Ispra (VA), Italy

ABSTRACT: Technological accidents and natural disasters have a major impact on society due to their potential to result in a loss of lives, damaging long-term consequences on the environment, the economy, and disruption of community life. A comprehensive analysis of the circumstances that led to a disaster yields valuable lessons that can contribute towards the prevention or mitigation of a future event. This paper introduces the Natural and Environmental Disaster Information Exchange System (NEDIES), which is one of the European Communities' systems for lessons learning and information exchange on the management of technological accidents and natural disasters. The paper describes the structure and functioning of NEDIES and briefly outlines the NEDIES methodology for lessons learning. As an example lessons learned on landslide disasters in Europe and guidelines for flash-flood prevention and mitigation are presented. The mainstreaming of lessons learned into disaster risk management practices and the most important fields of application of lessons learned are also discussed.

1 INTRODUCTION

The protection of the citizen and the environment in Europe faces a continuing challenge from a wide range of risks that arise from both natural and technological hazards. Therefore, the lessons learned from the systematic analysis of the evolution of past accidents and disasters and the circumstances that facilitated their occurrence are of paramount importance for future risk reduction and priority setting in terms of vulnerability management. The widespread application and publicizing of these lessons learned is another vital step towards combating the occurrence of undesirable events and, in particular, mitigating their consequences.

The Joint Research Centre (JRC) of the European Commission (EC) supports this goal by ensuring that the information available throughout Europe and beyond on the management of natural and technological disasters is systematically exploited and the lessons learned disseminated. More specifically, the Joint Research Centre maintains the Natural and Environmental Disaster Information Exchange System (NEDIES), whose aim is the generation and dissemination of lessons learned for the prevention, preparedness and response to natural disasters and to non-Seveso (European Communities, 1982; European Communities, 1996) technological accidents in support of the European Communities' action programme on Civil Protection (European Communities, 1999).

This article presents the structure and functioning of the European Communities' Natural and Environmental Disaster Information Exchange System and outlines the employed methodology for lessons learning using examples from the NEDIES knowledge base. Moreover, this paper discusses how lessons learned can be mainstreamed into risk-management practises, how useful recommendations can be addressed to the local and regional authorities involved in natural disasters management and how this system contributes to the development of a risk reduction culture.

2 THE NATURAL AND ENVIRONMENTAL DISASTER INFORMATION EXCHANGE SYSTEM

2.1 *Scope and objectives*

The increase in the number of disasters in Europe and the suffering and damage they caused has led the Member States of the European Union to strongly support the sharing of experience acquired in the management of these disasters, cooperation and mutual assistance. As a consequence, European Community cooperation in the field of civil protection was reinforced, eventually resulting in the launch of the NEDIES project whose main objectives are:

- To supply the EC, the Member State Authorities and other civil protection, rescue and emergency

services with validated information on past natural disasters and non-Seveso technological accidents and their management.
- To contribute to the protection of the citizens from disasters via the dissemination of targeted information on risk perception and awareness.
- To constitute a common repository of lessons learned from disasters in support of the development of effective prevention, preparedness and response policies and measures.
- To provide an international interdisciplinary platform for dialogue to facilitate the exchange of information between all the actors involved in disaster management.

Starting with lessons learned on the management of floods and earthquakes during the pilot phase of the project in the beginning of 1997, the scope of NEDIES has been extended in the years to follow to now comprise all major natural hazards, technological accidents not covered by the European Seveso directives (which regulate the process industry) (European Communities, 1982; European Communities, 1996), and multi-hazard events (e.g. natural-hazard triggered technological accidents, so-called Natech disasters). Although NEDIES is a European project and has initially focused on disasters that occurred in Europe, its scope is not restricted to European stakeholders only. It encourages users worldwide to contribute with input on lessons learned and/or to benefit from lessons learned generated by other users. Currently it is under investigation how NEDIES could be embedded within a global operative system to facilitate the worldwide exchange of disaster risk management data.

2.2 *The NEDIES knowledgebase*

Technically, NEDIES is a web site interfaced with an underlying knowledgebase. The web site provides the interested user with publicly available information on the disasters covered by the NEDIES objectives and permits the downloading of lessons-learned reports, guidelines and other documentation prepared by the NEDIES team. The knowledgebase supports the objectives of the project by storing and exploiting relevant data in a systematic way and by contributing information back into disaster-management practices in the form of lessons learned on past events.

The very core of the knowledgebase are natural and technological event (disaster) reports, compiled by civil-protection authorities and services, which contain validated and structured information on 1) the evolution of a disaster and its consequences, 2) the implemented prevention, mitigation and response measures, and 3) the lessons learned on the management of the event. Since part of this information is sometimes confidential, the knowledgebase can only be accessed via user registration and logon through the website. Any information provided by a registered user remains his or her property. As the owner of a report alone can access and modify its contents, data security is guaranteed. Users are also invited to supply document references pertaining to the described event, including references to books, journals, proceedings, reports or URLs that direct the interested user to other web sites of interest. Work is in progress to offer the users access to the full documents through NEDIES instead of only a bibliography.

In addition to the event reports NEDIES also contains a fully-implemented lessons-learning data model that offers the user the possibility to provide structured details on a given lesson learned. This includes information on the context and validity of a lesson, as well as actors and possible actions derived from it in view of the development of possible disaster reduction techniques. In support of this activity NEDIES is interfacing with the Disaster Reduction Hyperbase project with aims at collecting and disseminating community-based and implementation-oriented disaster risk reduction practices (Scheer & Ranguelov, 2008).

At present, NEDIES counts 383 registered users from national and international organisations, academia and the private sector that contribute to and benefit from currently 98 fully validated event descriptions and their accompanying lessons learned.

2.3 *The NEDIES methodology for lessons learning*

Irrespective of the disaster dimension (international, national, regional or local) the lessons-learning process comprises the following six major steps (Krausmann & Mushtaq, 2005):

- Investigation of a disaster
- Reporting of a disaster
- Data collection, in particular on the management of a disaster
- Data analysis and evaluation
- Generation of lessons
- Implementation and monitoring of lessons learned.

The work performed within the NEDIES project focuses on steps 3 to 6, with the main emphasis being on data collection, analysis, extracting lessons and their dissemination.

In the absence of the mandatory reporting of data pertaining to the evolution and management of a disaster, NEDIES collects information on past events using predefined templates that have been developed in close cooperation with authorities and experts (Table 1). This ensures that the users' needs and wishes are adequately addressed, thereby facilitating the practical application of the generated lessons. The voluntary reporting of information by designated bodies (e.g. civil-protection authorities and services) is the most effective means to collect reliable data. In the absence of this, data must be collected from the media or open literature sources, which is rather cumbersome and a significant effort has to be invested in corroborating the various data sources to guarantee the quality of the collected information. The data collection step aims at compiling into a suitable format the information from the original data sources for carrying out an in-depth analysis to detect patterns of accident or disaster causation, to identify adequate and inadequate (including missing) disaster-management policies and measures, and to pinpoint vulnerable risk receptors. The evaluation of the outcome of the data analysis using expert judgment or systematic tools is the final step in the data analysis process.

Table 1. The NEDIES lessons-learned template format.

GENERAL INFORMATION	
Starting date	
Duration	
Location(s) involved	
Administrative Unit (NUTS3 Eurostat classification, if available) or Region (Province)	
Country	
Consequences to persons **- Number of fatalities/ injured/homeless**	
Economic losses (in Euro) **- Material losses/ Response action costs**	
Prediction made **- Yes/No** **- Short comment (if any)**	
INFORMATION ON DISASTER MANAGEMENT AND LESSONS LEARNED	
Short description of event ***Prevention phase*** - Prevention measures - Lessons learned ***Preparedness phase*** - Preparedness measures - Lessons learned ***Response phase*** - Response actions - Lessons learned ***Information dissemination*** - Prior to event/During event/Following event - Lessons learned	

Lessons can be learned in all phases of disaster management, i.e. prevention, preparedness, response and recovery. Within the NEDIES activity mono- and cross-hazard types of analyses are undertaken for the former three disaster-management phases. Recovery aspects are not currently tackled within the system. Emphasis is also given to lessons learned on information dissemination to the public, which is an integral part of all three phases. The mono-hazard analysis aims at providing input to risk-management practices with respect to one type of hazard (e.g. landslides). The cross-hazard analysis compares and contrasts information across different hazards and investigates if a cross-fertilisation of lessons learned from one disaster-management field to another could provide added value. This type of analysis is applied to multi-hazard events, such as technological accidents triggered by a natural disaster, or occurrences with domino effects (e.g. a coastal or submarine landslide triggering a tsunami).

The implementation of lessons learned into everyday disaster risk management begins with the dissemination of the outcome of the performed analyses. In support of this task the NEDIES team prepares lessons-learned reports, guidelines and recommendations that are made available to all interested parties through conferences, workshops, publications and/or the NEDIES website. These documents can be targeted to responsible authorities, urban planners, the general public or any other stakeholder. The effective mainstreaming of lessons learned into disaster risk management, which is undertaken by decision makers at local, regional or national level, and the establishment of a monitoring mechanism that provides feedback on the adequacy of the adopted measures are the final steps in the implementation phase. It is foreseen that NEDIES, which currently focuses on the dissemination part of the implementation phase, will in the future also encompass the monitoring of new risk-reduction measures by developing dedicated indicators.

2.4 *Findings from case studies: Landslides and flash floods in Europe*

2.4.1 *Lessons learned from landslide disasters in Europe*

Landslides are responsible for significantly higher economic losses and casualties than is generally recognised. Destructive landslides are triggered not only by prolonged or heavy rainfall but also by other natural hazards such as earthquakes, flooding and volcanic eruptions, as well as by construction or other human activities. Moreover, in coastal zones landslides can trigger other natural hazards, such as tsunamis, and they can cause technological accidents in industrial areas. Recognising the need for concerted action to tackle landslide disaster management, the JRC organised a workshop within the NEDIES framework to get a better understanding of landslide disaster-management practices in Europe (Hervás, 2003). This workshop, which gathered representatives from authorities, civil protection services and other experts from EU Member States and Candidate Countries, constituted a platform for the sharing of expert knowledge and opinion and provided a source of validated information on past landslide disasters, the analysis of which by the NEDIES team led to the publication of lessons learned for coping with future landslides. As an example a selection of generic key lessons learned for the management of landslides is given in the following.

Lessons learned concerning landslide prevention measures

- Identify precipitation thresholds that enable the definition of three alert levels (attention, warning and alarm), especially for debris flows and mudflows triggering.
- Produce landslide hazard maps and preferably risk maps in residential and industrial areas scaled in accordance with local land-use maps.
- Issue specific legislation that requires the elaboration of natural hazard/risk maps and couple land-use decisions to the hazard or risk level present.
- Provide sufficient financial resources for geological and geotechnical surveys in landslide-prone areas and for keeping them updated.
- Set up landslide monitoring systems coupled with automatic warning and alarm systems connected to the headquarters of emergency intervention bodies in the case of major active landslides affecting populated areas or major lifelines.

Lessons learned concerning landslide preparedness measures

- Develop a specific emergency plan according to the type of potential landslide and risk level. For active landslides the plan must especially include evacuation measures and the enlargement of the risk area when substantial slope movement has been ascertained, even before damage occurs.
- Establish systems for real-time weather information at a local scale in areas where landslides are most often related to intense rainfall.
- Identify diversion routes for key roads subject to land sliding for use by emergency services.

Lessons learned concerning landslide response measures

- Identify and organise the civil-protection emergency control centres, the persons in charge of the centres and their various functions.
- Ensure that rescue and protection services are staffed with sufficiently trained and adequately equipped personnel.
- For landslides linked to rainstorms or floods utilise houses or other permanent structures as reception centres for evacuees rather than tents or caravans.

Lessons learned concerning the dissemination of information

- Provide regular and open information to residents of landslide hazard areas to raise awareness of the danger and to attain confidence in and promote collaboration with the rescue services.
- Issue general warnings through TV and radio or directly warn the public by fixed/mobile telephone using an automatic voice/SMS messaging system.
- Establish a notification centre and appoint an information manager to inform both the public and the media.

2.4.2 *Guidelines on flash-flood prevention and mitigation*

The vulnerability of our society to flash floods is continuously increasing, in particular due to the spreading of urban areas and critical technologies and infrastructure into mountain valleys and flood plains. The resulting colonisation of flood plains in search of building sites and more arable land frequently leads to a clash between the population's needs or economic interests and nature's forces. Flash floods strongly affect natural and man-made environments with often deplorable losses in terms of human lives and damage to property and infrastructures. As flash floods are difficult to predict, efforts on the management of flash-flood risk tend to be curative and focus on emergency response and recovery. Being firmly convinced that prevention and mitigation strategies should receive more attention, the NEDIES team prepared a set of non-prescriptive guidelines (Colombo et al. 2002) based on the results of three EC-funded pilot studies on flood management (Pichler et al. 1999; Dautrebande et al. 2000; Vignozzi et al. 1999), in fulfilment of a request from the EC. The aim of these guidelines is to support decision makers at the various levels in the national, regional and local organisations in their quest to deal with complex flash-flood phenomena.

The guidelines on flash-flood prevention and mitigation identify and extensively discuss both structural and non-structural measures to manage flash floods. Structural measures aim at controlling the hazard component of flash-flood risk. They include works and interventions for either controlling floods or reducing flood peak (by means of e.g. river training interventions including transversal and longitudinal protection works, water retention basins or catchment-wide adequate agriculture and forestry management practices and water control works). To this end, use of bio-engineering structural works should be promoted, where possible, so as to diminish negative effects on watercourse dynamics and environment (Schiechtl & Sten, 1997). Non-structural measures in turn affect the flash-flood vulnerability component (e.g. through land-use planning or the application of financial measures in support of individuals and local communities or monitoring and warning systems).

With a view to assisting decision makers in prioritising and implementing measures to cope with flash floods, the guidelines propose a general prioritisation scheme that is shown in Figure 1. In the first step the scope of the study is defined (step A), which is followed by the data collection (step B). The identification of possible measures to cope with flash floods and their grouping follow

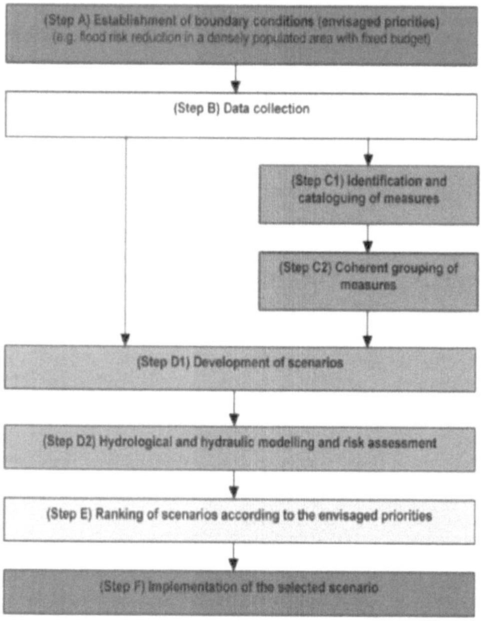

Figure 1. Schematic representation of the proposed steps for prioritising measures to be implemented in flood management.

in steps C1 and C2. After that, the development of scenarios (step D1), which serve as input to the models assessing the flash-flood risk in step D2, is repeated for various sets of measures including a reference simulation that does not foresee the implementation of any prevention or mitigation measures. The resulting scenarios are then ranked (step E) in accordance with the boundary conditions established at the outset of the study. As a final step the selected scenario is implemented (step F) within the legal framework at the desired geographical scale.

3 MAINSTREAMING LESSONS LEARNED INTO DISASTER RISK MANAGEMENT

The mainstreaming of lessons learned into disaster risk management is the last step and probably the final objective of the lessons-learning procedure, as described in Section 2.3. It is the phase where the concentrated experience gained from one or many disasters is transferred into risk-management practices, with improved risk-reduction measures as a result. Moreover, this is the phase that informs and supports the decisions to be taken by each decision or policy maker. What are, however, the needs of the decision maker in terms of the usefulness of lessons learned? For the decision maker a useful lesson is a comprehensive and validated conclusion of general applicability and value on how a particular disaster or type of disasters could have been prevented or mitigated. This lesson can take the form of a recommendation, guideline, good practice or bad practice to be avoided.

The whole range of disaster risk reduction decisions and actions can and should be informed by lessons learned from past events. Prevention and preparedness strategies and measures, the management of the emergency, and recovery practices need to be based upon past experience, as well as on systematic analyses using models and scientific tools. However, lessons learned have a clear advantage compared to analytical models because of their direct character, the fact that they are based on real events and the lack of simplifying assumptions. For that reason, if identified and given the appropriate attention, they receive a much higher acceptability, which facilitates their mainstreaming into disaster risk management practices.

The end user of lessons-learned studies is anyone who deals with the management of risks. This includes civil protection and planning authorities, regulators, emergency planners and responders, risk analysts and practitioners, insurers and the general public. Clearly, in order for the lessons to reach the various end-user categories and to be effective, they need to be formulated differently. But whatever form they may take, the lessons need to be clear, comprehensible and focused. Unrealistic and general statements (e.g. "if we could have stopped the hurricane we would have avoided the disaster") are of not much help for risk management. Moreover, it is very important for the risk manager to be provided with a clear definition of the conditions for which the lesson is valid.

The potential range of application of lessons learned is broad, covering the whole spectrum of disaster risk management. Of particular importance are:

1. The application to risk assessment,
2. The application to risk management,
3. The application to policymaking and
4. The contribution to the creation of a risk reduction culture.

Risk assessment techniques gain much from the experience gathered during real events. Lessons learned provide valuable information, data, and validation cases, while they define new needs for research and development. For instance, after a catastrophic earthquake in Puglia, Italy, in 2002 (Tozzi, 2005) the seismicity of many vulnerable areas has been reassessed and hazard maps have been reconstructed, based on a much more accurate analysis including new event information. Research needs for addressing combined hazards or Natech events were identified based on lessons learned from real multi-hazard events. Risk-management strategies and practices are also supported by lessons learned. In fact, in most cases they are the only source of information for judging whether a risk-reduction measure is adequate or not.

Traditionally, policymaking has been reactive rather than proactive, which means that the regulators tend to respond to an emergency rather than prevent it. In the aftermath of the Seveso (Mannan, 2005) and similar disasters, for instance, the Seveso Directive has been developed to control major accident hazards. As a consequence of the lessons learned from the above-mentioned earthquake in Italy, the application of Eurocodes (European Commission for Standardization) for the anti-seismic construction of buildings became mandatory in many areas in Italy. Last but not least, lessons learned can contribute to the development of a disaster risk reduction culture throughout the stakeholder community. The direct and objective character of lessons learned is an important element in this respect. Similarly, this also serves to improve communication amongst the different stakeholders.

Being the last step in the lessons-learning procedure, the task of mainstreaming lessons learned into disaster risk management policies and practices has not yet been exploited sufficiently within the NEDIES framework. This is foreseen for the future, when various stakeholder groups can be targeted and informed. This will also include the monitoring of implemented lessons learned by means of appropriate indicators to get feedback on their success or lack thereof. Without doubt, there are lessons to be learned from the dissemination and mainstreaming steps, too.

4 CONCLUSION

The exchange of information on the effectiveness of disaster-management practices and the widespread application of lessons learned across different hazards is gaining in importance due to the increase in inter-dependence and vulnerability of our societies. This is especially valid for multi-hazard events, such as Natech disasters, whose management would certainly profit greatly from an integrated approach to disaster risk reduction. NEDIES tackles these issues by providing a platform for information exchange between all stakeholders that share an interest in disaster risk management. This promotes the communication between civil-protection authorities and services and fosters the dialogue between the public and risk experts, thereby impacting on the stakeholders' perception of risk and eventually contributing to a better-targeted allocation of resources for risk reduction. In the frame of its activities NEDIES also identifies existing gaps in prevention, preparedness and response activities and proposes and publicises improvements, which may provide valuable input to decision-support systems.

The results of a lessons-learning exercise and consequently their usability are highly dependent on the quality, extent and accuracy of the data reported or collected. Since the data can be affected in every step of the methodology outlined above, minimum standards are required to guarantee the reliability of the utilised information. In addition, the availability and accessibility of data may be hampered by a lack of free-flowing information due to the confidentiality of the outcome of external investigations into the factors contributing to a disaster or the investigations' perceived mandate to assign blame. Consequently, lessons-learned type studies would greatly benefit from the mandatory reporting of the circumstances that led to a disaster and the measures undertaken to cope with it to a centralised location where the data can be collected, analysed and shared. For this to function effectively, targeted legislation is required.

REFERENCES

Colombo, A.G., Hervás, J. & Vetere Arellano, A.L. 2002. *Guidelines for flash-flood prevention and mitigation*. EUR Report 20386 EN, European Communities.

Dautrebande, S., Leenaars, J.G.B., Smitz, J.S. & Vanthournout, E. (eds.) 2000. *Pilot project for the definition of environment-friendly measures to reduce the risk for flash floods in the Geul river catchment* (Belgium and The Netherlands). Contract No. B4-3040/97/730/JNB/C4, European Commission.

European Committee for Standardisation, http://www.cenorm.be/cenorm/index.htm

European Communities, 1999, Council Decision 1999/847/EC of 9 December 1999 establishing a Community action programme in the field of civil protection (2000–2004), Official Journal of the European Communities, No. L327.

European Communities, 1997, Council Directive 96/82/EC of 9 December 1996 on the control of major-accident hazards involving dangerous substances, Official Journal of the European Communities, No. L10.

European Communities, 1982, Council Directive 82/501/EEC of 24 June 1982 on the major-accident hazards of certain industrial activities, Official Journal of the European Communities, No. L230.

Hervás, J. (ed.) 2003. *Lessons learned from landslide disasters in Europe*. EUR Report 20558 EN, European Communities.

Krausmann, E. & Mushtaq, F. 2005. A methodology for lessons learning – Experiences at the European level. submitted for publication in: *Measuring vulnerability and coping capacity to hazards of natural origin – Concepts and methods*. United Nations University.

Mannan, S. (ed.) 2005. *Lee's loss prevention in the process industries, Hazard identification, assessment and control*. 3rd Edition, Vol. 3, Elsevier.

NEDIES, European Commission, Natural and Environmental Disaster Information Exchange System, http://nedies.jrc.it

Pichler, F. et al. 1999. Ecologically oriented flood and erosion management in alpine river basins, Regional government of Carinthia (A) and Autonomous Province of Trento (I).

Scheer, S. & Ranguelov, B. 2008. *The development of templates capturing information regarding disaster reduction technologies*. In Proc.: Int. Disaster and Risk Conference IDRC Davos, Short abstracts, 25–29 August, 2008.

Schiechtl, H.M. & Stern, R. 1997. *Water Bioengineering Techniques for Watercourse Bank and Shoreline Protection*. Oxford: Blackwell Science.

Tozzi, M. 2005. *Catastrofi*. Rizzoli.

Vignozzi, G., Fratini, R., Nocentini, G., Bartalucci, L., Binazzi, P. et al. 1999. *PREMO98—Principles and guidelines for the environmental protection of drainage basins prone to flash floods (PREMO2)*. Contract No. B4-3040/97/734/JNB/C4, European Commission.

Risks Challenging Publics, Scientists and Governments – Menoni (ed)
© 2010 Taylor & Francis Group, London, ISBN 978-0-415-58072-4

Multi-criteria decision aid for nuclear emergency management: Problem structuring based on a stakeholders survey

C.O. Turcanu
Service de Mathématiques de la Gestion, Université Libre de Bruxelles, Bruxelles, Belgium

B. Carlé & F. Hardeman
Department Decision Strategy Research, Belgian Nuclear Research Centre, Belgium

ABSTRACT: Stakeholders' involvement in risk management and decision making has received increasing attention in the last years. In nuclear emergency management, decisions taken have firstly to deal with the radiological impact to people and environment. Multi-criteria decision aid offers a suitable methodological framework for risk-informed decision making, compatible with the democratic perspective. We argue that in order to be successful, the use of multi-criteria decision tools and methods has to be subject to a systematic interaction with the relevant stakeholders in all stages of the decision analysis, and especially in the problem structuring stage. The paper focuses on this process and illustrates its findings from a questionnaire conducted in the Belgian context and related to agricultural countermeasures for milk following a radioactive release to the environment. The points of view of the different stakeholders (decision makers, experts, production and processing sectors) are discussed and suggestions are given for incorporating this information into a decision-aid model.

1 INTRODUCTION

The management of nuclear emergencies involves complex decisions dealing with high stakes and potential far-reaching consequences. Since the hazard is due to radioactivity, the decisions taken aim firstly at reducing the radiological health risks to people and impact on the environment. In practice, there may be important drawbacks associated with potential countermeasures: additional health risks, high economic and environmental costs, anxiety or social disruption. Since the Chernobyl accident, the importance of non-radiological aspects received increasing attention in international research projects on off-site emergency management (French et al. 1992; French et al. 1993; Hämäläinen et al. 1998; Hämäläinen et al. 2000) and, more recently, in the European Commission project STRATEGY for site restoration (Howard et al. 2005) and the FARMING project for extending the stakeholders' involvement for countermeasures in the food chain (Nisbet et al. 2005).

The heterogeneous nature of the consequences associated with potential decisions, as well as the need to deal with conflicting objectives and value systems that come into play, highlight the multi-criteria decision aid methods as a suitable framework for promoting a good decision making process (Keeney & Raiffa, 1972), by helping the decision makers to better understand the problem they face, their own values and priorities and the different perspective of the other stakeholders (Belton & Stewart, 2002). Banville et al. (1998) point out that the systemic socio-political context of the decision making process must be sufficiently integrated in the models, strategies and techniques developed for decision support. The past use of formal decision aid techniques involving concerned stakeholders includes decision conferences (French et al. 1992; French et al. 1993), decision interviews (Hämäläinen et al. 2000) and, more recently, decision analysis workshops (Geldermann et al. 2005), all supported by multi-attribute utility/value methods. The findings revealed a general agreement that the problem structuring process contributes to an improved understanding and common problem definition (Hämäläinen et al. 2000). However, a certain reluctance manifested by stakeholders, notably at the level of modelling and aggregation of preferences, indicates that the envisaged process does not fully accommodate the needs (Carter et al. 2005).

In the current paper, we make an attempt to fill this gap by stimulating the interaction with real decision makers and relevant stakeholders from the initial stages of designing a decision aid model. This is carried out by means of a questionnaire, the results of which are presented and discussed. The motivation for this study lies in the fact that the type of input information and reasoning associated to a decision-aid method should be understood and accepted not only by the decision-makers/advisers, but also by the other stakeholders involved, or else the whole analysis might be rejected.

In the next Section some basics related to agricultural countermeasures for the management of radioactively contaminated milk are introduced.

The stakeholders' process employed in this study is described in Section 3. The case study used concerns the countermeasures for management of contaminated milk in the early to intermediate phases of nuclear emergency management. Milk was chosen given its importance in the food chain, the continuous nature of its production and the rapid transfer of some radionuclides to milk.

In Section 4, basic notions of multi-criteria decision aid are presented in parallel with the results from the questionnaire, which are discussed in detail. The perspectives of the stakeholders involved are illustrated and suggestions are made on how this information can be taken into account in a decision-aid framework, in order to reach a higher degree of acceptability of the decision aid tools.

Final conclusions are drawn and directions for further use are envisaged in the last Section.

2 AGRICULTURAL COUNTERMEASURES FOR THE MANAGEMENT OF CONTAMINATED MILK

Through the dispersion of radioactivity in the environment, depending on the scale and timing of the release, large quantities of feedstuff and foodstuff may become contaminated, as a result of direct deposition and later by root uptake. Based on available measurements and model predictions, and taking into account prescribed Food Intervention Levels, various food countermeasures can be taken. The first aim is to reduce the radiological risk for people consuming plant and animal products from the contaminated zones. In the later phases, countermeasures are also needed to restore as much as possible normal life in the affected regions, preferably allowing for unrestricted activities and production of clean produce.

In case of milk, the scale of the accident (Nisbet, 2002) is determined by the quantity of produce affected, the activity concentrations and the isotopic composition of the release. The possible countermeasures (Figure 1) range from food bans and clean feeding to radionuclide-specific countermeasures like administration of feed additives for reducing the radionuclides' transfer to milk (e.g. calcium concentrates for radiostrontium and AFCF concentrates for radiocaesium), or processing of milk to dairy products with low activity retention factors (e.g. butter).

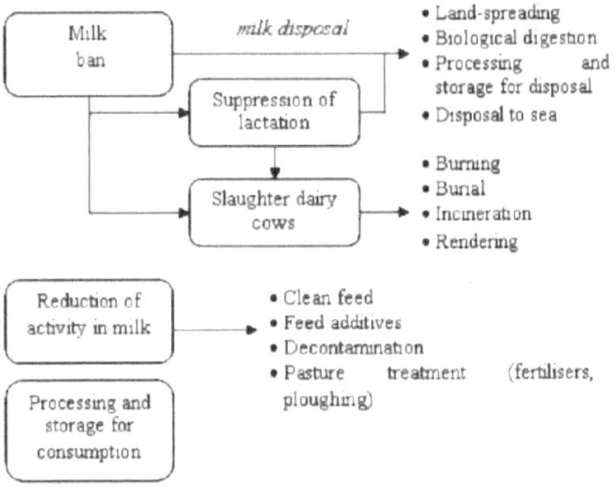

Figure 1. Countermeasures for management of contaminated milk.

3 STAKEHOLDERS' APPROACH

3.1 *Participants in the study*

Following Banville et al. (1998), the term stakeholder will hereafter designate an actor in the decision process who has a vested interest, either directly affecting the decision through the position and resources he/she controls (Savage et al. 1991), either being affected by the way the problem is formulated or solved, having something to win or to loose (Freeman, 1984), either simultaneously influencing and being influenced.

It has previously been emphasized (Vandecasteele et al. 2005) that agricultural countermeasures do not only have to be scientifically founded, and feasible from the technical and economic points of view, but also socially and psychologically acceptable by the concerned stakeholders.

Several representatives of the decision makers and their advisers have been selected to participate to the research, according to the responsibilities and competencies, as derived from the nuclear emergency plan (Moniteur Belge, 2003). In fact, the responsibility for the final decision bears modifications in the different phases of the emergency management: in the first hours to days it belongs to the Emergency Director (Minister of Interior or his delegate), after which it shifts towards the Federal Agency for Food Safety Control.

Other stakeholders, which do not take part actively in the decision process, but are affected by its consequences and whose preferences must be considered when taking a decision are the production, processing and retail sectors, the provincial crisis centres, the emergency workers, the waste management companies, various interest groups—e.g. environmental organizations or consumers interest groups- and last but not least the consumers or the general public. Drawing from the Belgian experience of the FARMING network (Vandecasteele et al. 2005), the final stakeholders group addressed in the interviews consisted of 18 persons having expertise, responsibilities or interests in the management of contaminated milk and being willing to participate in the study. This covered the following fields:

- radiation protection and emergency planning (practitioners and/or experts with consultant role in the federal crisis centre for radiological evaluations);
- radioecology (practitioners and/or experts with consultant role in the federal crisis centre for agricultural countermeasures);
- food chain control;
- decision making (Ministry of Interior);
- dairy industry;
- farmers association;
- communication;
- social science;
- public health (expert with consultant role in the federal crisis centre);
- provincial decision making and implementation of countermeasures;
- radioactive waste management.

In what regards the public opinion and public acceptability, in the early phase the best source of information is the experience gained in past (non-nuclear) crises of the food chain. Nevertheless, such issues should be re-iterated in the later phases, e.g. by making use of methods involving public participation (Bolinger & Pictet, 2003). Another stakeholder that could not be covered in the list of interviews concerns the environmental organisations. Since the problem addressed was a theoretical one, it was not considered by the organisations contacted as one of high priority. This role was nevertheless partly played by the radioecologists.

3.2 *Structure of the interviews*

As the study concerns a theoretical problem with which many of the persons interviewed were rather unfamiliar, the questionnaires were carried out on an individual basis. The individual meetings had two main parts:

- In order to facilitate the interaction, a short presentation was made, describing the general problem framework and including basic notions of radiation protection, a description of the types of agricultural countermeasures for contaminated milk, with some of their strong and weak points, and finally a presentation of the scope of formal decision aid;
- The presentation was followed by a discussion based on a questionnaire, as support in structuring the decision process for milk countermeasures for the early to intermediate phase.

The interviews were taken on an anonymous basis and do not necessarily represent an official point of view of the organisations to which the persons interviewed belonged to, but rather personal point of views.

4 RESULTS

Roy (1996) advises to construct the decision-aid process on four successive levels: i) definition of the object of decision and the type of recommendation needed; ii) consequence analysis and elaboration of criteria; iii) modelling of global preferences and operational approaches for aggregation of performances and iv) elaboration of a recommendation.

This paper tackles mostly issues related to the first two levels enumerated above, while the remaining two will be addressed in a future study. In the followings, basic notions from multi-criteria decision aid theory which were investigated in the questionnaire are introduced.

4.1 *Problem formulation*

4.1.1 *The set of potential actions*

The possible decisions (options, strategies, plans) that have to be explored during a decision process constitute the set of potential actions. This can be characterised either by an explicit list, or by a set of constraints.

The knowledge base created within the framework of the E.C. project STRATEGY (Sustainable Restoration and Long-Term Management of Contaminated Rural, Urban and Industrial Ecosystems), (Howard et al. 2005) provides the state of the art in site restoration strategies and waste disposal options. In case of contaminated milk, the optimal strategy must combine individual management options into a practicable strategy, involving both waste disposal, as well as waste minimisation (Nisbet et al. 2003).

Several types of combinations of countermeasures were proposed in the questionnaire:

- two or more countermeasures executed simultaneously (Brown et al. 2000), e.g. addition of AFCF to animal feed + processing milk to butter;
- variable timing and duration for countermeasures, e.g. for 10 days dispose of the contaminated milk, then process the milk to butter;
- fractionated treatment of production (Perny & Vanderpooten, 1998), e.g. process 50% of the contaminated milk to butter and dispose of the rest, in case there would be no market for this other 50%;
- any combination of the previous three types.

No objections were formulated against the first type of combination of countermeasures, but it was underlined that the consumer's acceptance could stand as a veto condition.

Concerning the second type of combination, most persons agreed that such an approach can certainly be useful in the preparation phase. The exceptions were persons with practical experience (dairy industry, radioactive waste management) and the communication specialist who suggested that in reality, decisions would probably be serial due to i) the situation being completely new and ii) in order to avoid an increase in uncertainty. From the producers' point of view however, it is important even in a climate of crisis, that the farmers receive a total overview of the countermeasures that will or can be taken. The possibility to switch on/off such a type of combination in a decision-aid model is probably a satisfactory compromise.

For the third type of combination, two issues were investigated, one regarding the zoning of the affected area, the other the adequacy of combining disposal of contaminated milk with processing for consumption. Except for some of the radiation-protection experts and the person from public health, the favoured option for zoning was on larger areas - with delimitation of the hot spots of activity concentration -, as opposed to making a finer differentiation on activity levels. From the dairy industry's point of view it is even important to make just one distinction: contaminated/non-contaminated milk; if separate collection is possible in Belgium, fractionate processing is problematic, since the milk has many flows and suffers many recombination processes until the final products. Regarding the consumption-disposal combinations, all persons interviewed expressed doubts that such a decision can be explained to the public, but some found it a reasonable way to deal with a limited market segment.

Another point of interest was the description of the set of potential actions: by enumeration of all possible countermeasure strategies, or based on a set of constraints processed by a decision aid model (e.g. timing or resource-related constraints). The latter was seen as interesting, but the following remarks were made:

- meaningful input from the experts should be used;
- care should be taken to the danger of relying exclusively on the model;
- the model must be tried out and sufficiently robust.

Using a computer model to generate potential actions may lead to an unreasonably large number of actions (Papamichail & French, 2000). Screening techniques could then be used to reduce this number (Walker, 1986, e.g. i) by identifying and retaining only the non-dominated actions, possibly by making use of simplified model that uses a smaller number of criteria, or ii) by setting bounds on the space of possible alternatives. Keeney (1980) suggests that criteria exhibiting large uncertainties should not be used for screening, while Anandalingam and Olsson (1989) propose a technique allowing to identify and eliminate the alternatives with unacceptable risk.

A recurrent issue in nuclear emergency management concerns the operational intervention levels (ICRP 63, 1993), in our case activity concentration levels in milk, starting from which countermeasures should be applied. The intervention principles formulated by the International Commission on Radiological Protection (ICRP 60, 1991) state that: i) the reduction radiation detriment should justify the harm and the costs of the intervention (including social costs) and ii) the benefit from the reduction in radiation detriment should be optimized with respect to the form, scale and duration of the intervention. The opinions in the questionnaire were rather convergent for activity levels above the legal intervention limits: countermeasures need to be taken, whenever operationally possible. For activity levels below the intervention levels many persons (including the one from public health) expressed the opinion that, depending on the situation, even if on a purely scientific basis it is not necessary to take actions, the public pressure may lead to this.

4.1.2 *The type of decision framework*

Depending on the way that decision aid is employed, various –possibly non-exclusive- types of decision frameworks be identified (Roy, 1996):

- choice framework, i.e. determination of the best option or the set of best options possible;
- sorting framework, i.e. classification of possible options based on given profiles;

- ranking framework, i.e. preference-based ordering of all possible options;
- description framework, i.e. methodological description and formalization of the set of possible options in a systematic manner.

The above types of decision frameworks can be further detailed to fit particular situations (Bana e Costa, 1993; Belton & Stewart, 2002). In the questionnaire, a fifth type of framework, the design one, was added to the four types listed above, i.e. identification/ construction of new alternatives that meet certain goals or aspirations (Belton & Stewart, 2002), in order to explicitly underline the creative potential.

The questionnaire showed that opinions were spread on the five types of decision frameworks which were subject to analysis, with pro's and contra's recorded for each of them. The ranking and the design frameworks counted favourable advice from the largest number of different categories of stakeholders. Ranking was supported, among others, by the decision makers at federal and provincial levels and the person from the farmers' union, provided that, as some stated it, the rationality behind it can be explained to the public. However, ranking assumes comparability between any two actions and some persons expressed doubts that this would be possible, due to the inherent, large, uncertainties. The design framework was supported, among others by the decision maker at federal level and the public health representative who underlined the need for model flexibility in what regards the sensitivity analysis. The description framework was favoured by fewer persons, but notably by the one from the food chain control.

The choice framework was notably supported by the person from the dairy industry and the provincial decision-maker, while the opponents (e.g. the person from food chain control) found this framework too limitative. In what regards the sorting framework, this was mainly favoured by radiation-protection experts, but the farmers' representative suggested that a classification of potential actions would probably not be understood by the farmers.

Some persons suggested that depending on the particular situation, one or another approach could become the most adequate.

4.2 *Evaluation criteria*

The consequences, effects, attributes of potential actions allow the actors involved in the decision process to build, justify and modify their preferences. In practice, these consequences are often vaguely known and weakly differentiated, therefore needing to be structured in a consistent family of evaluation criteria. The bottom-up approach starts by identifying the elementary consequences (Roy, 1996), i.e. those which are understandable, measurable and operational.

The elementary consequences identified are then used as a basis for building the family of evaluation criteria. This should satisfy at least some basic consistency properties (Roy, 1996):

- Exhaustiveness, i.e. all fundamental aspects of the consequences are taken into account (the family is complete);
- Cohesiveness, i.e. the partial preferences modelled by each criterion are consistent with the global preferences (monotonicity of preferences);
- Non-redundancy, i.e. the chosen family is minimal with respect to the previous two properties (no double-counting).

Furthermore, concision is recommended (Keeney, 1992), i.e. the number of criteria should be as restricted as possible. In practice, depending on their semantics, several consequences have to be grouped together provided that they correspond to non-contradictory points of view (Roy & Bouyssou, 1993).

Based on the existing literature in the domain (STRATEY fiches, Howard et al. 2005) and the outcome of the FARMING project (Vandecasteele et al. 2005), in which the rural management strategies were screened by several stakeholders, a list of primary points of view for our case study was created and used as a starting point for the discussion (Figure 2).

The stakeholders interviewed were asked to comment on these points of view and to amend the list, if found necessary. The discussions revealed a number of 34 potential consequences of potential actions, which could be interpreted as being exhaustive for the case study considered.

Some of these consequences were further discarded for being non-operational (i.e. not possible to gather sufficient information with reasonable time and effort), such as the long term economic

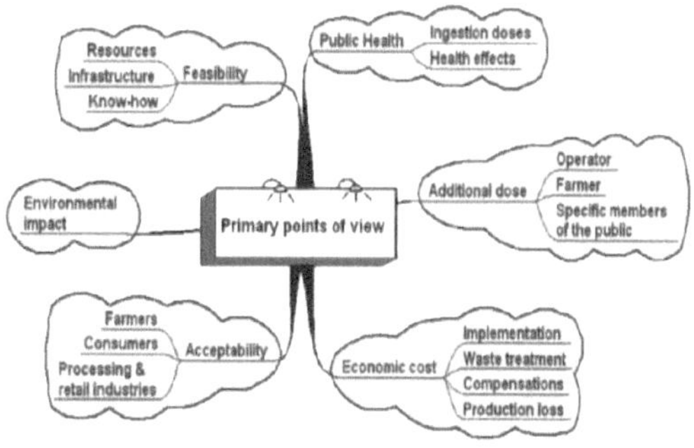

Figure 2. Primary points of view (initial list).

effects on the economy. Others turned out to be better suited to be modelled as screening thresholds for countermeasures instead of evaluation criteria (e.g. individual ingestion dose to the critical population group, the time needed for the countermeasure's implementation, the geographical area which is contaminated, the possibility to substitute the contaminated production with clean one, the effective collective or individual dose from ingestion after the application of the countermeasure).

Finally, from the remaining set of consequences, a number of 10 evaluation criteria were synthesized, as described in the following subsections. For each criterion an evaluation method is proposed: direct evaluation on a certain type of scale (e.g. ordinal, interval, etc) or relative evaluation, by pair-wise comparison between actions. Interdependencies were checked as well and, when appropriate, alternative attributes were proposed for the same criterion.

4.2.1 *Impact on public health*

More than one attribute was highlighted in connection with this criterion:

- Avertable effective collective dose from ingestion of dairy products;
- Avertable collective dose to thyroid from ingestion of dairy products;
- Stochastic health effects (e.g. lethal cancers, cancer incidences);
- Thyroid cancers (lethal or total incidences);
- Maximal individual effective dose from ingestion for the critical population group.

The choice of one attribute from the above should be made, depending on the situation. Some persons underlined that the ingestion dose is a more objective measure than the number of health effects due to the uncertainties related to the dose detriment conversion factors, but more difficult to be understood by non-experts.

One way to calculate the effective collective dose from ingestion when a strategy S is applied for T days after deposition is (see also Müller et al. 2003):

$$D_{ing}^{col}(S) = \sum_{k}\sum_{r} DCF_{ing}^{r} \sum_{t=0}^{T} PR(k,t|S) \cdot A^{r}(k,t|S)$$

where

$D_{ing}^{col}(S)$ = collective dose from ingestion under strategy S, [person·Sv];
k = index for the milk products;
r = index for the radionuclides released;
DCF_{ing}^{r} = ingestion dose factor for effective dose, for the chosen population group, for radionuclide r, [Sv/Bq];
$PR(k,t|S)$ = quantity of product k produced at time t, under strategy S, [kg];
$A^{r}(k,t|S)$ = activity concentration of radionuclide r in product k produced at time t, when good for consumption, under strategy $S[Bq/kg]$;

In case production data are not available, a summation can be made only with regards to milk.

The avertable dose in case of implementation of strategy S can thus be calculated as $D_{ing}^{col}(\varnothing) - D_{ing}^{col}(S)$, where the symbol $\varnothing$ stands for no countermeasure taken.

The use of collective dose has been criticized over the past years, e.g. when very low individual doses are summed over large periods of time, or when it hides an inequitable dose distribution over individuals. In our case however, the time period addressed regards the early to intermediate phases and the assumption is made that all products would be consumed somewhere, while the ultimate consumer is not known (see also ICRP 63, 1993, paragraph 90). On the other hand, the individual dose for the critical population group is assumed to act as a screening threshold. The use of the collective dose was also supported by the person from public health.

Several sources of uncertainty affect the calculation of health effects, starting from dose calculations (e.g. prediction of radionuclides' transfer to milk), to estimation of health risks based on ingestion doses (ICRP 60). While decision-makers would often settle for the experts' best estimate of ingestion doses, many of the persons interviewed agreed that the use of variable preference and indifference thresholds (Vincke, 1992), can be an adequate way to deal with uncertainty when comparing potential actions (e.g. values of the dose differing with less than 10% can be considered indifferent). Another proposal was to represent uncertain values by intervals of values.

We note the remark that in order to preserve the exhaustiveness of the chosen family of criteria, the calculated doses should be calculated for a sufficiently long time period, in order account for longer term benefits of a potential countermeasure (e.g., total elimination of contaminated milk by slaughter of dairy cows).

4.2.2 *Additional dose to workers*

This refers to the incremental dose to workers implementing the countermeasure strategy. Two attributes were proposed in connection with this criterion:

- Maximal worker's dose;
- Collective dose to workers, calculated based on the dose distribution.

In this case, it was difficult to envisage preference thresholds and a majority of people suggested to use this criterion only for screening, i.e. the legal limits for the yearly individual dose to workers should be complied with. In order to build a criterion with a point state indicator, one of the attributes above can be selected.

Wilkins et al. (2001), Marchant et al. (2002), Mercer et al. (2002), Nisbet et al. (2003) propose methodologies for estimating the worker's dose associated with various countermeasures for the management of contaminated milk.

4.2.3 *Feasibility of the food countermeasures implementation*

Feasibility comprises many aspects: qualitative, e.g. know-how, and quantitative, e.g. resources and infrastructure available. Some aspects have clearly the meaning of a constraint, while others indicate the degree of effort needed to implement a countermeasure. The majority of persons interviewed favoured a qualitative evaluation of feasibility as a comprehensive, but not very clearly stated concept. To make it operational, one can take a closer look at specific aspects for the strategies considered as "globally feasible" after a preliminary screening: i) technical feasibility of the food countermeasure strategy (excluding waste) and ii) feasibility of waste treatment. In the following we detail the first issue, while the latter will be treated in the next sub-section.

The availability of a resource s (e.g. man-power, additives, clean feedstuff, etc) for a strategy S can be expressed in percentages as:

$$\frac{\min(AN(s), N(s|S))}{N(s|S)} \cdot 100,$$

where

$N(s|S)$ = number of units of resource s needed for strategy S;

$AN(s)$ = number of available (or easy to acquire) units of resource s.

The aggregation of these values can be done for example using a "min" operator. Care has to be taken however from the exhaustiveness point of view, since, e.g. 50% availability of resources doesn't give any information about the capability to substitute for the remaining 50%. A solution can be to derive an overall qualitative score (e.g. ordinal scale) based on both qualitative and quantitative information.

4.2.4 *Feasibility of waste treatment*

This criterion refers mainly to the volume of waste, its activity levels (which fraction can be treated as non-radioactive waste?) and the availability of temporary storage capacities.

For the evaluation of actions with respect to this criterion, an assessment must be made in collaboration with the waste companies. A qualitative scale (ordinal scale) is preferred by the radioactive waste company. For exercise purposes, either a generic price or the total volume of waste can be used.

4.2.5 *Acceptability for farmers*

This criterion comprises the following aspects: direct costs for farmer (excluding compensations), disruption from normal agricultural practices, psychological degree of acceptance and agricultural impact.

In general, a global cost is calculated, in which the cost for the farmer may be marginal, while being on the other hand an important factor influencing the farmer's acceptability. Therefore it appears more adequate to separate the costs on categories of stakeholders.

An ordinal scale is better suited for this criterion and the best way to evaluate this criterion is to ask the farmers' representatives to give their opinion and either to rank accordingly the different strategies or to give a qualitative appreciation (preferred by the persons from the farmers' union). However, according to the results of the discussions the following preference order could be used for exercise purposes: {additives, processing for consumption, clean feedstuff} preferred to {landspreading, processing for disposal, biological digestion of milk} preferred to {suppression of lactation, slaughter of dairy cows}. The person from the farmer's union mentioned that the most important thing for the farmers is to have a guarantee that the production can continue, while the additional costs should be kept as low as possible.

4.2.6 *Acceptability for the dairy industry*

The contribution in assessing this criterion comes from the economic cost for the dairy industry, either from additional work, including cost of products that cannot be sold, or from decontamination of installations, etc. As above, the evaluation of potential actions with respect to this criterion should come from the dairy industry representatives themselves. The favoured evaluation methods were the use of qualitative labels and the ranking of potential actions, respectively.

It was underlined by the person from the dairy industry that the processing of contaminated milk for subsequent human consumption must be ensured for the case when consumers would not buy the final products (and therefore the retail industry would not accept them). More radical measures are preferred in general (e.g. disposal of all contaminated milk), provided they can ensure a faster return to normal life.

4.2.7 *Public acceptability*

In the early to intermediate phases there is a large uncertainty associated with this criterion. If results from public enquiries or consumption statistics are not available, an estimation based on the historical data of past crises in the food chain can be made.

Vandecasteele et al. (2005) point out that in Belgium the recent crises which affected the food chain (BSE, dioxin crisis, foot and mouth disease, fowl plague) increased the sensitivity and intolerance of consumers to any perceived degradation of food quality, and leading many times to very conservative decisions. They also suggest that there will probably be a strong pressure to discard and dispose of any suspicious food product.

A likely ranking of countermeasures could accordingly be assumed to be e.g. {food ban} preferred to {clean feedstuff} preferred to {feed additives} preferred to {processing, decontamination}. Alternatively, a qualitative evaluation on an ordinal scale can be used, if sufficient data are available to validate such an evaluation.

4.2.8 *Impact on the environment*

This criterion comprises two aspects: the severity of environmental damage and time needed for remediation. While there are norms for specific agricultural activities (e.g. land-spreading of milk) depending on the season, soil type, etc, the most suitable scale appears to be a qualitative one. A number of persons suggested that pair-wise comparison between alternatives –and eventually ranking- would be the most adequate ways of modelling the preferences with respect to this criterion.

4.2.9 *Economic cost*

This refers to the financial direct cost for the state, including costs of implementation (resources, labour, compensation paid to farmers or dairy industry) and the cost of waste disposal.

In the early phase it was emphasized as having lower importance than other criteria. Preference thresholds (variable) can also be envisaged for this criterion.

4.2.10 *Reversibility*

Reversibility was mentioned by only two persons, both with radiation protection background. It refers to the possibility to correct an inadequate decision in the light of new evidence. This criterion is recognized (Tyler, 1988) as being one of procedural fairness.

Clearly some countermeasures are irreversible (slaughter of dairy cows), other reversible, but with certain drawbacks (processing), the rest being easily reversible (food ban, addition of food additives, clean feeding).

4.2.11 *Other comments on evaluation criteria*

An additional criterion which was not included in the proposed list encompasses the socio-psychological impact on the public. This only becomes operational when sufficient data become available from public enquiries, medical statistics, etc. This criterion could be useful for the re-evaluation of long term countermeasures.

Two other issues were also analysed and discarded as candidate criteria due to violation of the cohesiveness property:

- Uncertainty of outcome of a countermeasure. This was mentioned by the person from the food safety control in connection with countermeasures like processing of milk to butter or cheese: even if the public accepts the countermeasure, it doesn't mean that it will actually buy these products. Therefore e.g. disposal of all contaminated milk is preferred to processing by the representative of the food chain control. However, the uncertainty of outcome may be taken into consideration by translating it into uncertainty of economic cost. One idea can be then to look for robust solutions (Rosenhead, 1980), which perform reasonably well even in the worse case scenario.
- The possibility to control the correct implementation of the decision. It can be regarded also as a qualitative aspect of feasibility, but also as uncertainty related to the estimation of doses and health effects due to the follow up of a countermeasure. This aspect can be translated into uncertainty in the health impact of the countermeasure.

4.3 *Specification of criteria importance*

In order to derive global preferences in the context of conflicting evaluation criteria, priorities have to be set, i.e. the relative importance of criteria has to be defined. Roy and Mousseau (1996) show that this notion can be interpreted differently, depending on the specific parameters and the type of aggregation used to model the global preferences.

In the questionnaire, four types of commonly used approaches were considered:

- substitution rates (tradeoffs) between criteria, i.e. scaling constants; these are used in fully compensatory methods such as the multi-attribute utility/value methods (Keeney & Raiffa, 1993) or some interactive methods.
- criteria weights as intrinsic importance coefficients; some methods using such parameters are ELECTRE (Roy, 1991) and PROMETHEE (Brans et al. 1984) use. In the ELECTRE type methods, these importance weights are combined with veto thresholds.
- complete preorder on the set of criteria (complete and transitive binary relation), i.e. a complete ranking with possible ties, like in e.g. the ORESTE (Roubens, 1981) and QUALIFLEX (Janssen et al. 1990) methods and the lexicographic aggregation.

- a preorder relation (reflexive and transitive binary relation), not necessarily complete, defined on subfamilies of criteria. Such an approach has been used, e.g. by Vincke (1999) to construct a robust ELECTRE-type.

The adequacy preference of the four methods was assessed by each of the person interviewed on a qualitative scale: strong opposition- weak opposition – neutrality – weak approval – strong approval. The results of the questionnaire revealed a preference directed towards the more qualitative approaches (via a relation defined on criteria or families of criteria), which met only two strong oppositions (public health and social science, for partial preorder on subsets of criteria) and, respectively, one weak opposition (radioecology, for complete preorder on criteria). It was noticed that the concept of weights as intrinsic importance coefficients proved harder to be understood, but still encountered a smaller number of strong opponents than the concept of weights associated with substitution rates. The main argument against the latter was based on ethical motivations, e.g. the difficulty to argue for a value trade-off between the number of expected health effects and the economic cost. On the other hand, the person from the public health drew attention that in real life, the nuclear problem could be just one item added to a big list of problems of other types and therefore priorities would have to be set for allocating resources. In line with this, some persons backed up the opinion that trade-off judgements could still be established for certain subgroups of criteria.

5 CONCLUSION

The results of the proposed questionnaire highlighted key elements of a decision aid framework.

Different definitions for the set of potential actions were studied, corresponding to certain types of combinations of individual countermeasures. The results show that for preparedness of emergency response it is worthwhile considering combinations of consecutive actions. In a real case, the focus is likely to be set for practical reasons on a single time interval.

The specification of the decision framework points mainly to ranking all possible actions, but a considerable amount of interest has also been expressed connected to the sorting and design frameworks, with an emphasis on the need for model flexibility.

A family of 10 criteria was synthesized, for which basic consistency properties, as well as different methods of assessing the relative importance of criteria were analysed.

The study has of course its limitations: the views of the persons interviewed are not official, but personal views on the problem and, as responsibilities change, the opinions may change as well. Despite inherent difficulties due to the multi-disciplinary character of the problem studied and the different background of the persons interviewed, the process was positively appreciated. The divergences observed in the various stakeholders' viewpoints - even inside the same category of stakeholders, e.g. radioecologists- substantiate the need for stakeholders' involvement from the first stages of multi-criteria decision-aid model design, in order to ensure a higher degree of acceptability of the methodologies and tools employed.

Further study needs to address the most suitable method for deriving the global preferences on actions and the development of a final recommendation, given the characteristics of the problem context.

ACKNOWLEDGEMENTS

The authors would like to thank all persons who willingly contributed with their knowledge and experience to our study, by expressing their point of view on the decision-making process and its outcome.

REFERENCES

Anandalingam, G. & Olsson, C.E. 1989. A multi-stage multi-attribute decision model for project selection. *European Journal of Operational Research* 43: 271–283.

Bana e Costa, C. 1993. Les problèmathiques dans le cadre d'aide à la décision. *Document du Lamsade no.74*, Université Paris Dauphine.

Banville, C., Landry, M., Martel, J-M. & Boulaire, C. 1998. A Stakeholder Approach to MCDA. *Systems Research* 15(1): 15–32.

Belton, V. & Stewart, T.J. 2002. *Multiple Criteria Decision Analysis: An integrated approach*. Dordrecht: Kluwer Academic Publishers.

Bolinger, D. & Pictet, J. 2003. Potential use of e-Democracy in MCDA Processes. Analysis on a Basis of a Swiss Case. *Journal of Multi-Criteria Decision Analysis* 12(2–3): 65–77.

Brans, J.P., Mareschal, B. & Vincke, P. 1984. PROMETHEE: a new family of outranking methods in multicriteria analysis. In J.P. Brans (ed.), *Operational Research, IFORS '84.*: 477–490. Amsterdam: North Holland.

Brown, J., Mansfield, P.A. & Smith, J.G. 2000. Modelling approach in the terrestrial late countermeasures module, LCMT within RODOS 4.0, RODOS Report *RODOS(WG3)-TN(99)-43* www.fzk.rodos.de.

Carter, E. & French, S. 2005. Nuclear Emergency Management in Europe: a Review of Approaches to Decision Making, *Proc. 2nd Int. Conf. on Information Systems for Crisis Response and Management: 247–259*, 18–20 April, Brussels, Belgium, ISBN 9076971099.

Freeman, R.E. 1984. *Strategic Management: A Stakeholder Approach*. Boston, MA: Pitman.

French, S., Kelly, G.N. & Morrey, M. 1992. Decision conferencing and the International Chernobyl Project. *Journal of Radiological Protection* 12: 17–28.

French, S., Walmod-Larsen, O. & Sinkko, K. 1993. Decision Conferencing on Countermeasures after a Large Nuclear Accident, *Risoe-R-676(EN) Report*. Roskilde, Denmark: Risoe National Laboratories.

Geldermann, J., Treitz, M., Bertsch, V. & Rentz, O. 2005. *Moderated Decision Support and Countermeasure Planning for Off-site Emergency Management*. In R. Loulou, J.-P. Waaub and G. Zaccour (ed.), *Energy and Environment: Modelling and Analysis*. Kluwer Academic Publishers.

Hämäläinen, R.P., Sinkko, K., Lindstedt, M.R.K., Amman, M. & Salo, A. 1998. RODOS and decision conferencing on early phase protective actions in Finland. *STUK-A 159 Report*, STUK - Radiation and Nuclear Safety Authority, Helsinki, Finland, ISBN 951-712-238-7.

Hämäläinen, R.P., Sinkko, K., Lindstedt, M.R.K., Amman, M. & Salo, A. 2000. Decision analysis interviews on protective actions in Finland supported by the RODOS system, *STUK-A 173 Report*, STUK - Radiation and Nuclear Safety Authority: Helsinki, Finland. ISBN 951-712-361-2.

Howard, B.J., Beresford, N.A., Nisbet, A., Cox, G., Oughton, D.H., Hunt, J., Alvarez, B., Andersson, K.G., Liland, A. & Voigt, G. 2005. The STRATEGY project: decision tools to aid sustainable restoration and long-term management of contaminated agricultural ecosystems. *Journal of Environmental Radioactivity* 83: 75–295.

ICRP Publication 60, 1991. Recommendations of the International Commission on Radiological Protection, 60, *Annals of the ICRP* 21(1–3).

ICRP Publication, 1993. Principles for Intervention for Protection of the Public in a Radiological Emergency, 63, *Annals of the ICRP* 22(4).

Janssen, R., Nijkamp, P. & Rietvelt, P. 1990. Qualitative multicriteria methods in the Netherlands, *Readings MCDA*, Ed. Carlos Bana e Costa. Berlin: Springer, Verlag.

Keeney, R.L. & Raiffa, H.A. 1972. Critique of formal analysis in public sector decision making. In A.W. Drake, R.L. Keeney and P.M. Morse (ed.), *Analysis of Public Systems:* 64–75. Cambridge, MA: MIT Press.

Keeney, R.L. 1980. *Siting Energy Facilities*. NY: Academic Press.

Keeney, R.L. 1992. *Value-focused Thinking: A path to creative decision-making*. Cambridge, MA: Harvard University Press.

Keeney, R.L. & Raiffa, H. 1993. *Decision with multiple objectives: preferences and value tradeoffs*. NY: John Wiley & Sons.

Marchant, J.K. & Nisbet, A.F. 2002. Management options for Food Production Systems Affected by a Nuclear Accident: Task 6: Landspreading as a Waste Disposal Option for Contaminated Milk, *NRPB-W11*, National Radiological Protection Board, Chilton, Didcot, U.K.

Mercer, J.A., Nisbet A.F. & Wilkins, B.T. 2002. Management options for Food Production Systems Affected by a Nuclear Accident: Task 4: Emergency monitoring and processing of milk, *NRPB-W15*, National Radiological Protection Board, Chilton, Didcot, U.K.

Moniteur Belge, 2003. *Plan d'Urgence Nucléaire et Radiologique pour le Territoire Belge*, 20.11.2003.

Müller, H., Gering, F. & Pröhl, G. 2003. Model description of the Terrestrial Food Chain and Dose Module in RODOS PV6.0, *RODOS(RA3)-TN(03)06*, www.rodos.fzk.de.

Nisbet, A.F. 2002. A strategy for management of milk contaminated as a result of a nuclear accident, *NRPB-W5*, National Radiological Protection Board, Chilton, Didcot, U.K.

Nisbet, A.F., Marchant, J.K., Woodman, R.F.M., Wilkins, B.T. & Mercer, J.A. 2003. Management options for food production systems affected by a nuclear accident. Task 7: Biological treatment of contaminated milk, *NRPB-W38*, National Radiological Protection Board, Chilton, Didcot, U.K.

Nisbet, A.F., Mercer, J.A., Rantavaara, J.A., Hanninen, R., Vandecasteele, C., Carlé, B., Hardeman, F., Ioannides, K.G., Papachristodoulou, C., Tzialla, C., Ollagnon, H., Jullien, T. & Pupin, V. 2005. Achievements, difficulties and future challenges for the FARMING network. *Journal of Environmental Radioactivity* 83: 263–274.

Papamichail, K.N. & French, S. 2000. Decision support in nuclear emergencies. *Journal of Hazardous Materials* 71: 321–342.

Perny, P. & Vanderpooten, D. 1998. An interactive multiobjective procedure for selecting medium-term countermeasures after nuclear accidents. *Journal of Multi-Criteria Decision Analysis* 7: 48–60.

Roubens, M. 1981. Preference relations on actions and criteria in multicriteria decision making. *European Journal of Operational Research* 10: 51–55.

Rosenhead, J. 1980. Planning under Uncertainty: II A Methodology for Robustness Analysis. *Journal of the Operatational Research Society* 31(4): 331–341.

Roy, B. 1991. The outranking approach and the foundations of ELECTRE methods. *Theory Decis.* 31: 49–73.

Roy, B. 1996. *Multicriteria Methodology for Decision Aiding*. Dordrecht: Kluwer Academic Publishers.

Roy, B. & Mousseau, V. 1996. A theoretical framework for analysing the notion of relative importance of criteria. *Journal of Multi-Criteria Decisions Analysis* 5: 145–159.

Savage, G.T., Nix, T.W., Whithead, C.J. & Blair, J.D. 1991. Strategies for assessing and managing organizational stakeholders. *Academy of Management Executive* 5: 61–75.

Tyler, T. 1988. What is procedural justice: Criteria use by citizens to assess the fairness of legal procedures. *Law and Society Review* 22: 103–135.

Vandecasteele, C., Hardeman, F., Pauwels, O., Bernaerts, M., Carlé, B. & Sombré, L. 2005. Attitude of a group of Belgian stakeholders towards proposed agricultural countermeasures after a radioactive contamination: synthesis of the discussions within the Belgian EC-FARMING group. *Journal of Environmental Radioactivity* 83: 319–332.

Vincke, P. 1999. Robust and neutral methods for aggregating preferences into an outranking relation. *European Journal of Operational Research* 112: 405–412.

Vincke, P. 1992. *Multicriteria Decision Aid*. Chichester: J. Wiley and Sons.

Walker, W. 1986. The Use of Screening in Policy Analysis. *Management* Science 32(4): 389–402.

Wilkins, B.T., Woodman, R.F.M., Nisbet A.F. & Mansfield, P.A. 2001. Management options for Food Production Systems Affected by a Nuclear Accident(5): Disposal of waste milk to sea, *NRPB*-R323, National Radiological Protection Board, Chilton, Didcot, U.K.

Risks Challenging Publics, Scientists and Governments – Menoni (ed)
 ISBN 978-0-415-58072-4

The Potenza Province emergency plan: The seismic vulnerability map of buildings for the construction of earthquake damage scenarios

A. Attolico
Head of Civil Protection Office, Province of Potenza, Potenza, Italy

A. Bixio & S. Pacifico
Consultant of Province of Potenza, Potenza, Italy

ABSTRACT: A large-scale evaluation of building vulnerability requires the knowledge of the territorial distribution of the structural typologies. In the 2004 edition of the Province of Potenza Civil Protection Plan, vulnerability data derived from ISTAT 1991 census were used, coming from a study published in 2001 by the Italian National Seismic Survey. In 2005 updating of the Plan, a deeper analysis of vulnerability distribution on the provincial territory has been carried out, starting from recognition of available vulnerability studies at the local or national scale. To complete the collected data, a specific investigation survey has been performed in many seismically exposed municipalities of the province, by means of a simple and quick method, named "*Guided Interview Protocol*". The vulnerability data, together with the population exposure, the geographical collocation, the damage probability matrices and the seismic attenuation laws, represent a unique database, from which it is possible to create damage scenarios, consequent to seismic event that could happen in future, by means of an apposite software procedure.

1 BUILDINGS VULNERABILITY STUDIES OVER THE PROVINCIAL TERRITORY

The seismic risk can be described as a function of the following three factors: hazard, vulnerability and exposure:

- the hazard is the probability that an earthquake with a specific intensity will occur in a given area in a determined return time;
- the vulnerability is the propensity of a building or a facility to suffer damages because of its intrinsic physical and functional characteristics, when an earthquake occurs;
- the exposure represents the population, goods and activities which will be directly or indirectly struck by the earthquake.

To evaluate the seismic vulnerability on a large scale, such as for a province or a big city, the only way to run is to employ simple descriptive instruments by considering the effects of historical earthquakes and statistical processing methods.

At this scale, testing the safety levels of buildings through structural computation models is not feasible, both for the difficulty of formulating simple and reliable models, and for the great amount of needed fieldwork. Hence, available methods must be based on few empirical parameters and their validation requires the observation of the damages caused by real earthquakes.

The vulnerability level of a building can be estimated through quick analyses, using information about design data, type of structures, typological and formal characteristics, mechanical properties of employed material, state of preservation and presence of critical elements or systems for the right seismic behaviour.

Assigning the entire buildings stock to vulnerability classes allows to put together buildings with a similar behaviour under the action of an earthquake; each vulnerability class is associated to a relationship between earthquake intensity and experienced damages.

At present, four classes (named A, B, C and D) are used in vulnerability studies, where "A" represents the highest vulnerability and "D" the lowest one.

In the study (A. Lucantoni et al., 2001) used as reference for the 2004 edition of the Provincial Civil Protection Plan, the method for assessment of building vulnerability, based on a statistical – typological approach, assigns buildings to three decreasing vulnerability classes A, B, C, fixed in MSK macroseismic scale.

The vulnerability data based on ISTAT (Italian Statistics Institute) 1991 census, used in that

Figure 1. Area of study: Province of Potenza (South Italy) with an extension of about 6,500 km², a population of about 400,000 inhabitants and an average population density of 60 in/km² (ISTAT 2001).

study, even though they are the only ones available at the moment for the whole national territory, have an inadequate reliability. So, in the 2004 edition of the Provincial Civil Protection Plan was already established to start a closer study on seismic vulnerability, in co-operation with DiSGG (Department of Structures) of Basilicata University (Masi et al. 2004). The study includes 28 municipalities having the following two requirements: being assigned to high seismicity zone (zone 1) by the Italian Seismic Rules and having experienced intensity equal or greater than VI MCS degree during the Irpinia – Basilicata earthquake, occurred on November 23rd 1980.

The vulnerability data concerning those 28 municipalities were obtained in the following way: in 11 of them, a thorough survey was made after the Potenza earthquakes of 1990 and 1991, using damage survey cards developed by GNDT (National Group for Defence from Earthquakes); in the other 17, the data were surveyed through the "*Guided Interview Protocol*".

In details, the *Guided Interview Protocol* (GIP) is a questionnaire submitted to a municipal technician or a local expert; through it, useful information to estimate the vulnerability distribution on the territory of a town can be obtained. By means of the *Guided Interview Protocol*, the "*built environment*" of investigated towns are divided into one or more subareas, in which buildings have homogeneous characteristics (*homogeneous compartments*). On this basis, it is possible to estimate the vulnerability distribution in the subarea and, consequently, in the whole territory (Figure 2).

In 2005 a Plan update was decided to perform a deeper analysis of the seismic vulnerability of buildings in the Province, so an analysis of all the available local or national applied studies and surveys on vulnerability in the provincial territory was started.

Besides the ISTAT derived data processed from National Seismic Survey and the study made in 2004 by DiSGG, the performed research made available the following vulnerability studies concerning the provincial territory:

- Evaluation of the vulnerability in 9 municipalities of the Upper Agri Valley;
- Microzoning of the regional territory affected by the September 9th 1998 earthquake, including hazard and vulnerability studies in 27 municipalities

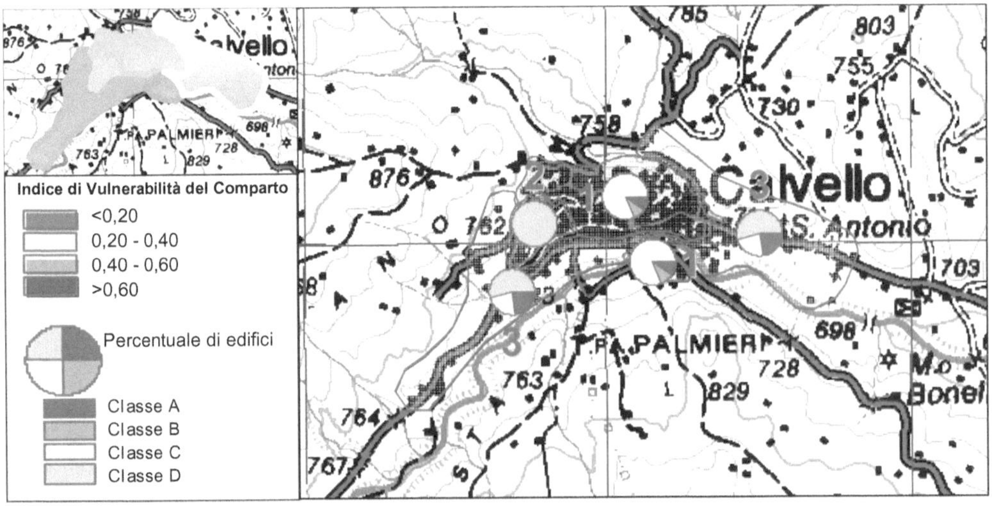

Figure 2. Estimate of the vulnerability in the homogeneous compartments of a municipality.

of "Pollino-Lagonegrese" area, in the southern part of the Province (Dolce et al., 2001);

- CNR – GNDT, Buildings vulnerability: a nationwide assessment of seismic vulnerability of ordinary buildings (edited by A. Bernardini, 2000), including three municipalities in the Province of Potenza.

To integrate the available data, a on-the-field vulnerability survey was performed in other 31 municipalities in the Province of Potenza, by means of the *Guided Interview Protocol.*

In this phase, for the residual 20 towns it was decided to use the statistical-derived data of National Seismic Survey (ISTAT 1991), updating them with the 2001 population data.

The map in Figure 3 shows the various vulnerability data sources.

To have a synthetic and effective indicator of the average vulnerability in the studied municipalities, starting from the four vulnerability classes (A, B, C, D) a *Vulnerability Index* **Iv** has been defined, having the following expression:

$$Iv = \sum_{1}^{n} \frac{V_i f_i}{n} \qquad \text{with} \quad 0 \leq Ie \leq 1 \qquad (1)$$

where:

- V_i is the *Vulnerability Level*, ranging from 0 to 3; it is defined as the numerical expression of the vulnerability class, according to the Table 1:
- f_i represents the relative frequency of each vulnerability level.

The sum is extended to n = 3 vulnerability levels >0. Therefore, Iv = 0 indicates that all the building in the considered area are classified in D class, while Iv = 1 means that all the buildings are assigned to A class.

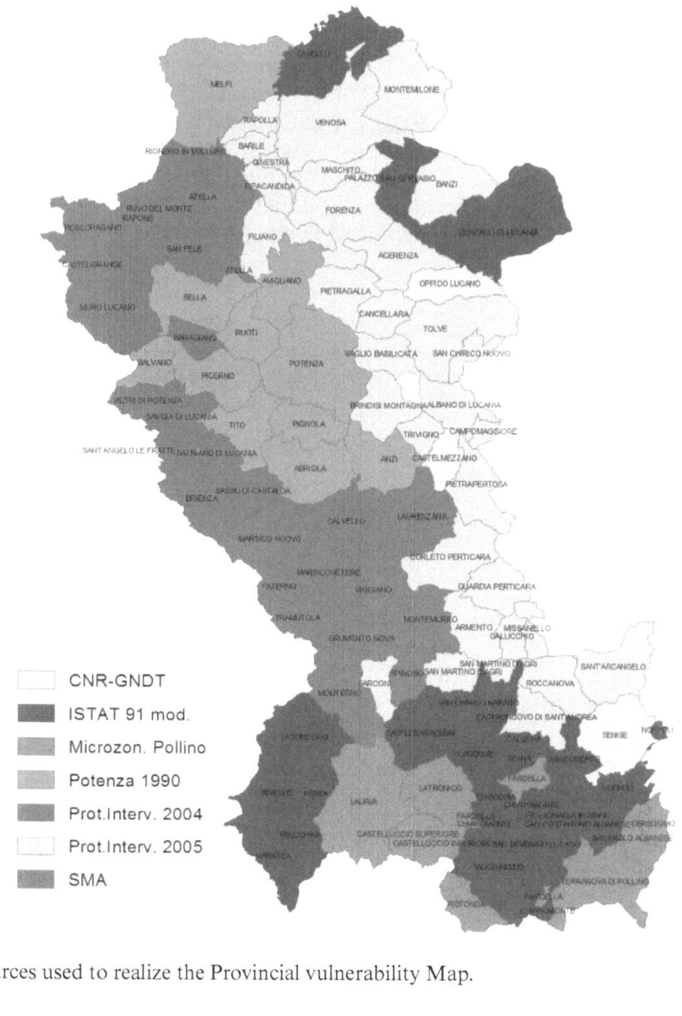

Figure 3. Data sources used to realize the Provincial vulnerability Map.

Table 1. Vulnerability class.

Vulnerability class	A	B	C	D
Vulnerability level V_i	3	2	1	0

Note that $Iv = 0$ doesn't mean zero-vulnerability: for the buildings assigned to "D" vulnerability class, corresponding to the antiseismic ones, some damages under the action of strong earthquakes are acceptable, but not collapse. So, $Iv = 0$ indicates the minimum, non-zero, vulnerability for a building.

In the same way, an *Exposure Index* **Ie** has been defined, to have an indication of the average exposure of the population. Its expression is similar to that of the vulnerability index:

$$Ie = \sum_{1}^{n} \frac{V_i fp_i}{n} \qquad \text{with} \quad 0 \leq Ie \leq 1 \qquad (2)$$

where:

- fp_i represents the relative frequency of the people living at each vulnerability level, as seen above;
- V_i and n are defined as above.

As in the previous case, $Ie = 0$ does not mean zero exposure for the population, but it indicates its minimum, non-zero, level.

Starting from the implementation of literature and surveyed data, maps with the vulnerability and exposure distribution on the provincial territory have been produced (Figure 4). The vulnerability has been represented in terms of its distribution in the four classes A, B, C, and D, overlaid to the representation of the vulnerability index; in a similar way, the exposure is represented through both the exposure index and the distribution of population in the four vulnerability classes.

The figure clearly shows that the towns with lower vulnerability and exposure indices are mostly localized in the North-West area of the Province: they were the most affected by the 1980 Irpinia-Basilicata earthquake and where the largest number of reparation and reconstruction interventions was made, thanks to the Reconstruction Law of 1981.

The vulnerability data show that the massive rehabilitation work (reparation, seismic strengthening, renovation, reconstruction) made after the 1980 earthquake reduced very significantly the seismic vulnerability of the private buildings.

It is also interesting to consider the data about people living in the various vulnerability classes: in 94 towns out of 100, the exposure index is lower than the vulnerability index; meaning that the population is, in general, less "vulnerable" than the buildings.

As a proof of this fact, Table 2 shows the distribution of the buildings and of the resident population in each vulnerability class. In particular, about 260,000 people, corresponding to

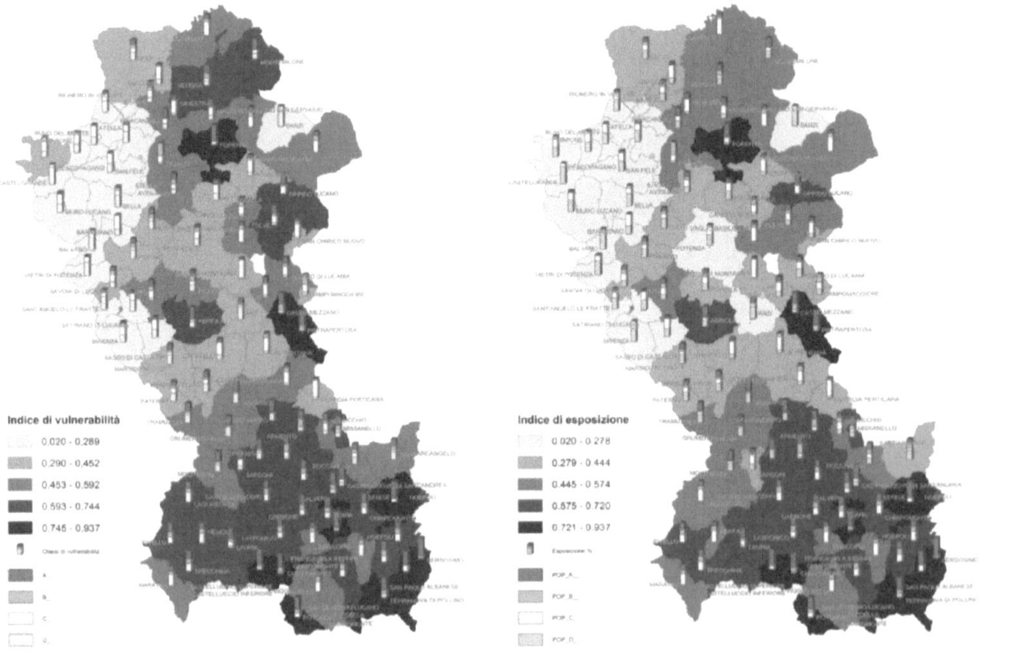

Figures 4 and 5. Provincial maps of vulnerability and exposure.

Table 2. Synthetic table of vulnerability and exposure distributions in the Province.

Vulnerability class		A	B	C	D
Buildings	N.	34475	16272	36534	28362
	%	29,8%	14,1%	31,6%	24,5%
Resident population	N.	88481	44411	135908	123524
	%	22,6%	11,3%	34,6%	31,5%

66% of the total population of the province, live in buildings belonging to less vulnerable classes ("C" and "D"), corresponding about to 56% of the total. Moreover, only 34% of the population lives in buildings assigned to the most vulnerable "A" and "B" classes, that are about 44% of the total.

It can be deduced that most people live in the least vulnerable houses. This also depends on the strong depopulation occurred, and still occurring, in ancient parts of the towns, where most of the high vulnerability buildings are localized.

Collecting the data about buildings vulnerability and population exposure, on the provincial territory, a unique database has been realized, containing the *provincial model of seismic attenuation* (Attolico, Harabaglia et al., 2005) and damage probability matrices.

From this database, by means of a purpose-built software procedure (Attolico, Bixio, 2005), it's possible to create damage scenarios of both simulated and historical seismic events involving the provincial territory, useful for civil protection planning (risk assessment and prevention) and management (emergency and post-emergency operations).

2 CONCLUSIONS

Until today, information about residential building stock vulnerability collected by the Provincial Administration of Potenza for civil protection activities (risk assessment and prevention, emergency planning and management) and implemented in the provincial GIS of Risk Exposed Elements come from:

- research and collection of available studies carried out by other local and national authorities and institutions in 21 municipalities;
- vulnerability data collected in 28 municipalities, thanks to a co-operation with DiSGG (Department of Structures) of Basilicata University, during the 2004 research activity for the development of the Civil Protection Provincial Plan; in 11 of them, they were collected by DiSGG by means of a thorough survey made after Potenza earthquakes of 1990 and 1991, using damage relief cards made by GNDT (National Group for Defence from Earthquakes); in the other 17, data were surveyed through the "*Guided Interview Protocol*";
- the on-the-field survey carried out by the provincial working group, during the 2005 updating of the Civil Protection Provincial Plan, by means of the Guided Interview Protocol, in other 31 municipalities;
- the statistical-derived data of National Seismic Survey (ISTAT 91), updated with the 2001 population data, in the residual 20 municipalities.

By analysing the collected data together with those about exposure, it is possible to say that, in the considered territory, population is less "vulnerable" than buildings.

In the next updates of the Plan, it will be possible to extend the survey by means of the Guided Interview Protocol to all the remaining municipalities, with an absolutely bearable economical and human resources commitment. This way, we could have up-to-date and reliable vulnerability data with coverage of 100% of the provincial territory.

REFERENCES

Medvedev, S.V. 1997. Seismic Intensity Scale M.S.K.-76 - Publ. Inst. Geophys. Pol. Ac. Sc. Varsaw.

Braga, F., Dolce, M. & Liberatore, D. 1982. *A statistical Study on Damaged Buildings and Ensuing Review of the MSK-76 Scale.* 8th European Conference on Earthquake Engineering, Athens.

Braga, F., Dolce, M. & Liberatore, D. 1987. *Rassegna critica dei metodi per la stima della vulnerabilità.* Proceedings of Third National Conference of Earthquake Engineering, Roma.

Dolce, M. 1996. *Seismic vulnerability evaluation and damage scenarios - US-Italian Workshop on Seismic Evaluation and Retrofit.* New York City: Columbia University.

Dolce, M., Lucia, C., Masi, A. & Vona, M. 1997. *Analisi delle tipologie strutturali della Basilicata ai fini di una valutazione di vulnerabilità.* 8th Anidis Conference, Taormina.

ESC Working Group "Macroseismic Scales", 1998. *European Macroseismic Scale 1998*, GeoForschungs - Zentrum Potsdam, Germany.

Grunthal, G., European Macroseismic Scale 1998. In Chaiers du Centre Europèèn de Géodynamique et de Séismologie, Volume 15, Luxembourg.

Dolce, M., Papa, F. & Zuccaro, G. 1999. *Protocollo d'intervista.*

Bernardini, A. (A cura di), 2000. *La vulnerabilità degli edifici: valutazione a scala nazionale della vulnerabilità sismica degli edifici ordinari*, CNR-Gruppo Nazionale per la Difesa dai Terremoti – Roma.

Lucantoni, A., Bosi, V., Bramerini, F., De Marco, R., Lo Presti, T., Naso, G. & Sabetta, F. 2001. *Il rischio sismico in Italia*, Servizio Sismico Nazionale – Roma.
Dolce, M., Marino, M., Masi, A. & Vona, M. 2001. Uno scenario di danno sismico per la città di Potenza, 10th Anidis Conference, Potenza – Matera.
Dolce, M. (coordinatore), 2001. *Rapporto finale della Convenzione "Microzonazione del territorio della Basilicata colpito dal terremoto del 9.9.98"*, Potenza.
Masi, A. (coordinatore) 2004. *Rapporto finale della Convenzione Provincia di Potenza – DiSGG "Raccolta ed elaborazione dati relativi alla vulnerabilità del costruito"*, Potenza.
U.O. Protezione Civile Provincia di Potenza, 2004. Piano Provinciale di Protezione Civile; Vol. 1: Programma Provinciale di Previsione e Prevenzione dei Rischi; Vol. 2: Piano Provinciale d'Emergenza, Edizione n. 1, Potenza, Italy.
Attolico, A. (coordinatore) 2005. *The Civil Protection System of Provincia di Potenza: interoperability between operational instruments for emergency prevention and management.* 14th SRA Europe Annual Meeting, Como, Italy.
Attolico, A. & Bixio, A. 2005. *Seismic Risk Assessment: a software procedure for the simulation of damage scenarios in emergency management.* 14th SRA Europe Annual Meeting, Como.
Attolico, A., Ciorciari, C., Harabaglia, P. & Rosa, A.B. 2005. *Seismic Risk Assessment: a numerical model for the construction of Event Scenarios suited for provincial territory of Potenza.* 14th SRA Europe Annual Meeting, Como.

Risks Challenging Publics, Scientists and Governments – Menoni (ed)
© 2010 Taylor & Francis Group, London, ISBN 978-0-415-58072-4

Seismic risk assessment: A software procedure for the simulation of damage scenarios in emergency management

A. Attolico
Head of Civil Protection Office, Province of Potenza, Potenza, Italy

A. Bixio
Consultant of Province of Potenza, Potenza, Italy

ABSTRACT: The paper describes the methodology to produce event scenarios based on both historical analysis and knowledge of the physical characteristics of the involved phenomena. A *Damage Scenario* is obtained by overlapping an event scenario with risk-exposed territorial elements. Reliability of the damage estimation given by scenarios is strictly related to an accurate modelling of all the different systems in an area. Based on the seismic vulnerability data on the whole provincial territory, a software procedure has been realized, which makes possible to obtain a first estimate of victims and damage of goods, only a few minutes after a seismic event. The procedure has been implemented in the Provincial GIS of Risk Exposed Elements, as an effective and simple instrument to manage the first moments of a seismic emergency. As a case study, a simulation of the 1857 Basilicata Earthquake (M = 6.8) was performed to know the effects it would cause if should it happen today.

1 VULNERABILITY AND EXPOSURE

In 2005, to update of the Provincial Civil Protection Plan, a recognition of available data about the seismic vulnerability of buildings in the provincial territory was perfor med. The collected vulnerability data are listed below:

- Study of buildings vulnerability in 28 municipalities of the Potenza province: 11 of them were based on post seismic relief in 1990 and 17 were studied through the "*Guided Interview Protocol*" (Dolce, Papa & Zuccaro, 1999), within a cooperation agreement between the Province of Potenza and DiSGG (Structures Department) of University of Basilicata (Masi et al. 2004);
- Study of seismic vulnerability in 9 municipalities of Upper Agri Valley (cooperation agreement between Basilicata Region and SMA S.p.A.);
- Cooperation agreement between Basilicata Region and DiSGG for the Microzoning of the regional territory affected by the September 9th 1998 earthquake, including hazard and vulnerability studies in 27 municipalities of "Pollino-Lagonegrese" area (Dolce et al. 2001);
- CNR—GNDT, Buildings vulnerability: a nationwide assessment of seismic vulnerability of ordinary buildings (edited by A. Bernardini, 2000), including three municipalities in the Potenza province;
- ISTAT (Italian Institute of Statistics) 1991 census data processed by the National Seismic Survey to obtain seismic vulnerability information (Lucantoni et al. 2001).

To integrate the available data, within the 2005 update of the Civil Protection Provincial Plan, a seismic vulnerability survey was performed in other 31 municipalities in Potenza province, by means of the *Guided Interview Protocol.*

In this phase, for the residual 20 municipalities it was decided to use the statistical-derived data of National Seismic Survey, but updating it with the 2001 population data.

According to acquired data, a provincial map showing vulnerability index and distribution in four vulnerability classes (A, B, C, and D) was created. In the same way, a map of exposure was produced too, showing the exposure index and the population distribution in the four vulnerability classes (Attolico, Bixio & Pacifico, 2005).

2 THE CONSTRUCTION OF DAMAGE SCENARIOS

To assess the damages caused by an earthquake, the study of scenarios has gradually come up to the traditional risk analyses. By using them we can simulate the consequence of a seismic event with specified features, occurring on a specific territory. According to the results of the scenario, we can calibrate quite adequate programs for damages assessment and prevention.

To build a damage scenario we first need to proceed to the realization of an event scenario. In this study, the "*Provincial Attenuation Model*" (Attolico et al. 2005), was used to evaluate the felt intensity on the territory struck by an earthquake. This model is obtained by the analysis of macroseismic data of all the historical events from CFTI and DOM catalogues that reported observation on the Province of Potenza territory.

The model is defined for epicentral intensities I_0 ranging from MCS VI to XI with isoseismals as low as MCS V (Table 1). The whole model was coded in the Provincial *GIS of the Risk Exposed Elements*, so to be able to build a real-time scenario as soon as the National Civil Protection Department communicates the instrumental epicentral location and magnitude.

From the intensity maps of the various assessed seismic events, it is possible to define the distribution of damage levels on buildings, on a probabilistic basis and for each vulnerability class, through the "*Damage Probability Matrices*" (DPM) shown in Figure 2 (Braga et al. 1982; Dolce et al. 2001).

Table 1. Attenuation relation in km, obtained for the Potenza area and surroundings (Attolico et al. 2005).

	I_0 XI	I_0 X	I_0 IX	I_0 VIII	I_0 VII	I_0 VI
XI	4.2					
X	13.6	8.0				
IX	23.3	16.2	6.9			
VIII	54.3	33.5	11.9	5.5		
VII	93.8	61.0	27.5	15.1	7.7	
VI	153.7	98.7	46.3	22.9	15.1	10.2
V	161.4	135.0	83.7	38.9	31.3	18.4

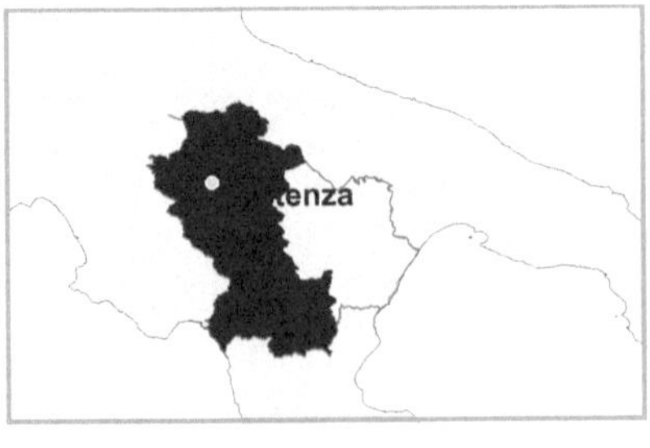

Figure 1. Area of study: Province of Potenza (South Italy) with an extension of about 6,500 km^2, a population of about 400,000 inhabitants and an average population density of 60 in/km^2 (ISTAT 2001).

The considered damage levels, as defined by MSK and EMS macroseismic scales, are listed below:

- 0: no damage
- 1: negligible—light damage
- 2: moderate damage
- 3: strong damage
- 4: very strong damage
- 5: destruction

By using the DPMs we have an estimate of the number and the damage level of involved buildings, in those localities where the felt intensity is above the damage threshold, namely Is ≥ VI MSK. The assumption made in the construction of damage scenarios is $I_{MSK} = I_{MCS}$.

Figure 3 shows a scheme of the procedure used to build the damage scenarios.

Specifically, the damage to buildings was assessed as shown below:

1. buildings with D5 damage level are considered *collapsed*;
2. buildings with D4 damage level plus 40% of those with D3 damage level are considered *condemned*;
3. buildings with D2 damage level plus residual part of those with D3 damage level are considered *damaged*.

As concerns the effects on the population, according to the number of people living in buildings belonging to the various classes of vulnerability, such as the damage levels they have experienced, we can easily obtain an estimate of the *peoples involved in collapses* and the number of *homeless* peoples.

Since the effects of an earthquake on people strongly depend on a series of parameters, such as the instant in which the event happens, the activities of population, the rapidity of aids, the percentage of survivors, etc, while realising a damage scenario is very difficult to estimate reliable numbers, unless

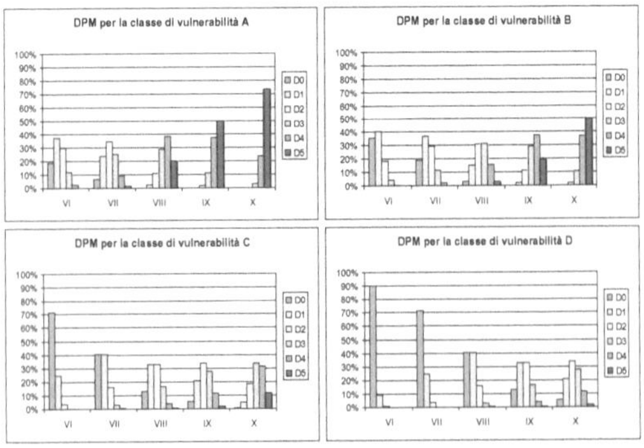

Figure 2. Charts of Damage Probability Matrices (DPM).

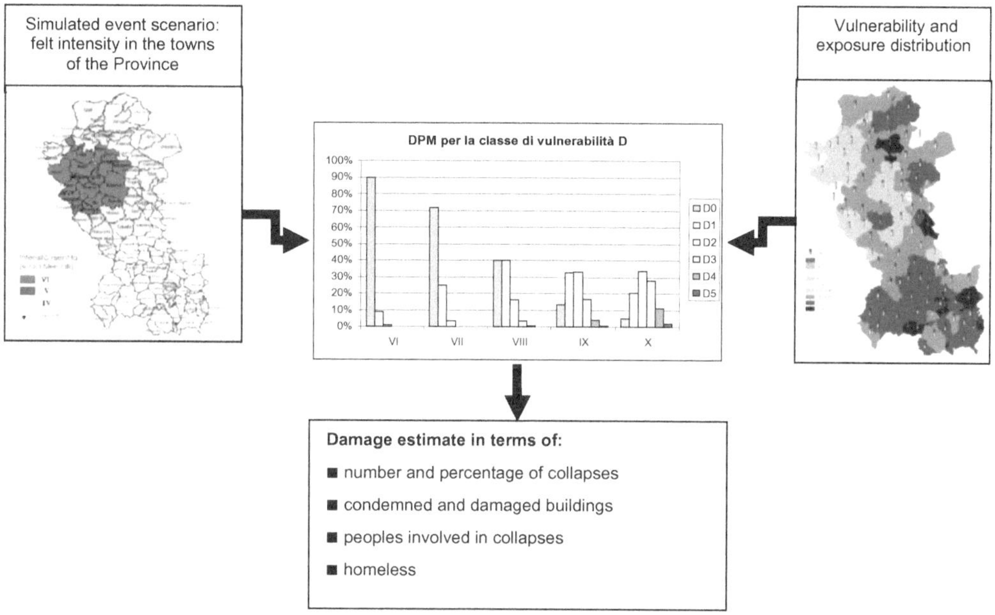

Figure 3. Construction of a damage scenario.

experiencing enough reliable studies on the population dynamics during the day. Thus, the effects on population were estimated only in terms of *homeless* peoples and peoples *involved in collapses*.

Besides the number of collapsed, condemned and damaged buildings, a *Damage Index* **Id** was used, to obtain a synthetic representation of the damage level. The index is defined as below (Masi et al. 2004):

$$Id = \sum_{1}^{n} \frac{D_i f_i}{n} \quad \text{with} \quad 0 \le Id \le 1 \tag{1}$$

where $\mathbf{D_i}$ represents the **ith** damage level, ranging from 1 to 5, and $\mathbf{f_i}$ is the relative frequency of every damage level.

The sum is extended to n = 5 damage levels > 0. Therefore, $Id = 0$ indicates the absence of any damage, while $Id = 1$ means that all the buildings in the considered area collapse (D5 damage level).

In the same way, for every town in the scenario a *Population Damage Index* **Idp** was defined as follows:

$$Idp = \sum_{1}^{n} \frac{D_i fp_i}{n} \quad \text{with} \quad 0 \le Idp \le \tag{2}$$

where $\mathbf{D_i}$ represents the $\mathbf{i^{th}}$ damage level, ranging from 1 to 5, and $\mathbf{fp_i}$ is the relative frequency of people living in buildings undergoing the various damage levels.

$Idp = 0$ means that the whole population lives in buildings with no damage, and $Idp = 1$ indicates that the whole population lives in collapsed buildings (D5 damage level).

3 THE SOFTWARE PROCEDURE TO CREATE DAMAGE SCENARIOS

A simple software procedure (running in Excel® environment and called "Scenari") to be used in the Operating Room has been realized to create *event* and *damage* scenarios, in order to allow a quick estimate of the predicted damages subsequent to a relevant seismic event, namely with $I_0 \ge$ VI MSK.

For relevant earthquakes (over the damage threshold), the procedure allows to create an event and damage scenario both as a tabular report and as a ready-to-view ArcMap® DBF file, by using the previously shown buildings vulnerability data and the *provincial attenuation model*.

The worksheet is easily manageable from the start menu page (Figure 4), by means of a series of buttons, whereby we can enter all the implemented functions. In particular, the first action to do for the operator is to insert the geographical coordinates and the magnitude of the earthquake, to calculate the corresponding damage scenario.

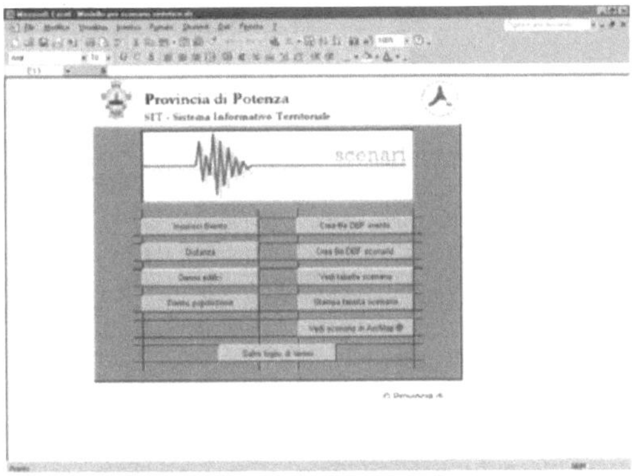

Figure 4. Start menu page of the worksheet "scenario".

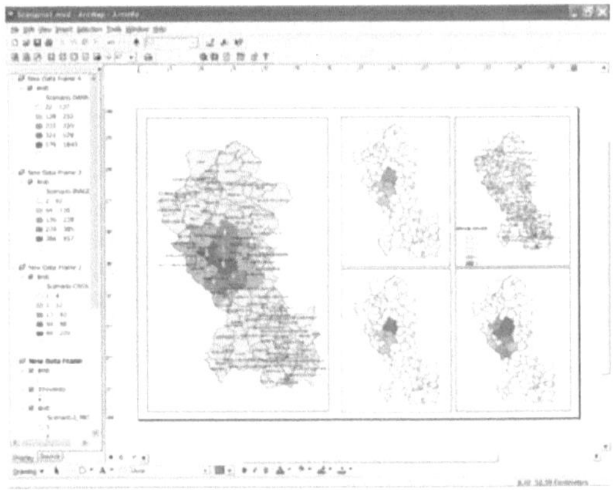

Figure 5. ArcMap® display with the map of the scenario.

After that, the worksheet calculates the kilometric distances between epicentre and all the municipalities in the province and the subsequent macroseismic felt intensities, by using the *provincial attenuation model.*

For all the municipalities where the felt intensity is upper than the damage threshold, the worksheet calculates the percentage of buildings and population for each damage level, and creates two DBF files, containing respectively the epicentre coordinates and the damage scenario, ready to be seen as a map representation in ArcMap® (Figure 5).

Finally, the procedure allows to view and print a synthetic tabular report of the calculated damage scenario (Figure 6). In particular, the report shows the data listed below:

- denomination, magnitude and epicentre intensity of the earthquake;
- number of municipalities with $I_r \geq VI_{MSK;}$
- total number of the buildings interested by the earthquake;
- total population interested by the earthquake;
- total estimated number of the collapsed, condemned and damaged buildings;
- total estimated number of homeless and people involved in collapses;

and, for every municipality:

- felt intensity;
- number of building;
- population;
- damage index *Id*;
- population damage index *Idp*;
- estimated number of the collapsed, condemned and damaged buildings;
- estimated number of homeless and people involved in collapses.

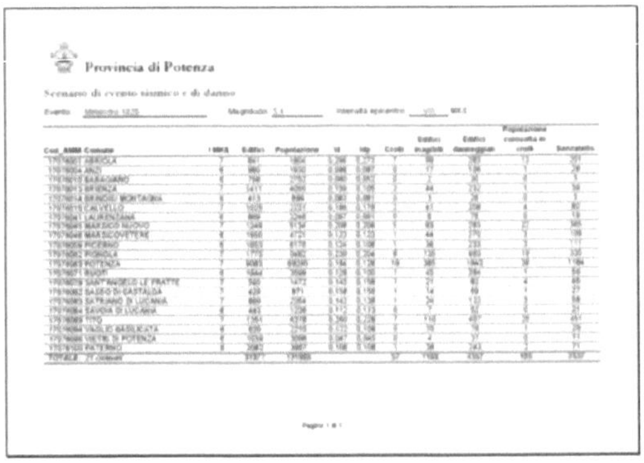

Figure 6. Synthetic tabular report of the damage scenario.

By using this procedure, in only few minutes after the earthquake, the Province of Potenza is able to have a first estimate of the struck localities and the damages to buildings and peoples, useful for the calibration of the aids.

4 A SIMULATION OF THE BASILICATA EARTHQUAKE OF DECEMBER 16TH, 1857

The earthquake occurred on December 16th, 1857 was probably one of the strongest which interested Basilicata in a recent time, with a magnitude of 6.8 Richter degrees. Its epicentre was localised in the Agri Valley, and it interested all the localities of the Province of Potenza and many other localities in the Southern Italy.

From the historical observations, in two municipalities next to the epicentre the felt intensity was XI MCS; in thirteen towns it was about X MCS and in other seven it was IX MCS. The victims were over 9,000 (actually 9,000 is the official datum; unofficially the victims were more than 15,000) and there were tens of thousand wounded peoples.

As a case study, a *damage* scenario has been built to have a simulation of the damages which an earthquake like that would cause if it could happen today.

Figure 7 shows the *event* scenario with the felt intensities, calculated by using the software procedure previously shown.

According to the created damage scenario, it would have a macroseismic intensity above the damage threshold in all the 100 towns in the province, interesting a population of about 400,000 peoples. Its effects would be catastrophic: there would be about 5,400 collapsed, 14,500 condemned and 23,600 damaged buildings; homeless peoples would be over 38,000 and about 8,500 people would be involved in buildings collapses, being potential victims.

Figure 8 shows the damage to buildings, represented by the Damage Index Id and the percentage of collapses, condemned and damaged buildings. In Figures 9, 10 and 11 the damage to population is represented respectively by the Population Damage Index Idp, the number of people living in collapsed buildings and the number of homeless peoples.

The strongest damages would be located in the middle part of the Province, with strong effects also in localities quite far from the epicentre. As an example, in the major city, Potenza, far 33 km from the epicentre, there would be about 670 people involved in collapses and 6,300 homeless, out of a population of about 70,000 inhabitants.

5 CONCLUSIONS

The software procedure, called "*scenari*", is a simple and quick tool for the construction of event and damage scenarios when an earthquake with intensity above the damage threshold ($I_0 \geq VI\ MSK$) will occur, striking areas in the Province of Potenza.

Its characteristic development as an Excel® automatized worksheet makes the training time dedicated to the use of the procedure extremely short.

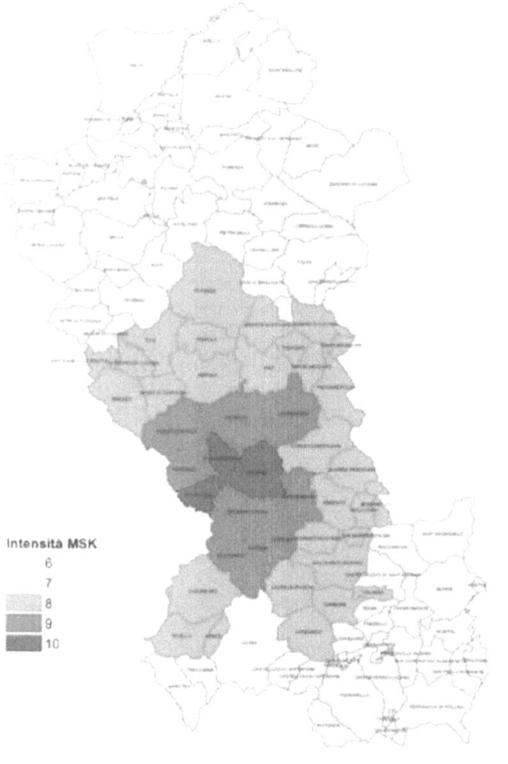

Figure 7. 1857 Basilicata earthquake: simulated event scenario.

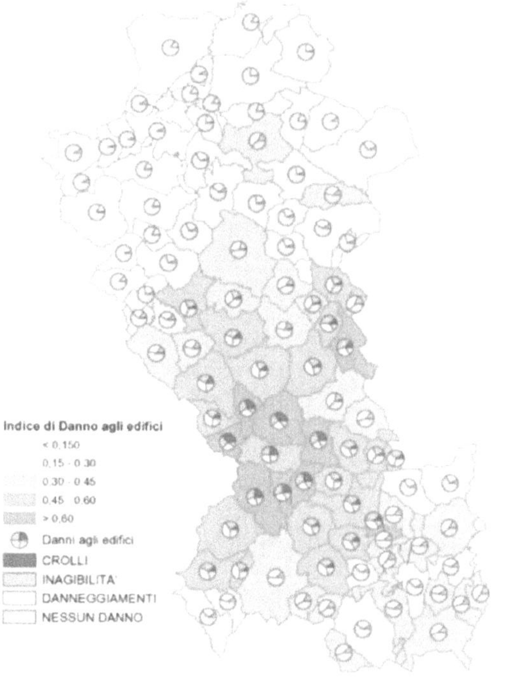

Figure 8. Simulated 1857 Basilicata Earthquake – M = 6.8: Damage to buildings.

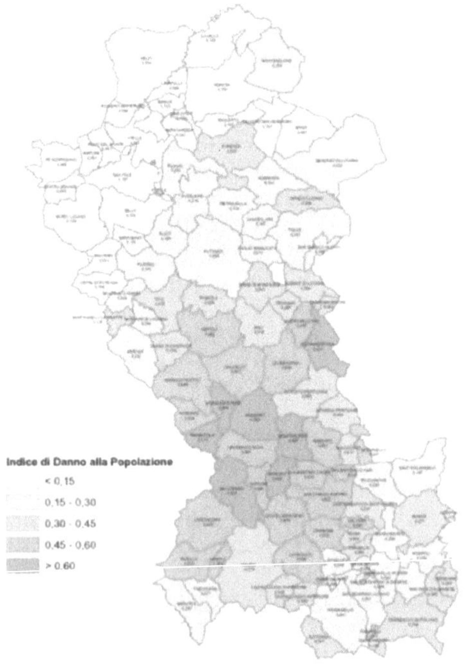

Figure 9. Simulated 1857 Basilicata Earthquake – M = 6.8: Damage to Population.

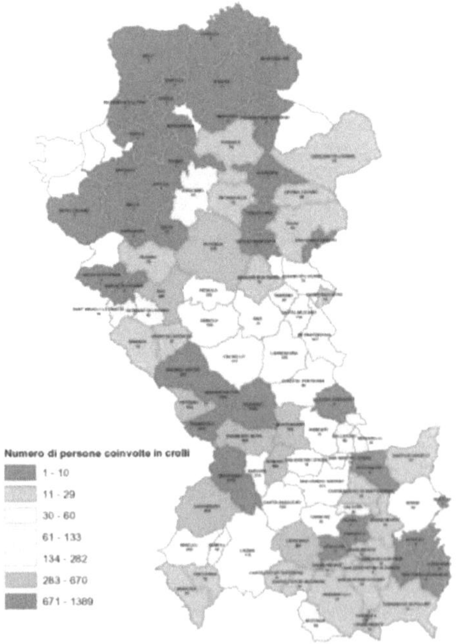

Figure 10. Simulated 1857 Basilicata Earthquake – M = 6.8: peoples living in collapsed buildings.

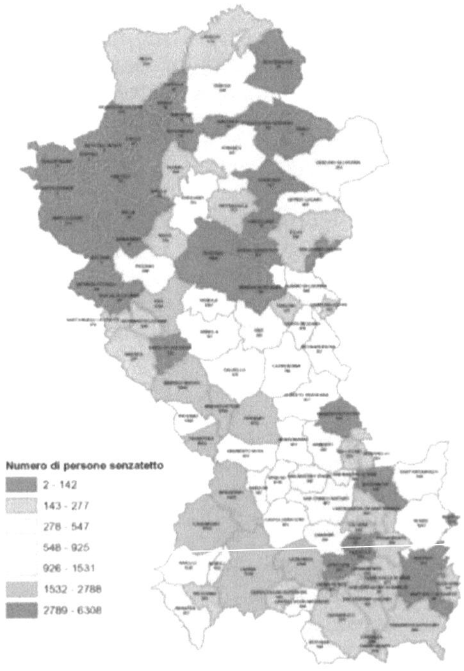

Figure 11. Simulated 1857 Basilicata Earthquake – M = 6.8: homeless peoples.

The Province of Potenza, within its role in coordination of aids after a relevant seismic event, is able to have, in few minutes, a well detailed picture of the expected damage. Even if it is only a first estimate, yet it allows not to operate in the dark or with indirect and fragmentary information.

Finally, the procedure is adaptable to other territories too, simply modifying the geographical (coordinates of the towns) and the seismic risk data (vulnerability of buildings, lifelines, infrastructures, population, etc. and exposure of lives and properties). Moreover, it is possible to implement various attenuation laws, both generical and specific for a territory.

REFERENCES

Attolico, A. (coordinatore) 2005. *The Civil Protection System of Provincia di Potenza: interoperability between operational instruments for emergency prevention and management.* 14th SRA Europe Annual Meeting, Como, Italy.

Attolico, A., Bixio, A. & Pacifico, S. 2005. *The Emergency Plan of the Province of Potenza: The seismic vulnerability map of buildings for the construction earthquake damage scenarios.* 14th SRA Europe Annual Meeting, Como.

Attolico, A., Ciorciari, C., Harabaglia, P. & Rosa, A.B. 2005. *Seismic Risk Assessment: a numerical model for the construction of Event Scenarios suited for provincial territory of Potenza.* 14th SRA Europe Annual Meeting, Como.

Bernardini, A. (A cura di), 2000. *La vulnerabilità degli edifici: valutazione a scala nazionale della vulnerabilità sismica degli edifici ordinari.* Roma: CNR-Gruppo Nazionale per la Difesa dai Terremoti.

Braga, F., Dolce, M. & Liberatore, D. 1982. *A statistical Study on Damaged Buildings and Ensuing Review of the MSK-76 Scale.* 8th European Conference on Earthquake Engineering, Athens.

Braga, F., Dolce, M. & Liberatore, D. 1987. *Rassegna critica dei metodi per la stima della vulnerabilità.* Proceedings of Third National Conference of Earthquake Engineering, Roma.

Dolce, M. (coordinatore) 2001. *Rapporto finale della Convenzione "Microzonazione del territorio della Basilicata colpito dal terremoto del 9.9.98".* Potenza.

Dolce, M., Marino, M., Masi, A. & Vona, M. 2001. *Uno scenario di danno sismico per la città di Potenza.* 10th Anidis Conference, Potenza—Matera.

Dolce, M., Marino, M., Masi, A. & Vona, M. 2001. *Uno scenario di danno sismico per la città di Potenza.* Proceedings of 10th Anidis Conference, Potenza—Matera.

Dolce, M., Papa, F. & Zuccaro, G. 1999. Protocollo d'intervista.

ESC Working Group "Macroseismic Scales" 1998. *European Macroseismic Scale.* GeoForschungs, Zentrum Potsdam, Germany.

Lucantoni, A., Bosi, V., Bramerini, F., De Marco, R., Lo Presti, T., Naso, G. & Sabetta, F. 2001. *Il rischio sismico in Italia.* Roma: Servizio Sismico Nazionale.

Masi, A. (coordinatore) 2004. *Rapporto finale della Convenzione Provincia di Potenza—DiSGG "Raccolta ed elaborazione dati relativi alla vulnerabilità del costruito".* Potenza.

Medvedev, S.V. 1997. *Seismic Intensity Scale M.S.K.-76.* Varsaw: Publ. Inst. Geophys. Pol. Ac. Sc.

U.O. Protezione Civile Provincia di Potenza, 2004. Pia*no Provinciale di Protezione Civile; Vol. 1: Programma Provinciale di Previsione e Prevenzione dei Rischi; Vol. 2: Piano Provinciale d'Emergenza.* Edizione n.1. Potenza, Italy.

Risks Challenging Publics, Scientists and Governments – Menoni (ed)
© 2010 Taylor & Francis Group, London, ISBN 978-0-415-58072-4

Pilot project for a global safety plan on economic areas

C. Fiévez, D. Dupuis, X. Dugnoille & C. Delvosalle
Faculté Polytechnique de Mons, Major Risk Research Centre, Mons, Belgium

ABSTRACT: An economic area, near the town of Ghislenghien (Belgium), experienced, during summer 2004, a major accident on a high pressure gas pipeline. The whole region was deeply affected and the local authorities decided to make a pilot project, in cooperation with the district development company (IDETA), the Major Risk Research Centre and a safety specialised company. The aim is to set up a Global Safety Plan to promote the safety of the town. In order to achieve the objectives of the project, firstly companies were informed about the project. Then all the data have to be collected and indexed and the risks analyzed according to the identified dangers. Finally, the emergency plans was evaluated and coordinated in order to determine the possible scenarios and to deduce the reaction instructions, both internal and external.

1 INTRODUCTION

An economic area, near the town of Ghislenghien (Belgium), experienced, during summer 2004, a major accident on a high pressure gas pipeline, causing 24 fatalities and more than 100 injuries. The whole region was deeply affected by this disaster. Following this tragedy, the local authorities decided to make every effort to increase more and more the safety level in the economic area.

A pilot project was thus started by the local authorities, in cooperation with the district development company (IDETA), the Major Risk Research Centre ("Faculté Polytechnique de Mons") and a safety specialised company ("Ingenierie et sécurité sprl"). The aim of this project is to set up a Global Safety Plan in order to promote the safety in the vicinity of the town, especially in the economic area where the pipeline accident took place.

The pilot project began in November 2004. It is based on a methodological tool aiming at the development of a Global Prevention Intervention Plan for economic areas in general. Within the framework of the project, the economic area of Ghislenghien was chosen as pilot case.

This economic area is about ten kilometres on the south-east side of the town of Ath, two kilometres south of the motorway "A8" Lille – Brussels. About seventy companies are located there. The activities and the size of the companies are very different from one to another, ranging from small family business (less than 10 people) up to large leading companies (more than 300 people). It's a kind of economic area for which few global safety measures are usually taken although it gathers many various hazards.

This paper describes, in general terms, the Global Safety Plan through the methodological tool, the Global Prevention Plan, the Global Intervention Plan, the different phases and the collaborations required for the completion of the project. Finally, the post-project management is described, aiming at preserving the safety level reached.

2 THE GLOBAL SAFETY PLAN IN GENERAL

2.1 *The methodological tool*

A methodological tool was used to develop the Global Prevention Plan and the Intervention Plan for economic areas. The basis of this tool is a hierarchical set of bound files and cards. This set will constitute a data base included in the Global Prevention Plan.

All the stakeholders of the industrial area are concerned, which means industrial companies, rail/road transport networks and also supply networks (for gas, oil, water, electricity).

Each stakeholder is described in a file called "description file" providing detailed information about the considered stakeholder. For example, the following data are found in the "description files" for industrial companies:

> Name of the company, address, phone numbers, number of employees, contact people name and phone, maps with the localisation of the main hazardous equipment, available rescue material, fire exits, description of the activities through tables gathering hazardous installation and substances, Material Safety Data Sheets, means of access,

"Internal Emergency Plan" if it exists, vicinity (neighbouring industrial companies, transport networks and residential areas) and finally identified hazards.

The "description files" for the transport and supply networks include the characteristics of each of them and maps locating roads, rails, rivers, pipes and cables as well as identified hazards.

Based on the identification of the hazards, accident scenarios are built and used to write "accident scenarios files", namely including the description of the accidents and their effects.

The "description files" and "accident scenarios files" are the reference to write the "internal reaction cards" and "rescue reaction instructions".

For each identified scenario, the "internal reaction cards" will gather information needed to manage the accident with the means available inside the company. The "internal reaction cards" will thus be linked with the "internal emergency plan". The main actors of the emergency plan will have their own card in order to know what they have to do in such an accidental situation.

If the accident cannot be managed within the limits of the "internal emergency plan", a link will be established with the "external emergency plan", constituted of "rescue reaction instructions" for the various rescue services (fire brigade, police, health services, and civil protection)

The "rescue reactions instructions" contain many pieces of advice for the intervention of the various rescue services, depending on the events identified in the "accident scenario files" (for example if there is a risk of toxic fumes in case of fire, the reactions card will specify a safety zone) as well as rescue means which can be shared and procedure to inform the neighbours (people's contact, phone, ...).

The instruction will match the vocabulary and rules frequently used by the rescue service. The words will be chosen to avoid misunderstanding and to assure a quick comprehension of the described actions. The methodological tool is illustrated on Figure 1.

2.2 *The global prevention plan*

The Global Prevention Plan is a data base composed of the "description files" and the "accidents scenarios files". This data base should be as comprehensive as possible.

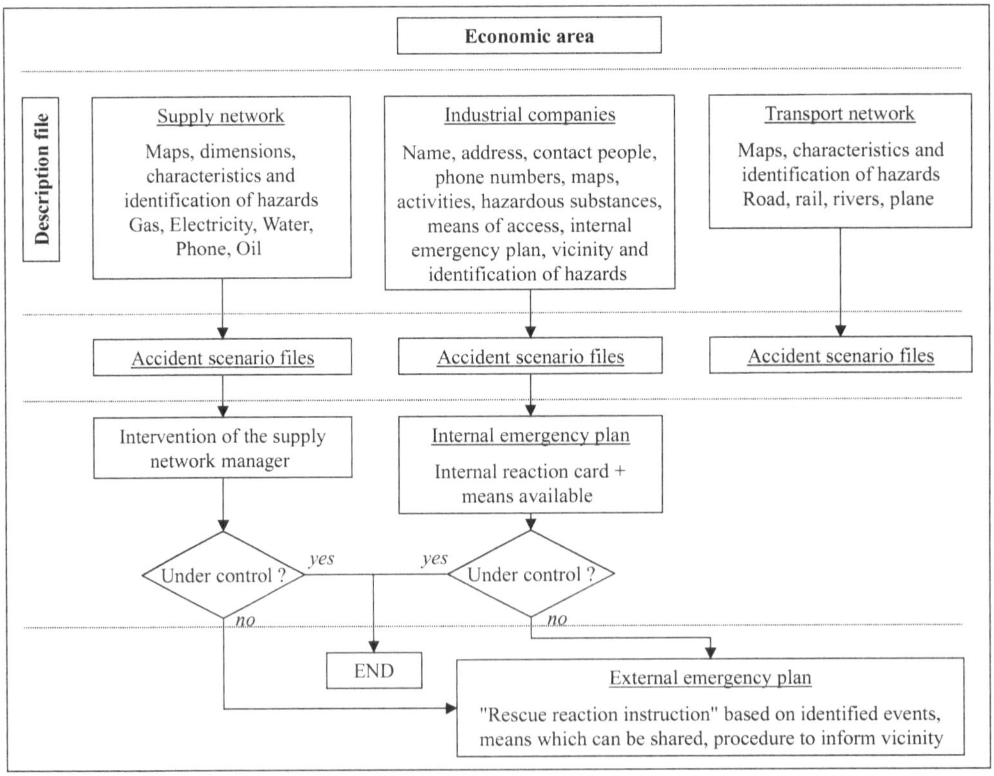

Figure 1. Methodological tool.

The database is a crucial information tool for the rescue services. With the "description files" and the "accident scenarios files", rescue services find essential data for their intervention (where are main and alternative gates, how many people are present on site, where are located hazardous substances, is there radioactive sources in the company, what are the possible evolutions of the early phase incident, etc.). In the same way, they have information about the availability of the road network (for example, if there is a temporary diversion) and the hazards that some supply networks can involve in some cases.

In this preventive part of the project, there are two other very important aspects to preserve and to support the high safety level reached in the economic areas.

The first one is to analyse particularly the safety aspects when a new licence to operate concerns the industrial area (new company, works, building site, etc.). In Belgium, each application for a licence to operate must be introduced to the local authorities and must gather a set of information requested by the law. In general, this information is a good outline of the hazards introduced by the extension or the new project. Through the deliverance of the permit, local authorities can act in order to maintain the high safety level in the economic area, for example by requiring special operating measures.

Secondly, the Global Safety Plan will also include a "safety charter", a document in which all partners will commit themselves to give priority to safety. This state of mind of the stakeholders should encourage a high level of safety in the industrial area.

2.3 *The global intervention plan*

The Global Intervention Plan will gather all information needed to manage an accident. The information will be listed in the "internal reaction cards" and "rescue reaction instructions".

If the consequences of the accident can be controlled only with the means available inside the company, the "internal reactions cards" are sufficient to manage the accident.

In other cases, a rescue service must be called to manage the accident. In this case, the "Rescue reaction instructions" are used by the various rescue services. The actors are informed about:

- The needed equipment, depending on the accident scenario,
- The additional hazards which can occur during the operation.

Their actions can thus be thought in advance to limit the consequences of the events and to avoid some new accident.

According to the identified scenario, safety zones will be defined to inform the rescue service about the safety measure which must be taken in these safety zones. For example, if a toxic release is possible in a scenario, the "Rescue reactions instructions" will define the different safety zones inside which people will be evacuated or confined in their building.

3 GLOBAL SAFETY PLAN: APPLICATION

In the frame of the pilot project, in order to achieve the objectives pursued, concrete actions must be carried out. These actions are divided in different phases.

3.1 *Information phase*

The first step is to inform all the stakeholders of the industrial area about the project and to explain them that their help is required to obtain the required data.

3.2 *Data gathering phase*

The data gathering phase is obviously the longest one because some information about the stakeholders is not directly available and must be searched for. For example, a lot of companies are small and employ few people, so have not an efficient structures to deal quickly with the project. Thus a lot of time can be spent for the exchange of information.

3.2.1 *Industrial companies*

In a first time, contact people in each industrial company are asked to send general information.

Secondly, a visit is necessary on the site of the companies. This visit provides data about the activities, the location of the equipment, the hazards and the kind of emergency means available inside the company.

A report is then written but often this document is incomplete just after the visit. A mail is then sent to obtain answers to outstanding questions.

When all the data needed are collected, they are indexed to fill in the datasheets and a first selection of hazards can be realised. The description file is completed at this time.

3.2.2 *Supply networks*

It is necessary to know exactly where are equipment used for supply networks like pipes and cables.

The managers of the network are asked to send maps and needed information. The location given on the maps are checked on site with a responsible people of the supply network because it is possible that the equipment is not exactly at the location shown on the map. Another encountered difficulty is the various scales of the maps.

3.2.3 *Transport networks*

The needed information is to be collected about transport network (number of cars, lorries, trains, boats, planes, frequency of hazardous substances transportation, etc.).

The data are collected among different departments managing the transport networks.

3.3 *Analysis and evaluation phase*

3.3.1 *At industrial company level*

Equipment and storage are chosen depending on the hazard that they represent.

The chosen equipment and storage are analysed in detail in order to build the possible scenarios and thus the "accident scenarios files". The scenarios will be compared with the planned actions envisaged in the "internal emergency plan" if it exists inside the company.

If the planned actions do not correspond with the scenarios, they are reviewed in cooperation with the company.

The "internal reaction files" are thus written at the end of this phase.

3.3.2 *At industrial area level*

The hazards identified on the transport and supply networks must be analysed and integrated in the evaluation of the possible scenarios in the economic area.

The neighbourhood of each company is considered like a possible source of hazard and domino effects are also analysed in order to obtain the most complete description of all risks to which the various rescue service can be exposed.

At the end of this phase, the "rescue reactions instructions" will be written.

At this time, in the pilot project the "rescue reaction instruction" begin to be written in close cooperation between the work team and the rescue service.

3.4 *Post-project management phase*

The methodological tool and thus the Global Safety Plan is built on a data base gathering a lot of information that can change more or less quickly with time.

The data base must be regularly updated in order to avoid that some information become false. According to changes made on the data base, the "internal reaction files" and "reactions instructions" have to be adapted.

This phase will take place when the Global Safety Plan is completed, namely through the request introduced to the local authorities when a new company arrive in the economic area or just when changes occur in a company.

The cooperation of the contact people in the company will obviously remain needed for some kinds of information (for example: number of employees, address and phone of responsible people, ...).

3.5 *Collaborations*

Throughout the realisation of the project, an intense cooperation is needed between the work team (Faculté Polytechnique de Mons), the stakeholders and the rescue services. The different necessary actions and cooperation are represented in the flow chart in Figure 2.

4 SPECIFITY OF THE PROJECT

The methodological tool used to build the Global Safety Plan is completely created in the frame of this pilot project. The Global Safety Plan on the economic area of Ghislenghien will be the first in Belgium.

An important result of the project will surely be the preventive communication which gives to the rescue services a detailed knowledge of each stakeholder and helps to enhance safety.

The high safety level expected from this project will give a new positive image of the economic area and of the industrial companies.

5 CONCLUSION

The Global Safety Plan is based on a methodological tool aiming at the development of a Global Prevention Plan and a Global Intervention Plan for economic areas in general. Within the framework of this pilot project, the economic area of Ghislenghien was chosen as a pilot case.

The methodological tool is in fact a hierarchical set of bound files and cards. Each stakeholder of the industrial area is described by a "description file" gathering all the information required. According to the identification of the hazards, "accidents scenarios files" are written, including namely the description of the possible accidents and their effects.

For each identified scenario, the "internal reaction cards" will gather the information needed to manage the accident with the means available inside the company. If the accident cannot be managed within the limits of the "internal emergency plan", a link will be established with the "external emergency plan", constituted of "rescue reaction instructions" for the various rescue services.

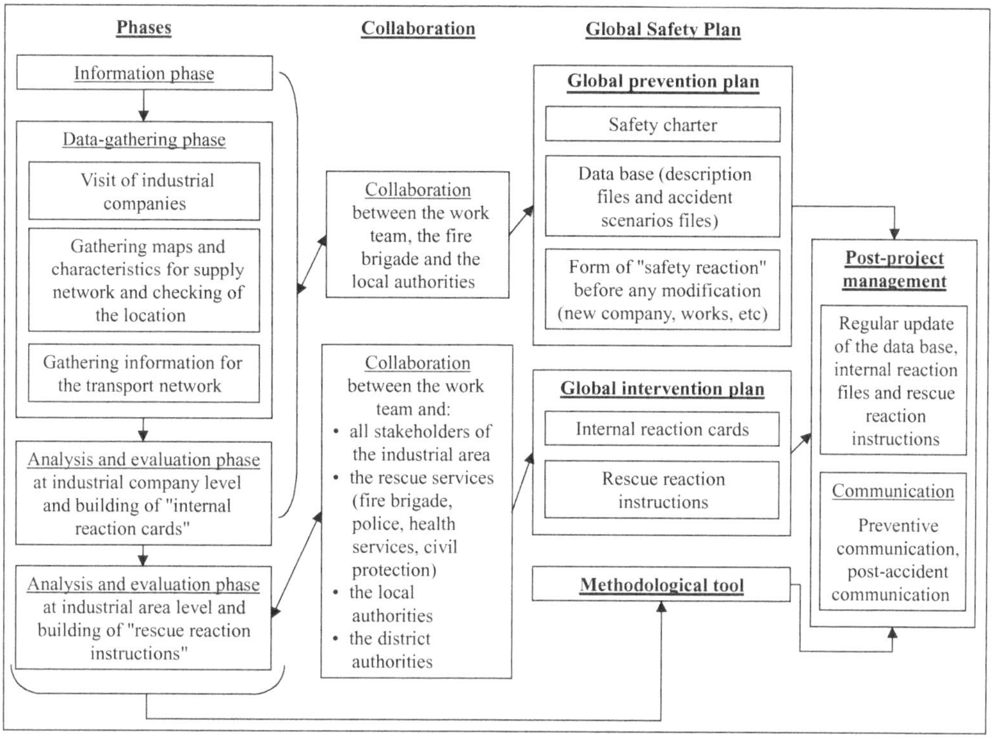

Figure 2. Flow chart gathering the phases and collaborations in order to build the Global Safety Plan and to maintain it.

The concrete actions such as the information of the stakeholders about the project, the collection of the needed data, the hazard assessment and also a close cooperation between the stakeholders, the rescue services and the local authorities will help to develop the Global Prevention Plan and the Global Intervention Plan.

The Global Prevention Plan relies on a data base composed of "description files" and "accidents scenarios files". It will also include a "safety charter", a document in which all partners will commit themselves to give priority to safety.

Through all the actions and cooperation explained here above, this project aim to avoid all unknown factors or misunderstandings when an intervention is necessary. By this way it is hoped to prevent accidents, to limit the consequences of an accident and increase the efficient communication between all the actors of the economic area. The end of the first Global Safety Plan is expected in March 2006.

Environment and public health:
Looking for new risk assessment tools

Risks Challenging Publics, Scientists and Governments – Menoni (ed)
© 2010 Taylor & Francis Group, London, ISBN 978-0-415-58072-4

A framework for decision support on HSE regulations

T. Aven, F. Asche & P. Lindoe
University of Stavanger, Norway

A. Toft
Norwegian Petroleum Safety Authority

H.S. Wiencke
University of Stavanger, Norway

ABSTRACT: This paper presents and discusses a framework for decision support on Health, Safety and Environment (HSE) regulations. The framework covers interests parties, main concerns, and the structure of the decision-making process itself, including problem identification, assessments of alternatives, and the decision-maker's review and judgement. Compared to existing frameworks, we clarify the decision processes' dependencies on the chosen risk perspective. This applies in particular to the way the evaluations of the various effects of the decisions should be carried out. The evaluations are based on classifications according to the dimension of possible consequences and associated uncertainties (risk), as well as manageability. The framework is developed for the petroleum activities in Norway, but the framework is to large extent general and should be of interest also outside the petroleum sector.

1 INTRODUCTION

How should the authorities regulate Health, Safety and Environment (HSE) in the petroleum sector? This is a question frequently being addressed by parties both within and outside the petroleum industry, and is an issue which the regulating authorities have a continuous focus on. HSE regulation of the petroleum activities is difficult as there are many conflicting goals and concerns to be taken into account. In addition, the effects of HSE regulation are not easily measured. For example, a possible regulating decision could give confidence that some potential accidents are avoided, but the uncertainties would be large and hence it would be difficult to demonstrate that the overall effect of the decision would be cost-effective. To evaluate the costs of an accident, it is not sufficient to reflect the direct costs – we also have to include the long term effects, for example related to loss of reputation. If an HSE measure is implemented, it would also often have other effects than reducing the danger of loss of lives and damage to environment. For example, if a requirement to HSE management is put in to force, this requirement would not just have HSE effects but could also give improved quality of the work and decision processes in general. Furthermore, the costs of HSE measures are in many cases difficult to assess as the motivation for incorporating a specific property or feature of a system often is a result of considerations of many concerns. Balancing the different concerns, means finding the appropriate weights for valuing accident risk reduction, seen in relation to other attributes such as costs. But what is the value of reduced uncertainty related to possible future events and losses?

In 2004 the Norwegian Petroleum Safety Authority (PSA) initiated a project to address these issues, with the aim of obtaining a better understanding of what should be the justification for HSE regulation and what concerns that should be taken into account, and how these should be evaluated and weighted (SEROS 2004). This paper summarises some of the main findings of this project.

The Norwegian authorities' initiatives on improving the quality of decisions on HSE regulation is a part of an international trend. We refer to works by OECD and EU, and in particular contributions from UK and the UK Health and Safety Executive (UK HSE), see e.g. HSE (2001, 2004). Inspired by the initiatives taken in UK on these issues, the PSA's ambition is to establish a guide on how to perform consequence assessments related to decisions on HSE regulation. Many of the ideas, principles and methods presented and discussed in the UK HSE reports are obviously applicable also for the Norwegian regulation regime, but we see the need for a rethinking and further development for some of these. In general terms, we would highlight the same aspects as UK HSE. Any policy intervention, and its enforcement, should meet the following principles which the Better Regulation Task Force devised in 1997:

- *Proportionality.* Regulators should only intervene when necessary. Remedies should be appropriate to the risk posed, and costs identified and minimised.
- *Accountability.* Regulators must be able to justify decisions, and be subject to public scrutiny.
- *Consistency.* Government rules and standards must be joined up and implemented fairly.
- *Transparency.* Regulators should be open, and keep regulations simple and user-friendly.
- *Targeting.* Regulation should be focused on the problem, and minimise side effects.

These principles are useful for measuring and improving the quality of regulation and its enforcement, setting the context for dialogue between stakeholders and government. But on a more detailed and specific level it is not clear what is the best way of structuring and guiding decisions on HSE regulations. This paper's ambition is to contribute to further understanding of these issues and provide some new ideas and perspectives. This is obtained by introducing a framework for structuring the decision-making process, including problem identification, assessments of alternatives, and the decision-maker's review and judgement. Emphasis is put on showing the decision processes' dependencies on the chosen risk perspective. In the Norwegian regulations risk is defined as the combination of probability and consequences, but in line with HSE (2002) and Cabinet Office (2002), a broad perspective on risk is adopted in practice reflecting that risk is more about uncertainties than science based probability calculations.

The paper is organised as follows. In section 2 we present the framework and in section 3 the structure used for assessing the performance of the various HSE decision alternatives. The structure is based on Aven et al. (2006) and Sandøy et al. (2005) and we refer to these papers for discussions of the structure in relation to other approaches and standards. Section 3 addresses in general terms the problem of measuring and valuing the various effects of HSE regulation. Section 4 gives some final remarks. In the appendix we include a review and discussion of different perspectives on risk, as a background for the approach adopted for the proposed framework.

2 THE FRAMEWORK AND CLASSIFICATION OF DECISION SITUATIONS

2.1 *The framework*

Our starting point is the regulating authorities having identified problem areas (areas for improvement) related to HSE, such as poor HSE results, lack of implementation of a risk reduction policy, the use of new technology, etc.

Having identified the problem areas, the authorities consider implementing different means and measures that can be used to solve these and obtain a desired improvement. Such means and measures include dialogue with specific companies to resolve a problem, modifying the existing regulation for the industry, and an increase in inspections activities etc. The challenge for the regulating authorities is to select the most appropriate way of handling this process. This paper presents a framework that guides the authorities in this task by formulating a structure for a systematic and holistic decision process, which includes stakeholder interests and main societal concerns. Figure 1 illustrates the principles of the proposed framework. The Figure shows the stakeholders influence the main concerns and the decision process, and further that the main concerns influence the decision process. In the following we describe the framework and its elements in more detail.

2.2 *Main concerns*

Main concerns refers to expectations from society, political signals, visions and ideal goals that should be used as guidelines for the development of the petroleum industry. These concerns will develop over time as society develops. However, at a certain point in time these concerns will be fixed, and would influence the decision process. It is therefore important to know whether there is a common understanding of what these concerns mean and how they should be incorporated into the decision process. Even though the main concerns are to be considered as constraints, they might be challenged as part of the decision process. The need for continuous improvement and development and the fact that expectations from society, political signals, visions and ideal goals may to some degree be conflicting, mean careful reflection of these concerns in the decision process.

A variety of concerns may be grouped according to how they relate to the stakeholders. Examples are; political guiding rules, "Man-Technology-Organisation" issues (MTO), application of the precautionary principle, societal concerns regarding major accidents, working environment and participation of workers in decision making.

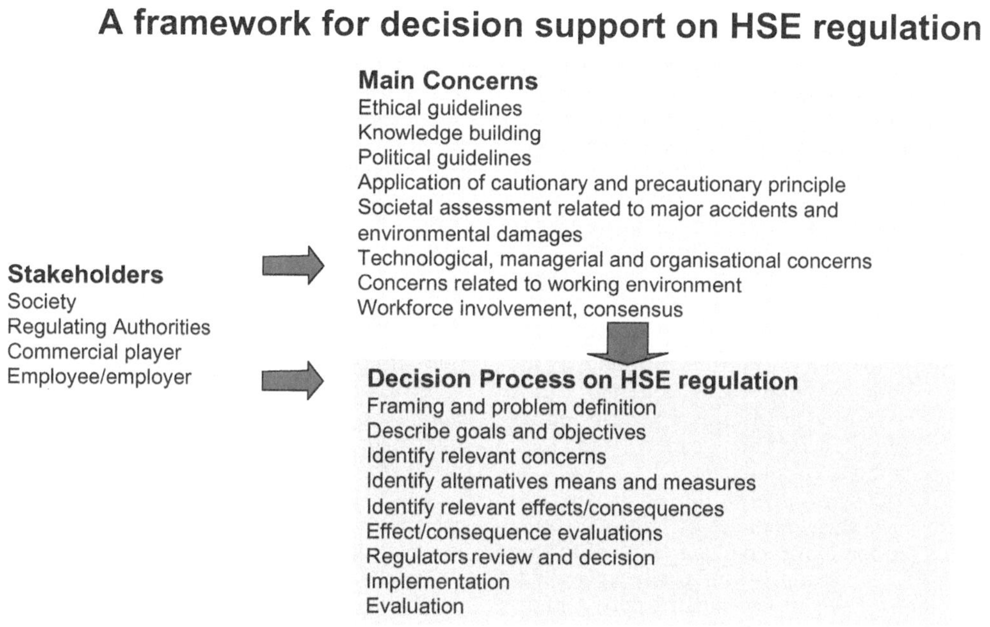

Figure 1. A framework for decision support on HSE regulation.

Other concerns are general and may be related to all kind of stakeholders as ethical issues and relevant knowledge. The following paragraphs give a brief presentation of some of the major concerns included in the framework:

2.2.1 *General concerns*

Ethical concern is embedded in all kind of decisions and it should be openly reflected in a framework on HSE regulation. In such a reflection it should be distinguished between *ethics* as a theoretical reflection on right and wrong, e.g. a "moral theory" and *moral* as the norms and values that may guide us in our daily actions. In such actions, norms can be defined as *standards* or pre-described patterns of action, which is based on individual or collective values.

Furthermore there should be a distinction between *juridical and legal* requirement/ norms/ standards, which is anchored in rules and regulations, and *moral* requirement/norms/ standards based on the choice of individuals or groups. To be right in a legal sense may be different from being right in a moral sense because the basis for the assessment is different. The issues of whistle blowing can illustrate a practical example of the difference between legal and moral norms and standards. A whistle blower leaking information of dangerous procedures or products may legally act in conflict with requirement/norms/standards of the firm and the action may be seen as a lack of loyalty in that perspective. However, stakeholders outside the firm may assess the act as moral and honourable.

There exist a vast body of knowledge regarding ethics and moral theory that may be taken into consideration. The table below gives an indication of how various moral theories may be applied towards different risk issues.

Knowledge is a necessary base for safe operation embedded in the new principles of "self-regulation" as a "*regime imposed from outside, in which part of the regulatory process is delegated to the organisation, but under condition which specify what must be regulated internally and how the outside world will check that this has taken place.*"

Table 1. For a further discussion of how ethical concerns relate to risk and safety issues (Erstal & Aven, 2005).

Level of operation	Main concern
Political authorities	Political guidance Use of precautionary principles
Local communities	Social assessment of major fatalities
Industry/Enterprises	Robust socio-technical design
Workplace	Active participation by workers Occupational Health and Safety

(Hopkins & Hale 2002, 1–2). This obligation to control activities according to criteria specified by the organisation itself depends on knowledge and mutual understanding among actors at different levels – from the legal, political level and to self-regulation in industrial relation at company level.

2.2.2 *Specific concerns*

In the table below stakeholders are grouped in three categories depending of the level they are operating. The highest level is the society at large, including local communities, which are affected, positively or negatively by the location of the industry, possible conflicting interest towards other industry (as fishery), risk of pollution, etc. The second level is the industry with enterprises as operators, entrepreneurs and suppliers. The third level is the local workplace with industrial relations.

Political guidance is how the politicians and authorities give their guidance or "signals" to the Regulator. In the Norwegian context the Storting (the Norwegian parliament) and Cabinet give guidance to the Norwegian Petroleum Safety Authority and the industry in the White papers. Here some issues are repeated and highlighted. Firstly the petroleum industry is described as a leading industry which continuously invests in knowledge and improvement by learning from best practise. "A levelling or decline in HSE performance is not in line with such objectives" the papers state. Secondly, the introduction of the "Zero-Philosophy" is seen as a milestone regarding attitude and behaviour in the industry. Thirdly the obligation of implementing international rules and regulation is stated. Such guidance contributes to priorities and focus areas for the regulating authorities, both in revising and developing the regulations, the inspection and instructing or guiding the industry.

The precautionary principle is rooted in Swedish and German environmental regulation from the end of 1960s and was highlighted in the Rio Declaration (1992): "*In order to protect the environment, the precautionary approach shall be widely applied by States according to their capabilities. Where there are threats of serious or irreversible damage, lack of full scientific certainty shall not be used as a reason for postponing cost-effective measures to prevent environmental degradation.*" The idea behind the principle is that if there is a lack of scientific certainty what will become the consequences of an action, then the action should not be carried out. After being applied regarding the external environment, the principle is more widely used. However, it is not unambiguous and the use of the principle is controversial (Löfstedt, 2003; Sandin, 1999).

In addition, the concept "cautionary principle" is useful, which says that in the face of uncertainty, *caution* should be a ruling principle. A leakage may occur, there are uncertainties, and we design for this by introducing safety barriers.

Societal concern regarding major accidents and environmental damages is related to the risks of catastrophes, affecting the whole industry and its reputation. The Alexander Kielland (1980) and Piper Alpha (1988) disasters are important points of references, where 123 and 167 persons were killed, respectively. The concern relates to the issue of what an acceptable risk level is, which is a thoroughly discussed issue in the literature, see e.g. HSE (2001) and Aven & Vinnem (2005) and the references therein. We also refer to the Risk level Norwegian sector project, see Vinnem et al. (2002) and Aven (2003, p.122), which bases its approach on a societal perspective related to major hazard risks, incorporating broad processes based on evaluations of risk perception, risk indicators, risk assessments, as well as experts panels and groups of representatives from the various interested parties.

The enterprise is responsible for a robust socio-technical design. In the design process the stakeholders should be aware of conflicting interest and how they should be solved. The regulations from the authorities frames the organisational and technological choice, and the enterprises assess

Table 2. Risk issue.

Risk issue	Moral theory	Represented by	Main concern
Risk exposure	Compulsory ethics Utilitarism	Immanuel Kant John S. Mill	Categorical imperative Maximise the utility for most
Acceptable risk level	Theory of Justice Discourse ethic	John Rawl Jürgen Habermas	Moral intuition The power of the best argument
Occupation Health and Safety	Ethics of nearness	Emmanuel Levinas	Ethics is rooted in the immediate relationship to "the Other"

the consequences and specify the criteria of risk acceptance (Frick et al. 2000).

The "Nordic Model" of a healthy working life (Kettunen, 1998) presupposes an active participation of the workers in job design, running operation and risk assessment. It is also embedded in the Working life Act and safety regulation. A major concern from authorities in the development of new regulation is to have a continuous focus on the role of workers and their unions as an educated and motivated resource to improve safety work.

2.3 *Stakeholders*

The petroleum industry in Norway is of significant importance for the national income, but also for employment, industrial and commercial development and competence development. At the same time the petroleum activities means a significant risk exposure to people and environment. Therefore interest groups, organisations and individuals are engaged and involved in the development of the petroleum industry, the frame conditions and the regulations the industry operates under.

In the framework emphasis is on three main categorises of stakeholders:

- Commercial stakeholders
- Employees
- Stakeholders representing the interest of the society.

The stakeholders may be presented in a hierarchical order with the political arenas and actors at the top and the work place with management, safety staff and workers at the bottom (Rasmussen, 1997).

Stakeholders representing the interests of the society are also focusing on areas outside the petroleum industry. This can be interests related to sustainable development of the fish population, clean ocean and air, tourist industry etc. Activities effecting the society and the environment usually engage stakeholders not directly involved in the industry. An important task for the regulating authority is to ensure that the interests of the whole society are properly taken care of. Key issues for the regulating authority is to balance the interests of different stakeholders in the society and the parties directly involved in the activities.

2.4 *The decision process on HSE regulation*

The following main steps are included in the proposed decision process on HSE regulation:

- Framing and problem definition
- Description of goals and objectives
- Identification of relevant concerns
- Evaluation of alternatives means and measures
- Identification of relevant effects/consequences
- Effect/consequence evaluations
- Governmental review and decision
- Implementation
- Evaluation.

In the framework we have discussed each of these steps and presented key issues that should be included in the evaluation (SEROS 2004). The key issues are summarized and presented as a checklist. The discussion and the checklist are to be used as a guideline through the decision process. An example is presented in the following textbox, the checklist "Description of goals and objectives".

The purpose of the proposed decision process is to establish a basis for the decision, that is well-structured and documented and at the same time gives the decision maker a good understanding of the pros and cons of the different alternatives.

Table 3. Key issues, description of goals and objectives.

KEY ISSUES – DESCRIPTION OF GOALS AND OBJECTIVES

- What should be changed?
- What factors should be considered constraints (eg. operational cost)?
- Are goals and objectives in line with the main concerns, i.e. expectations from society, political signals, visions and ideal goals?
- Timeframe; are changes needed immediately or is this a process that need to take some time?
- ...

An important part of the decision process is related to the evaluation of effects on the stakeholders of the different alternatives. This framework proposes a classification system were the alternatives are classified based on a high level evaluation of the different effects of the proposed alternative according to the dimensions risk and manageability. The classification system is presented in the following section.

By risk we understand the combination of possible consequences and uncertainties, where the uncertainties are expressed by probabilities when found appropriate. This perspective on risk is discussed in the appendix. It means a broad perspective on risk, acknowledging that risk cannot be distinguished from the context it is a part of, the aspects that are addressed, those who assess the risk, the methods and tools used, etc. Adopting such a perspective the consequence assessments for the HSE regulations need to reflect this, by

- focusing on different actors' analyses and assessments of risk

- what is acceptable risk and the need for risk reduction cannot be determined just by reference to the results of risk analyses
- risk perception has a role to play to guide decision makers; professional risk analysts do not have the exclusive right to describe risk.

3 ASSESSMENTS OF EFFECTS / CONSEQUENCES

For a specified HMS regulation measure, say M, we assess the consequences or effects of this measure seen in relation to the defined attributes (HSE, costs, reputation, etc.). Hence we first need to identify the relevant attributes (X_1, X_2, ...), and then assess the consequences of the measure M on these attributes. These assessments could involve qualitative or quantitative analysis. Regardless of the level of quantification, the assessments need to consider both what are expected consequences as well as uncertainties related to the possible consequences. In most cases of HSE regulation decisions the uncertainties would be large. In line with the adopted perspective on risk, the framework recommends a structure for the assessment according to the following scheme:

1. Identify the relevant attributes (HSE, costs, reputation, alignment with main concerns, ...)
2. Are the potential consequences of the measure M significant? The important point here is the *potential*. If there is a limited potential, further assessments are not required.
3. What are the assigned expected consequences, i.e. $E[X_i]$ given the available knowledge and assumptions?
4. Are there special features of the possible consequences? In addition to assessing the consequences on the quantities X_i, some aspects of the possible consequences might need special attention. Examples may for example be
   - the temporal extension,
   - aspects of the consequences that could cause social mobilization, i.e. violation of individual, social or cultural interests and values generating social conflicts and psychological reactions by individuals and groups who feel afflicted by the risk consequences.

   A system based on the scheme developed by Renn & Klinke (2002) is recommended, see Sandøy et al. (2005).
5. Are the large uncertainties related to the underlying phenomena, and do experts have different views on critical aspects? The aim is to identify factors that could lead to consequences X_i far from the expected consequences $E[X_i]$. A system for describing and characterising the associated uncertainties are presented in Sandøy et al. (2005). This system reflects features such as the current knowledge and understanding about the underlying phenomena and the systems being studied, the complexity of technology, the level of predictability, the experts' competence, and the vulnerability of the system. If a quantitative analysis is performed, the uncertainties are expressed by probability distributions.
6. The level of manageability during project execution. To what extent is it possible to control and reduce the uncertainties, and obtain desired outcomes? Some risks are more manageable than others, meaning that the potential for reducing the risk is larger for some risks compared to others. By proper uncertainty and safety management, we seek to obtain desirable consequences. The expected values and the probabilistic assessments performed in the risk analyses, provide predictions for the future, but what become the outcomes can be influenced. This leads to considerations on for example how to run processes reducing risks (uncertainties) and how to deal with human and organisational factors and obtain a good HES culture. Again we refer to Sandøy et al. (2005).

A matrix form has been developed that is used for each measure M, that reflects these aspects, and for each attribute (for example number of fatalities, number of injuries, societal costs, company costs, reduction in value added, etc.) the following points are addressed:

- potential of attribute (e.g. a certain number of fatalities)
- predictions of attribute (e.g. zero fatalities)
- expected value (e.g. 0.1 fatalities)
- probability distribution (e.g. expressing a probability of a "major accident")
- risk description on a "lower level" (e.g. prediction of number of leaks, expected number of leaks, etc.)
- uncertainty factors
- manageability factors
- aspects of the consequences.

These assessments provide a basis for making a decision. It is possible to introduce various type of scoring systems to integrate this information about the consequences of the measure M. Such a scoring system would mean some form of weighting of the various aspects, and would of course be subject to discussion as it is primarily value based. To simplify the decision-making process and obtain transparency in the process there is a need for some level of structure but we cannot replace the difficult value judgments by a mechanical procedure.

To what extent the analyses and evaluations provide clear recommendations on what decision alternative to choose would be measure dependent. Strict mechanical procedures for transforming the results of the analyses and evaluations to a decision cannot be justified. The analyses need to be evaluated in the light of the premises, assumptions and limitations of these analyses. The analyses are based on background information that must be reviewed together with the results of the analyses. The analyses and evaluations provide decision support – not a decision. When evaluating the decision support the decision maker needs to consider a number of issues, including

- Is the decision-making process managed and documented according to the decision principles and strategies formulated?
- What is the ranking of the alternatives based on the analyses and evaluations? What assumptions are the analyses and evaluations, and the ranking based on? What are the limitations of the analyses and evaluations?
- Are there concerns not taken into account in these analyses and evaluations?
- Are all relevant stakeholders taken into account? Would different weights of some stakeholders affect the conclusion?
- Robustness in the decision. What is required to change the decision?

It is often a requirement that the decision support should be available and the decision traceable and transparent, which means a documentation of which elements that have been given weight in the decision. Such a requirement does not apply in general, the point is that the process is in line with the decision principles and strategies defined. The authorities may not consider it desirable to trace all weights given to the various attributes. There will always be a trade-off between accountability, transparency and cost-effectiveness in the decision process.

The above approach does not exclude the use of cost-benefit analyses as a tool for supporting the decision-making. However, in most cases, such analyses need to be supported by information as discussed above. Cost-benefit analysis is not an objective procedure for ranking alternatives. The analyses have limitations and are based on a number of assumptions and presumptions, and their uses are based not only on scientific knowledge, but also on value judgments involving ethical, strategic and political concerns. Some of the challenges related to these tools are: assessment of uncertainties and assignment of probabilities, determination of appropriate values for quantities such as a statistical life and the discount rate, to distinguish between objective knowledge and subjective judgements, treatment of uncertainties and the way of dealing with intangibles. This discussion on the use of cost-benefit analyses have been going on for many years, see e.g. Ackerman & Heinzerling (2004) and Aven (2003).

We conclude that risk analyses, cost-benefit analyses and related type of analyses provide support for decision making, leaving the decision makers to apply decision processes outside the direct applications of the analyses. Managerial reviews and judgements are important elements in the decision process. It is not desirable to develop tools that prescribe or dictate the decision. That would mean a too mechanical approach to decision making and would fail to recognise the important role of management to perform difficult value judgments involving uncertainty. Given the methodological and technical challenges related to risk and cost-benefit analyses, such analyses should only be given status as a way of systemizing information. Sensitivity analysis should be seen as an important part of these analyses.

4 CONCLUDING REMARKS

We have presented a decision framework for HES regulations. This framework can be used together with existing standards on risk and safety management, such as AN/NZS 4360 (2004) and ISO (2005), as the framework follows standard procedures for risk and safety management. However, there are some features of the framework that extend beyond these standards, and these features represent the main contribution of the framework and this paper:

- The structure of the framework presented in Section 2
- The adoption of a perspective on risk in which risk is seen as the combination of possible consequences and associated uncertainties
- A system for assessing the alternatives with respect to the dimensions possible consequences, uncertainties and manageability.

The possible consequences of HSE regulation measures are often difficult to assess. The uncertainties are significant, and there are different views. To be able to structure the decision process, the proposed framework would provide guidance, but the difficult value judgments cannot be avoided. The use of cost-benefit analysis and other decision tools does not replace management review and judgements. In the face of uncertainties, *caution* should be a ruling principle. The level of caution, and cautionary, is primarily a management issue, and not science.

REFERENCES

Ackerman, F. & Heinzerling, L. 2004. *Prizeless.* NY: The New Press.

AS/NZS 4360, 2004. *Australian/New Zealand Standard: Risk management.*

Aven, T. 2003. *Foundations of Risk Analysis – A Knowledge and Decision-oriented Perspective.* New York: John Wiley & Sons Ltd.

Aven, T. & Kristensen, V. 2005. Perspectives on risk - Review and discussion of the basis for establishing a unified and holistic approach. *Reliability Engineering and System Safety* 90: 1–14.

Aven, T., Vinnem, J.E. & Wiencke, H.S. 2006. *A decision framework for HES management. Reliability Engineering and System Safety.*

Bedford, T. & Cooke, R.M. 2001. *Probabilistic Risk Analysis.* Cambridge: Cambridge University Press.

Cabinet Office, 2002. *Risk: improving government's capability to handle risk and uncertainty.* Strategy unit report, UK.

Cox, L.A. Jr. 2002. *Risk Analysis. Foundations, Models and Methods.* Boston: Kluwer Academic Publisher.

Douglas, E.J. 1983. *Managerial Economics: Theory, Practice and Problems,* 2nd ed. New Jersey: Prentice Hall.

Erstal, G. & Aven, T. 2005. On the ethical framework for decision making, Paper submitted for publication.

Frick, K., Langaa Jensen, P., Quilan, M. & Wilthagen, T. (eds.) 2000. *Systematic occupational health and safety management.* Amsterdam: Elsevier Science, Pergamon.

Hopkin, A. & Hale, A. 2002. Issues in the regulation of safety: Setting the scene, in B. Kirwan, A. Hale & A. Hopkins (eds.), *Changing Regulation.* Amsterdam: Pergamon, Elsevier Science.

ISO, 2002. *Risk Management vocabulary.* International Organization for Standardization ISO/IEC guide 73.

ISO, 2005. *Risk Management – General Guidelines for Principles and implementation of Risk management.* ISO/TMB/WG Risk Management, Discussion draft. 2005-06-22.

Kasperson, R.E. 1992. The social amplification of risk: Process in developing an interrative framework, in S. Krimsky & D. Golding (eds.), *Social Theories of risk:* 153–178. Westport: Praeger.

Kettunen, P. 1998. *Globalisation and the criteria of "us" – a historical perspective on the discussion of the Nordic model and new challenges, in Global Redefining of Working Life.* Copenhagen: Nordic Council of Ministers

Löfstedt, R.E. 2003. The precautionary principle: Risk, regulation and politics. *Trans IchemE* 81: 36–43.

Okrent, D. & Pidgeon, N. (eds.) 1998. Risk assessment vs. risk perception. *Special issue of Reliability Engineering and System Safety* 59: 1–59

Pidgeon, N.F. & Beattie, J. 1998. The psychology of risk and uncertainty. In P. Calow (ed.), *Handbook of Environmental Risk Assessment and Management*: 289–318. London: Blackwell Science.

Rasmussen, J. 1997. Risk management in a dynamic society: A modelling problem. *Safety Science* 27(2/3): 183–213.

Renn, O. & Klinke, A. 2002. A New approach to risk evaluation and management: Risk-based precaution-based and discourse-based strategies. *Risk Analysis* 22(6): 1071–1094.

Sandin, P. 1999. Dimensions of the precautionary principle. *Human and Ecological Risk Assessment* 5: 889–907.

Sandøy, M., Aven, T. & Ford, D. 2005. On integrating risk perspectives in project management. *Risk Management: an International Journal* 7: 7–21.

SEROS, 2004. *Important issues related to the authorities' regulation of HSE in activities exposed to large scale risks, with focus on the petroleum sector* (In Norwegian). SEROS report 91886, University of Stavanger/Rogaland Research.

Vinnem, J.E., Aven, T., Husebø, T., Seljelid, J. & Tveit, O.J. 2005. Major hazard risk indicators for monitoring of trends in the Norwegian offshore petroleum sector. *Reliability Engineering and System Safety,* available online 12 September 2005.

Risks Challenging Publics, Scientists and Governments – Menoni (ed)
© 2010 Taylor & Francis Group, London, ISBN 978-0-415-58072-4

Nutritional balance, the key to reducing the risks to the health of school children

R. Nelson
Communication Department, CAFRE, Loughry Campus, Cookstown, UK

D. McCleery & N. Windrum
Food Science Department (Food Microbiology), Queen's University Belfast, Belfast, UK

J. Phelan, A. Markey & R. Sweetnam
Department of Agribusiness, Extension and Rural Development, University College Dublin, Ireland

ABSTRACT: Dietary habits developed during pre-adolescence have the potential to be replicated in later life. These habits may pose a number of risks to human health. This includes conditions associated with obesity and nutritional disorders. It is widely recognised that the best diet is one that is balanced according to guidelines related to the five main food groups. The aim of this study was to determine the current situation regarding the consumption patterns of school children aged 9–11 across the island of Ireland. To meet this aim, 2861 school children completed food diaries, which recorded the consumption of all food and drink over a 24-hour period. Results showed that 83% of diets were classified as very poor or worse. Substitution of one or two portions of fatty/sugary foods by fruit and vegetables dramatically improved the situation and are recommended as a means to encourage both short term and long-term improvements.

1 INTRODUCTION

The risks associated with poor dietary intake of many people have been widely reported in both scientific literature and the wider media (Gregory et al. 2000). The outcomes associated with these risks generally fall into two categories – those that may lead to overweight and obese conditions, and those, which are more chronic in nature and are related to nutritional disorders. The former has received the most attention recently as the incidence of obesity levels in the population as a whole has risen significantly in the past two decades (Anon, 2003; Campbell, 2004; Lowell, 2004), and the risks associated with obesity have been related to a number of health problems. These include type 2 diabetes, heart disease, joint problems and cancer (Dietz, 1998; McPherson et al. 1995; Warwick et al. 1999). Although energy balance is central to weight loss or gain, the consumption of energy dense food (those high in fat and sugar) is seen as a major contributing factor. This is especially so in situations were there is a reduction of energy expenditure as a result of changes in leisure activities (Anderson et al. 1999; Anon, 1996; Prentice & Jebb, 1995). Excess energy consumed, which leads to accumulation of fat, is a major worry especially if it is deposited around the waistline (Chan et al. 1994).

Whilst the problems associated with obesity are currently receiving widespread coverage, it is also recognised that long-term health effects can also be influenced by the composition of the diet. A wide range of different foods in the diet has also been shown to provide a beneficial and protective effect to the individual (Heausman & Mellentin, 2001). As a result, dietary recommendations have been provided to the consumer to help combat diseases and disorders such as coronary heart disease, anaemia and colon cancer (Cooke, 2004). This has included recommendations to use whole grains wherever possible, select mono saturated oils, and limit the intake of sodium in the form of salt (Anon, 2003). Consequently the diet is seen as of primary importance and key to the maintenance of long-term health.

1.1 *The concept of a balanced diet*

It is generally agreed among health professionals that a diet, which is not balanced nutritionally, may contribute to either acute or chronic health problems Eurodiet Project Steering Committee (2001). From this perspective a number of models have been designed to represent the features of a balanced diet, usually in the form of pictorial guides such as segmented plates or pyramids of foods representing each of five food groups (carbohydrates, proteins, dairy, fruit & vegetables and fat/sugar). Although each model may have different ways to present this message they all have the following common features, i.e. the best diet is one that is based on carbohydrates (breads, potatoes and other cereals) and is rich in fruit and vegetables. It should also include moderate amounts of milk and dairy products, meat, fish or meat/milk alternatives, and limited amounts of food containing fat and sugar (Hunt et al. 2004; Teobald, 2004; Beauchamp & Moran, 1984).

From these broad descriptions the following recommended daily intake of foods from each of the food groups are advised:

- Four to six portions of carbohydrates
- At least five portions of fruit and vegetables
- Two portions of dairy
- No more than two portions of protein
- No more than two portions of food rich in fat & sugar

Using these guidelines concerns have been raised that people are consuming in particular a lower amount than recommended of fruit and vegetables per day, whereas the consumption of foods rich in fat and sugar are above that recommended. This as a consequence leads to an unbalanced daily diet (Vereecken & Maes, 2000).

1.2 *Development of eating habits*

Eating and consumption habits are mainly developed in childhood as a result of the interaction of two factors, (i) a genetic predisposition (ii) moderated by the immediate experiences associated with the child's environment (Longbottom, 2002; Hursti, 1999; Mela, 1999; Ton Nu et al. 1996). Genetically there is evidence to show that there is an innate desire to eat foods that will immediately provide energy and satiate hunger (Rozin, 1990). Foods that are high in fat and sugar fall into this category. The advantage of these foods is that the energy stored within them also has the potential to be stored for times of need, typically observed with the laying down of adipose tissue. The individual can then use this as an energy source when food is not so plentiful. Genetic diversity does lead to differences in individuals, which may lead to variations in food desires. However the early years environment of the child moderates the genetic predisposition. For example there is evidence that the consumption of foods high in fat are also influenced by the immediate environmental surroundings leading again to differences in consumption between groups of people (Murcott, 2002). This results in the observed wide-ranging differences in diet seen between families across the world, and also reflects the geographical availability of foods as to how they are prepared and presented.

Consequently, as consumption patterns may be formed early in life (and can then be potentially replicated in adulthood), this has implications for the long-term health of individuals. The aim of this study therefore was to determine the nutritional balance of primary school children's diet across the island of Ireland. This would identify the current situation regarding consumption patterns and pinpoint specific areas of potential risk.

2 METHOD

Data from epidemiological and microbiological investigations are often difficult to extrapolate to real life situations and as such it is difficult to identify actual behaviour (Redmond & Griffith, 2003). Also methods employing direct observation are often socially unacceptable and it has been shown that those people being observed tend to change their behaviour under scrutiny (Anderson & Bell, 2000). For this reason it was decided to use a food diary approach. Primary school children were selected to record the appropriate data as from the researchers' experience this group are less likely to convey a positive image that may be perceived as correct, rather than directly report that which may be thought of as undesirable. The schools involved were supportive as the skills associated in the completion of the diaries complemented the requirements set for the primary school curriculum, in both Northern Ireland and the Republic of Ireland.

Seventy-eight primary schools were visited individually across Ireland on three consecutive weeks during 2004 and completed a food diary detailing all food and drink consumed over a one-day period. A total of 2861 pupils aged between 9–11 fully completed their diaries, 1624 (57%) from NI and 1437 (43%) from ROI. One thousand nine hundred and forty five (68%) pupils came from urban schools whilst 916 (32%) pupils come from rural schools. Those pupils who had brought a home packed lunch to school recorded each food present in their lunch box. This applied to 2549 pupils, 631 from NI and 1918 from the ROI. This split reflected the difference in meal provisions within the two jurisdictions.

Each diary consisted of separate sections, and each meal occasion was recalled from the previous day up until the time of the visit. This provided information for breakfast, lunch (or lunch box), evening meal, and snacks taken during the day. Each diary was then coded according to the food consumed within each of the five food groups based upon the classification previously described. Portion sizes were estimated for each entry based upon discussion with each pupil. Analyses were based upon the frequency, amounts consumed and the timing of when the food was eaten during this 24-hour period.

3 RESULTS

The numbers of portions of each of the five food groups consumed in one day are presented in Figures 1–5. The consumption patterns for carbohydrate, protein and dairy foods are as expected. However over 25% of children did not eat any portions of fruit and vegetable within the one day period recorded, whereas 4% had eaten the recommended amounts of fruit and vegetables. The consumption of foods high in fat or sugar also indicates that there are a high percentage of children (33%) who are consuming more than 4 portions per day.

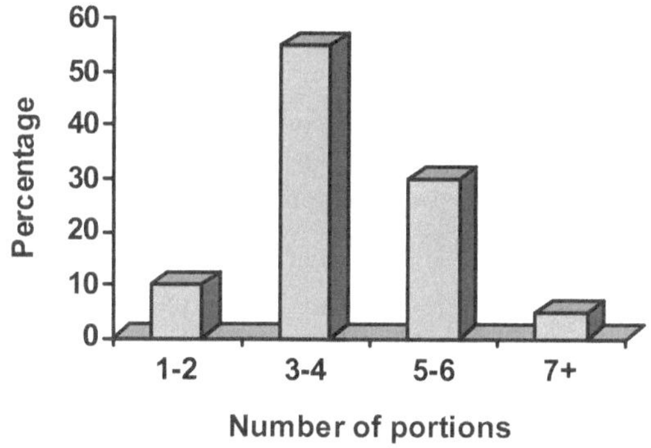

Figure 1. Number of portions of carbohydrate consumed over a one-day period (n = 2861).

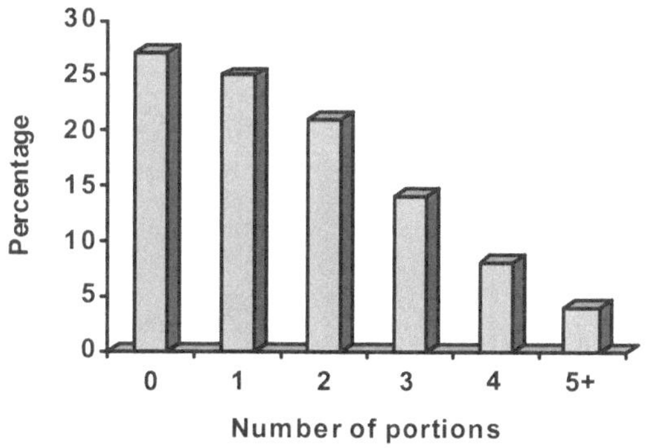

Figure 2. Number of portions of fruit and vegetables consumed over a one-day period (n = 2861).

In order to determine the occasions when food from each of the five food groups was consumed, they were allocated to one of four meal occasions, breakfast, lunch, evening meal or snacks. These results are presented in Figure 6.

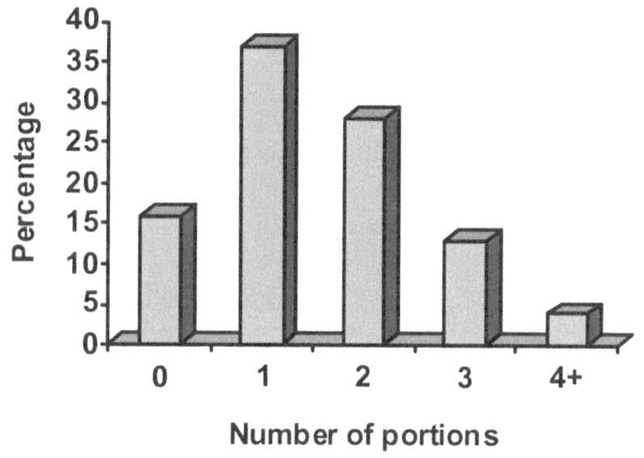

Figure 3. Number of portions of dairy consumed over a one-day period (n = 2861).

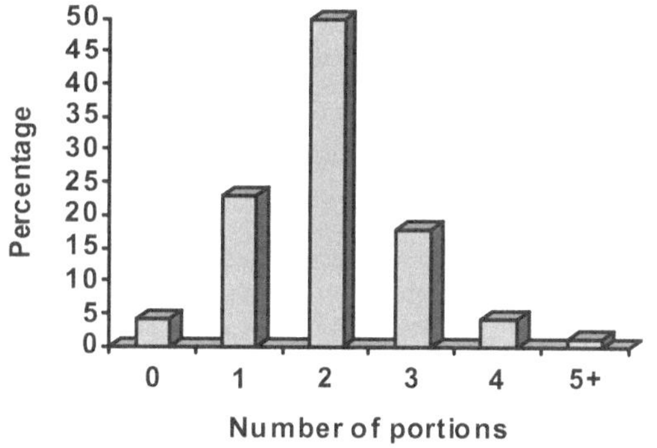

Figure 4. Number of portions of protein consumed over a one-day period (n = 2861).

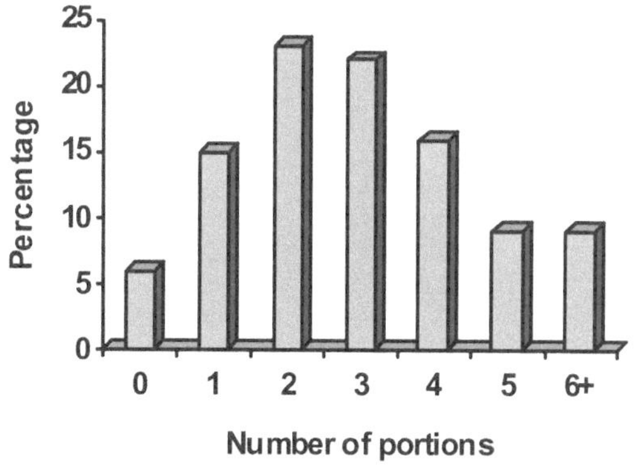

Figure 5. Number of portions of foods high in fat and sugar consumed over a one-day period (n = 2861).

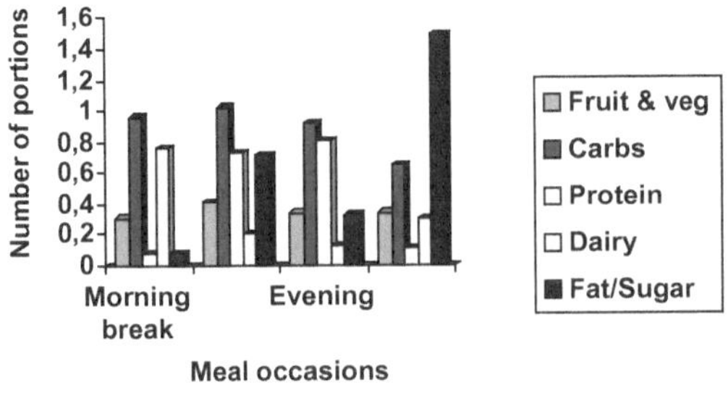

Figure 6. Meal occasions and the consumption of food (n = 2861).

The consumption patterns for carbohydrate, fruit and vegetables, dairy and protein foods (if perhaps not optimum) are as expected. However the timing associated with consumption of foods high in fat and sugar (mainly items such as confectionary and crisps) reveals that this takes place mainly during snacking. It is this area that potentially has the greatest effect on the overall daily balance of the diet.

In order to determine how balanced a daily diet was a daily balance score was calculated. The optimum consumption of each food group was determined from the recommendations and given a score of 1. If a pupil consumed the optimum amount of food from each group then they gained a score of five. Consumption amounts outside this ideal were then scored on a sliding scale according to each group and an additive score for each pupil calculated. This took into account under- and over-consumption in each of the food groups. The daily balance score for the children in this study ranged from 5 to 21, normally distributed around a mean of 10.68 with a standard deviation of 2.197 (n = 2,265, a reduction on previous due to some incomplete and/or missing data sets). The results of this can be observed in Figure 7.

Figure 7. Distribution of the daily balance score.

Scores from this distribution were then assigned into 5 groups perfect, good/adequate, poor, very poor and those showing serious concern. These allocations were based upon equal groupings of the raw data scores. The results of this indicate that 0.4% of primary school children's diet during the day of the visit would be classified as perfect with 85% classified as either poor, very poor or showing serious concern.

4 SENSITIVTY ANALYSIS

To determine the effects of specific improvements to the diet a substitution exercise was carried out. The balance score was recalculated taking into account the effect of substituting one, and then two portions of foods containing high amounts of fat or sugar with an equal number of portions of fruit and vegetables. The results of this hypothetical substitution in comparison to the original data can be seen in Figure 8 below.

From this substitution calculation an improvement in the balance of the diet can be observed. One substitution and the percentage of diets classified as poor to serious fell from 85% to 61%, and with two substitutions it fell again to 33%. With two substitutions, the diets showing serious concern would have been only 0.4%.

Two factors, urban/rural and north/south were also investigated to see if they affected the consumption patterns of the five food groups as well as their effects on the overall balance score. The results of these analyses are presented in Table 1 below.

Some statistically significant results were obtained highlighting differences between the geographical area and region, especially between Northern Ireland and the Republic of Ireland with respect to balance. A significant difference was observed for fruit and vegetable consumption where children in the Northern Ireland consumed a higher amount than that in the Republic of Ire-

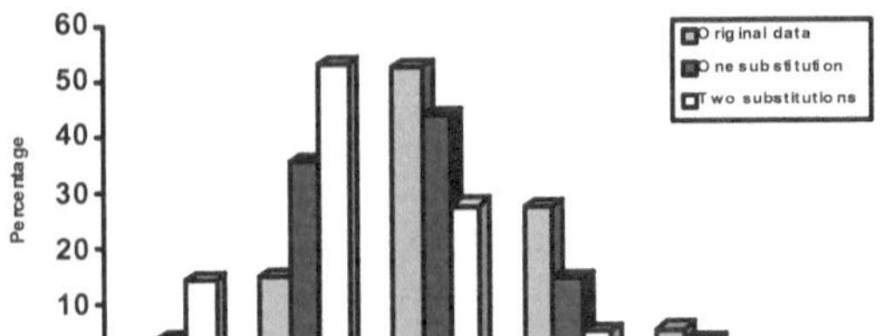

Figure 8. Original distribution of balanced diets – and the effect of substitution of fat/sugar foods with fruit/vegetables.

Table 1. The effect of geographical area and region on the consumption of the five food groups and the daily balance score.

Number of portions of	N	Mean Score	SD	t-value	df	Probability $p =$
Carbohydrates						
Urban	1567	4.01	1.35	-4.011	2263	**0.001**
Rural	698	4.27	1.49			
North	1017	4.26	1.46	5.209	2263	**0.001**
South	1248	3.95	1.33			
Fruit and Vegetables						
Urban	1567	1.59	1.39	-2.196	2263	**0.028**
Rural	698	1.73	1.532			
North	1017	1.72	1.53	2.489	2263	**0.013**
South	1248	1.57	1.35			
Dairy						
Urban	1567	1.55	1.12	-1.837	2263	0.066
Rural	698	1.04	1.19			
North	1017	1.57	1.20	-0.142	2263	0.887
South	1248	1.58	1.09			
Protein						
Urban	1567	2.0	0.93	-0.249	2263	0.803
Rural	698	2.0	0.93			
North	1017	2.0	0.91	0.03	2263	0.976
South	1248	2.0	0.94			
Fat and Sugar						
Urban	1567	2.94	1.76	-1.036	2263	0.3
Rural	698	3.02	1.82			
North	1017	3.31	1.95	8.504	2263	**0.001**
South	1248	2.68	1.57			
Balance Score						
Urban	1567	10.69	2.16	0.331	2263	0.741
Rural	698	10.60	2.29			
North	1017	10.95	2.37	5.213	2263	**0.001**
South	1248	10.47	2.02			

land. This was however offset by the significant difference in the consumption of foods high in fats and sugar. This indicated that school children in the Republic of Ireland consume less foods of this nature, which then contributes to an overall better balance score. This difference may reflect differences in school policy in the two regions and access to snack foods from outlets and shops.

5 CONCLUSION AND RECOMMENDATIONS

Snacking behaviour is seen to be an important contributor to the unbalanced nature of the diet. Currently only a small percentage of primary school children's daily diet can be described as balanced. This however can be changed with substituting foods that are high in fat/sugar with fruit and vegetables.

As the majority of meals consumed by primary school children are prepared by adults (either at home or in the school) and as dietary habits are formed early in childhood, then adults have a primary responsibility for the proportion of balanced meal occasions for the children under their control.

However the emphasis should be on ensuring long-term changes in the dietary habits of children. In this respect the development of healthier snacking behaviour is key to improving the overall diet. Whenever possible children should be encouraged to snack on fruit. This has the potential to develop good habits that will then persist into adulthood. To assist in this, fruit and vegetables need to become more available and more convenient, as well as portraying a 'cool' and desirable image. This requires a co-ordinated approach by educators and government in the development of school- and community-based curricula. The effects of this can be observed when differences when diets of children from Northern Ireland and Republic of Ireland are analysed, which represents two different jurisdictions. Each has a different policy that affects the final balance of the diets of the school children. This also needs reinforcing through good diet choices by the whole family during meal occasions.

Finally, the sustainability of any proposed interventions must be evaluated on a regular basis and models that reflect the nature of the behaviour desired. As the balanced diet is the concept to be promoted, and eating is essential, it is dietary change that is required as opposed to a cessation of any particular practice.

REFERENCES

Anderson, A.S. & Bell, A. 2000. The impact of a dietary intervention in a community based breakfast club: nutrient intake and measurement issues. *The proceedings of the Nutrition Society* 59, 25A.

Anderson, R.E., Waddon, T.A., Bartlett, S.J., Zemel, B., Verde, T.J. & Franckowaik, S.C. 1999. Effects of lifestyle activity v.s. structured aerobic exercise in obese women. *Journal of American Medical Association* 281: 335–340.

Anon, 1996. US Department of Health and Human Services (PHS) (1996). *Physical activity and health. A report of the Surgeon General (Executive Summary)*. Pittsburg, P.A: Superintendent of Documents.

Anon, 2003. Childhood Obesity September 2003. *Postnote report from the Parliamentary Office of Science and Technology* Millbank, London.

Beauchamp, G.K. & Moran, M. 1984. Acceptance of sweet taste preference in human infants. *Appetite* 3: 139–52.

Campbell, J.W. 2004. Obesity in Children – facts and fallacies. *European Journal of Lipid Science Technology* 106: 334–339.

Chan, J.M., Rimm, E.B. & Colditz, G.A. 1994. Obesity, fat distribution and weight gain as risk factors for clinical diabetes in men. *Diabetes Care* 17: 961–969.

Cooke, L. 2004. The development and modification of children's eating habits. *Nutrition Bulletin* 29: 31–35.

Dietz, W.H. 1998. Heath consequences of obesity in youth: Childhood predictors of adult disease. *Pediatrics* 101: 518–525.

Eurodiet Project Steering Committee 2001. Eurodiet Core Report. Nutrition and diet for healthy Lifestyles in Europe. *Public Health Nutrition* 4: 256–273.

Gregory, J., Lowe, S., Bate, C.J., Prentice, A., Jackson, L.V., Smillers, G., Wenlock, R. & Fairon, M. 2000. *National Diet and Nutrition Survey: young people aged 4 to 18 years. Volume 1: Findings.* London: The Stationery Office.

Healthy Eating: A whole diet approach 2003. British Nutrition Foundation.

Heausman, M. & Mellentin, J. 2001. *The Functional Food Revolution – Healthy People, Healthy Profits*: 3–197. UK: Earth Scan Publications Limited.

Hunt, P., Strong, M. & Poulter, J. 2004. Evaluating a new food selection guide. *Nutrition Bulletin* 29: 19–25.

Hursti, U.K. 1999. Factors influencing children's food choice *Ann. Med.* 31 Suppl. 1: 26–32.

Longbottom, P.J., Wrieden, W.L. & Pine, C.M. 2002. Is there a relationship between the food intakes of Scottish 5½ - 8½ year olds and those of their mothers? *Journal of Human Nutrition and Dietetics* 15(4): 271–279.

Lowell, J. 2004. The Food Industry and its impact upon increasing global obesity: A case study: *British Food Journal* 103(3): 238–248.

McPherson, R.S., Montgomery, D.H. & Nichamen, M.Z. 1995. Nutritional status of children: what do we know? *Journal of Nutrition Education* 27: 225–234.

Mela, D.J. 1999. Food Choice and Intake: The Human Factor. Proceeding of the Nutrition Society 58: 513–521.

Murcott, A. 2002. Nutrition inequalities. A note on sociological approaches. *European Journal of Public Health* 12: 203–207.

Prentice, A.M. & Jebb, S.A. 1995. Obesity in Britain: Gluttony or Sloth. *British Medical Journal.* 311: 437–439.

Redmond E.C. & Griffith, C.J. 2003. Consumer food handling in the home: A review of Food Safety Studies. *Journal of Food Protection* 66(1): 130–161.

Rozin, P. 1990. Acquisition of stable food preferences. *Nutr. Rev.* 48: 106–13.

Teobald, H.E. 2004. A whole diet approach to healthy eating. *Nutrition Bulletin* 29: 44–49.

Ton Nu, C., MacLeod, P. & Barthelemy, J. 1996. Effects of age and gender on adolescents' food habits and preferences. *Food Quality and Preference* 7(3–4): 251–262.

Vereecken, C. & Maes, L. 2000. *Eating habits, dental care and dieting*. In Health and Health behaviour among young people: Health Behaviour in school-aged children: A WHO Cross-National Study (HBSC) International Report. WHO Policy Series Health Policy for Children and Adolescents Issue 1. Eds. C. Currie, K. Hurrelmann, W. Settertobutle, R. Smith, & J. Todd. Copenhagen: WHO Regional Office for Europe.

Warwick, McIlveen & Strugnell, 1999. Food choices of 9–17 year olds in Northern Ireland – influences and challenges. *Nutritional Food Sciences* 5: 229–236.

Risks Challenging Publics, Scientists and Governments – Menoni (ed)
© 2010 Taylor & Francis Group, London, ISBN 978-0-415-58072-4

Release of toxic substances into the atmosphere: Forecast, prevention and protection of the human health Organizational model for the management of Civil Protection interventions

G. Evangelista, G. Francioni & E. Tonelli
Tecnomediterranea Sr, Rome, Italy

ABSTRACT: This project aims to provide an effective instrument for the organization of Civil Protection interventions following major accidents that provoke the release of toxic substances into the atmosphere. Supported by a solid normative base it shows an overview of the current Italian situation: not only a detailed description of the type of factories situated on the territory but also all instruments, institutions, persons and means to manage the Civil Protection interventions are presented. The key concept of planning and management of emergency cases is to try to preview possible events and to let a flexibility margin for the unforeseeable. The procedures have to become automatic but the answer model to the emergency must be sufficiently flexible. In the disaster management in order to plan the emergency, we have used "scenes" that provides a picture of what can happen to in case of the disaster and how the latter can be faced. An example is presented as an application of the mentioned method.

1 SCOPES AND GOALS

The main project targets are firstly the industrial risk analysis through critical estimation of the available information according to the law in force, safety analysis, risk areas location. Secondly, the definition of an organizational model of Civil Protection and the management of the emergency interventions in the incidental event of a release of toxic substances (liquids, vapours, fumes, dusts, fogs or the like) into the atmosphere are of interest.

2 MANAGEMENT REGULATION OF INDUSTRIAL RISK

As a starting point we have to consider the principal laws on the sector of industrial risk. The first law was the directive 82/501/CE, following the accident that happened on 1976 in the ICMESA factory in Seveso (Milan, Italy); it was emanated on 1982, named Seveso I and accepted in Italy on 1988 through D.P.R. 175. This European Directive evolved through the following years, until the Directive 96/82/CE, called Seveso II, was emanated in 1996 and accepted in Italy in 1999 through D.Lgs. 334. Currently, this decree is the most important law in force: It lists the factory classification with respect to the typology and the used quantity of toxic substances (C, B, A2 and A1 category), the Public Administrations involved in emergency interventions (Environment Ministry, Districts, Provinces, Municipalities, Fire Department etc) and the respective tasks, as well as, the control and vigilance systems in Italy through inspections and technical investigations. This decree has been followed from a series of decrees as DM 9/8/2000 (safety management system), DM 19/3/2001 (fire prevention), DM 9/5/2001 (urbanization control), DM 16/5/2001 (industrial and oil ports) and last but not least DPCM 25/2/2005 (guidelines for Emergency External Plan development).

After another catastrophic accident happened in Tolosa in 2001 (28 dead, 2000 wounded and 22000 evacuates) the European Community has emanated the directive 2003/105/CE called Seveso III; it accepted in Italy in 2005 through D.Lgs. 238.

In short this directive modify the application field with extension to some fields not comprised in the previous norm, it insert information for the population of all public structures and review the limits of the substances for the directive applications.

3 INDUSTRIAL RISK MAP IN ITALY

For constructing an industrial risk map of Italy, the data of a census, which identifies every 6 month the factories bound to the articles 6 and 8 of the Decree Legislative 334/99, was used. The analysis of the development process of industrial risk map began with the determination of the factories under considerable accident risk due to particular industrial activities. Also, the location of these factories by GIS, the kind of activity, the installation typology in every factory, the typology and quantity of the substances worked with, the possible incidental events, the vulnerable structures around the factory (hospitals, schools, commercial centre, offices, railway, roads, ports, airports protected areas, rivers etc) and the possible domino effects had to be identified. The data for these specifications, updated on 30 April 2005, were taken from ISPESL (National Authority for Health and Safety at Work), APAT (Environmental Protection Agency and Technical Services), the Environmental Ministry and ISTAT (National Institute of Statistics). In Italy, 1116 industrial factories bound to the decree legislative 334 were identified, concentrated mainly in the North, i.e. in Lombardia (about 23%), Piemonte (10%), Emilia Romagna (10%) and Veneto (8%) Other regions with important factory concentrations are Lazio, Campania and Sicily. The same analysis allowed to recognize the prevalent typologies of activities: storage of mineral oil (26%), chemical and petrochemical installations (25%) and storage of liquid gas (23%); lastly we have shown all Italian factories with name, location, substances worked and respective quantities.

Table 1. Factories distribution on Italian territory subject to the decree legislative 334/99.

Region	ART 6	ART 8	TOT
Abruzzo	13	9	22
Aosta	3	2	5
Basilicata	4	3	7
Calabria	5	7	12
Campania	49	27	76
Emilia Romagna	58	48	106
Friuli Venezia Giulia	19	12	31
Lazio	47	37	84
Liguria	18	17	35
Lombardia	147	113	260
Marche	9	5	14
Molise	1	4	5
p.a. di Bolzano e Alto Adige	8	3	11
p.a. di Trento	6	2	8
Piemonte	74	34	108
Puglia	27	20	47
Sardegna	22	23	45
Sicilia	35	34	69
Toscana	37	25	62
Umbria	10	7	17
Veneto	56	36	92
	648	468	1116

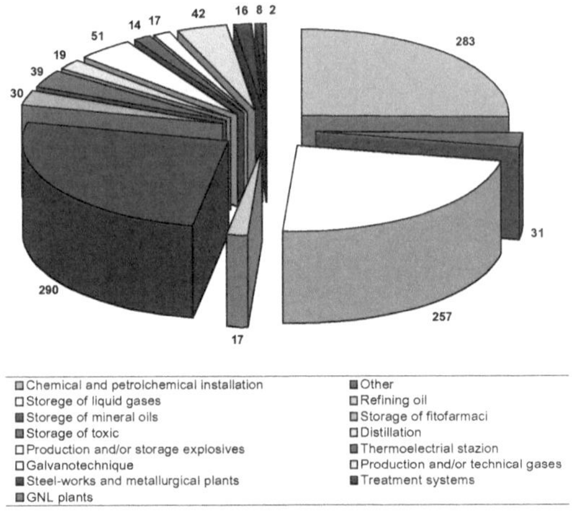

Figure 1. Prevalent typologies of activities in Italy.

4 RISK ASSESSMENT

In order to determine the risk areas, we started from an elementary definition of risk as a product of event frequency (F) by a magnitude of consequences (M). Figure 2 shows that the curves iso-risk (R1, R2, R3) assume a hyperbolic trend on the Cartesian plane: the decrease of event frequency is obtained through interventions of prevention whereas the decrease of magnitude of the consequences can be obtained through interventions of protection.

Risk analysis in industrial contexts consists of four integrated processes: identifying underlying sources of risk, determining the pathways by which such risks can materialize, estimating the

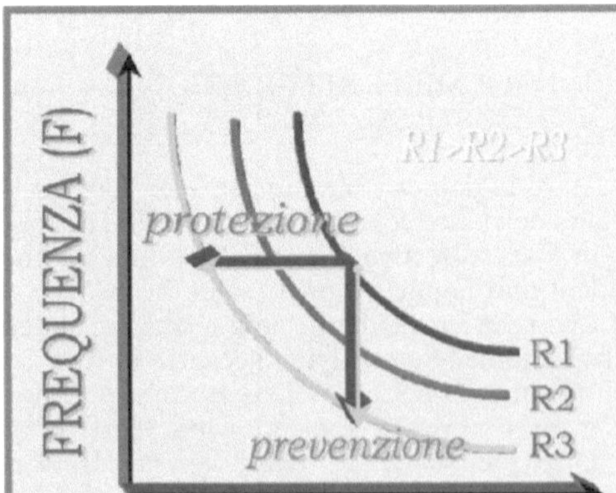

Figure 2. Curves of iso-risk and the actions of prevention and protection.

possible consequences of these risks under various scenarios and providing the means for mitigating and coping with these consequences.

The risks deriving from the production and use of dangerous substances can be of three types: risks connected to the substances (properties of the used substances), process risks (kind of work) and layout risks (system configuration); all these elements determine the typologies of the different incident events (release of toxic substances, fires, explosion and bursts).

There are multiple techniques of risk assessment: HAZOP (Hazard and Operability Analysis) in which the typologies of breakdown are estimated by the logical cause-consequence-remedy, FMEA (Failure Mode and Effect Analysis) which proceeds to the analysis of the system in inductive and qualitative way taking in consideration every single member of the system. Besides these qualitative methods, there are quantity ones that use a frequencies-consequences matrix. In this matrix, every possible accident is associated with a category, in relation to the event frequency and the gravity of the consequences: unacceptable (must be reduced in short time), undesirable (must be reduced in the long term), acceptable with verification control systems and acceptable without ties. Otherwise an inductive analysis of the emergency cases can be executed by event tree. In this system, the several possible sequences of propagation and arrest of the accident are the result of a predefined initiator event which will represent the departure point of the tree.

The diffusion of toxic substances deriving from incidental events have been simulated through the development of algorithms and mathematical equations which take into consideration the concentration, the conveyance and the diffusion of toxic substances into the atmosphere. The decision which model to use depends on many factors, such as scale space-time, orographic and meteorological conditions, source and substance typology. Among the most used models we have analytic models "at plume" (e.g. ISC3 and AERMOD) characterized by simplicity, not many meteorological input and steady space-time, three-dimensional models "at puff" (e.g. CALPUFF) characterized by more complexity, emission and parameters instability, three-dimensional Lagrange's models at particles through simulation with pseudo-particles and three-dimensional Eulero's models "at grid" that consist in integration of differential equation of mass conservation for each pollutant.

For every model, several methods can be applied, depending on type and requirements of desired output, atmospheric phenomenon (conveyance, deposition, reaction), land characteristics (urban or rural), available resources (human, economics, hardware) and type and quantity of available input data.

The evaluation of risk areas around the factory begins with defining the vulnerable elements (people and structures) through the PROBIT model. The next step is to calculate the damage probability through different functions for every incidental event depending on specific characteristic parameters: for example for toxic substances $P = a + b \cdot \ln(C^n \cdot t)$ where P is death probability, t is exposure duration in second, C is the contaminant concentration intensity and a, b and n are three constants depending on injurious substances.

Among the various methods applied for the calculation of the zones to risk (TNO, IAEA and quick method) the last one is the most frequently used in Italy and has been introduced from the guidelines of the 1994 and dawned by DPCM 25 February 2005. As input, it uses several parameters (physical state and quantity of the dangerous substance, type of incident event) and sure impact and damage distances are determined through tables and simple formulas. The distances are obtained in standard conditions of atmospheric stability (D5 in Pasquill-Gifford classification). In order to determinate the distances in various atmospheric conditions, corrective factors must be used.

5 PLANNING AND MANAGEMENT OF EMERGENCY

The planning of emergency consists in the identification of the responsible of planning activity, the drafting of emergency plan and management of emergency. It was defined for the first time in Italy by law n. 225 of 24 February 1992 and it has developed different characteristics on national, regional, provincial and municipal level.

The primary objective of the emergency planning is constituted by the definition of the "risk scenes" and therefore by the necessity of operating structures coordination: fire department, volunteers, structures of command and control, ways or routes of escape, shelters, sanitary areas etc. The fundamental complement of the risk scene is represented by the "participation model", that is the activity directory that constitute the operating procedure: it is a job tool adjusted on hypothetical situation based on acquaintances of the moment, renewable and that can be reviewed.

With the aim to face the incident event, the Decree Legislative 334/99 previews that for the plants re-entering in article 8 emergency plans are predisposed in order to reduce and to mitigate the accident consequences and to protect the workers and the population. There are two types of plans: the Internal Emergency Plan (PEI) and the External Emergency Plan (PEE).

5.1 *Internal Emergency Plan (PEI)*

This plan is developed by the factory's manager to face the important accident effects inside the factory. It describes who starts the emergency procedures, coordinates on site operations and cooperates with PEE responsible. Also, it specifies measures to be adopted (alarm system and behaviour regulations), equipment and resources available and instructions to train workers. Its targets are to control and circumscribe the accident (to minimize the effect and to restrict damages for people and environment), to plan the restoration operations of the environment after the accident, and to promulgate instructions (in order to start PEE in time and to train the workers).

5.2 *External Emergency Plan (PEE)*

The aim of the External Emergency Plan is to mitigate the damages and to protect the population from an important incident whose harmful effects fall back to the outside of the plant. This site specific plan is compatible with the already existing emergency plan (e.g. PEI). Its main objectives are: to contain and control incidents, as well as to minimize the effects and to limit damage to people, the environment and property, to implement the measures necessary to protect individuals and the environment from the effects of major accidents, to communicate the necessary information to the public and to the services or authorities concerned in the area, to provide for the restoration and clean up of the environment following the incidental event.

In Italy, the Department of Civil Protection has predisposed through the DPCM 25-2-2005 the elaboration and the modernization of External Emergency Plans for the factories to risk of major accident of which to art. 8 of D. Lgs 334. This law modernize the previous ones of 1994 and it is made up of: general part that it is devoted to site description, incident scene scenario that describes the event typology and make out risk areas (sure impact, damage and attention zone), intervention model (through emergency support functions, organization and procedures, alarm systems and levels, communications and management post-emergency), population information and cartographical section. The performance of this plan demands the participation and the involvement of many authorities: Red Cross, Fire Department, Local Sanitary System, Voluntary Service, Region, Police forces and ARPA (Environment Regional Protection Agency).

6 OPERATIVE AND SANITARY STRUCTURES

The main operative structure is represented by the unified operating centre (SOU) that coordinates all the activities concerning the emergency management, in ordinary periods (monitoring of events) and in emergency cases (fast and coordinate intervention of people and equipments). All the operative centres are coordinated by the SOU There are different structures for each emergency level: situations centre (CESI), direction of command and control (DICOMAC), unified air operations centre (COAU) and operative centre sea emergency (COEM) are involved in national emergency; unified regional operative room (SUR) and unified permanent operative room (SOUP) act in regional emergency cases; centre coordination aids (CCS), 15 support functions and mixed operative centre (COM) operate in provincial emergency situations as well as municipal operative centre and 9 support functions for municipal emergencies.

The sanitary aids are responsible for first aid and sanitary assistance, public health interventions and psychological and social assistance. Their activity is divided into two parts: in the preliminary phase, before the incidental event, they be devoted to land knowledge, events expected, the creation of the risk scene and of their resources, in the event phase they be devoted to alarm phase and sanitary emergency management. They operate with local sanitary systems, hospitals, Red Cross, voluntary and professional associations of sanitary area.

6.1 *Case study*

Finally, we would like to draw the attention to the main points of a practical case concerning the management of the emergency deriving by major accident. As an example, the case of two factories (one distillery and one GPL plant) located on Perugia's surroundings are presented now. The first step of these plan is to describe the position of the factories, their activities, dealt configuration of their systems, substances and their hazardous characteristics and possible damages to the atmosphere. The next step is the calculation firstly of the zone of sure impact characterized by elevated numbers of dead (the circular zone consists in a radius of 643 m, with its centre in the plant and more than 1350 individuals involved), secondly of the damage zone, characterized from "beginning death" and/or no reversible lesions (radius of 814 m and more than 650 individuals involved) and thirdly of the zone of attention with reversible lesions (radius of 1412 m and 3.400 individuals involved). These areas are illustrated in figure 3. As a third step the various main elements exposed to risk (e.g. population and historical, architectonic and environmental estates) and their organization in emergency case (e.g. traffic regulation, industrial and handicraft takeovers) should be localised.

The area to be evacuated would have to be divided into two sectors in order to assure a safe possibility to leave the contaminated area without impediment to help transport means. In order to plan the emergency, the following elements should be located on a map: the areas with high people concentration, the average routes aid, the population escape ways and one collection centre inside the area to be evacuated to bring individuals out of the area and to act as an information point, as well as three sorting centres, located outside the above described attention area where the staff could carry out the census of the individuals, and three aid centres close to the sorting centres to provide first aid and to assure a continuous link to the hospitals (Figure 3).

All the previous dispositions are integrated with of additional measures as prevention and protection corporate structures, security systems of the plants (antifire system, atmospheric surveys, system centers of block), external protection structures (sanitary structures, arrangements for the control of the air quality, average evacuation routes), and signalling incidents inside and outside the factory.

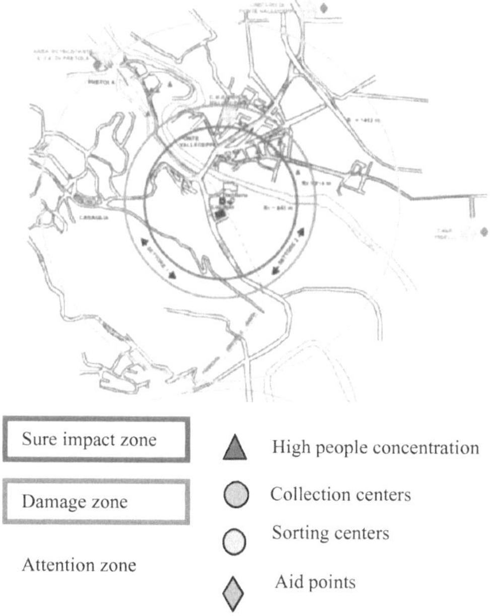

Figure 3. Example of organizational model for the management of Civil Protection interventions (Comune di Perugia—U.O. Ambiente e Protezione Civile).

7 CONCLUSION

We conclude that the development of an organizational model for the management of Civil Protection consists in three main steps. The first one is the territory characterization as aggregation of potential accidental events and connection among land, population, structure and planning. The second one is the quantitative analysis of risk in order to transform all available input into a cartography. Here a continuous revision is necessary that defines the characteristics of the damage areas as event typology and as procedure of spreading. Lastly, the planning of emergency as effective and efficient executive instrument for the Civil Protection interventions and settlement of any specific subject or authorities involved is needed.

REFERENCES

AIDIC, 2003. Lazio, Commissione Salute Sicurezza Ambiente e Qualità: "Rischi di Incidente Rilevante", agosto 2003.

Anzà, La Spada, Platania: Linee guida per la formazione del Piano per il risanamento ambientale ed il rilancio economico del Comprensorio del Mela—Commissione provinciale per la tutela dell'ambiente e la lotta contro l'inquinamento "Il rischio industriale e tecnologico", 11/06/2005.

APAT e Ministero dell'Ambiente e della Tutela del Territorio, 2002. "Mappatura del rischio industriale in Italia", Rapporto 22.

APAT, 2003. "Linee guida per lo svolgimento delle verifiche ispettive sui sistemi di gestione della sicurezza in impianti a rischio di incidente rilevante", Manuali e Linee Guida 23.

APAT, 2003. "Valutazione dell'impatto sull'ambiente degli incidenti rilevanti", Rapporto 36.

Astori, Boscaino, Caparresi, Capponi, Cerritelli et al. 2004. "Proposta di metodologia speditiva per la valutazione delle conseguenze ambientali di incidenti rilevanti" convegno VGR Pisa.

Berardi, S., Lazzaro 2001. "Supporto tecnico-scientifico alle attività di rilievo dei danni della vulnerabilità nonché per le analisi di rischio occorrenti per la corretta attività di ricostruzione – Attività svolta presso gli uffici della Regione in Ancona".

Bucci, Castrista, Contemori & Manning 1999. Atti del seminario inquinamento ed aziende a rischio: la compatibilità nel 2000 "Decreto Legislativo n. 334/99 "Severo bis": verifica degli adempimenti in relazione alla presenza di sostanze pericolose", 14/10/1999.

Colletta, C., Manzo, R. & Spaziante, A. 2001. Ministero delle Infrastrutture e dei Trasporti "Pianificazione del territorio e rischio tecnologico, il DM 9 maggio 2001: Il controllo dell'urbanizzazione nelle aree a rischio d'incidente rilevante", approvato con D.M. 382 del 17/12/2001.

Colletta, C. & Manzo, R. 2001. Ministero delle Infrastrutture e dei Trasporti: "Governo del territorio e rischio tecnologico: metodologie di intervento ed esperienze

di attuazione del DM 9 maggio 2001" approvato con D.M. prot.789/2003.
Comune di Perugia, 2000. U.O Ambiente e Protezione Civile: "Piano di emergenza esterna per gli stabilimenti industriali a rischio di incidente rilevante – Industrie Distillerie Di Lorenzo e Liquigas".
Comune di Perugia, 2000. Assessorato alla Protezione Civile: "Norme comportamentali per la popolazione in caso di incidenti rilevanti negli stabilimenti Liquigas e Distillerie G. Di Lorenzo a Ponte Valleceppi".
Comune di Perugia. Assessorato alla Protezione Civile - U.O Ambiente e Protezione Civile: "Piano Comunale di Protezione Civile".
Dierna, S. 2005. "Criticità ambientale e rischio tecnologico – Lo stato dell'ambiente e le condizioni di rischio naturale ed antropico", pubblicazione del 27/01/2005.
DPC in collaborazione con il Ministero degli Interni: "Il metodo Augustus".
DPC, 2000. Servizio Sismico Nazionale - Regione Emilia Romagna, Provincia di Modena: "Il sistema di raccolta dati a supporto della pianificazione comunale e provinciale di emergenza", Dicembre 2000.
DPC, 2003. Servizio Rischio Ambientale e Sanitario, Presidenza del Consiglio dei Ministri: "Criteri di massima sulla dotazione di farmaci e dispositivi medici di un Posto Medico Avanzato di II livello utilizzabile in caso di catastrofe", GU n°139 del 25 agosto 2003.
DPC, 2004. Regione Liguria: "Linee guida per la pianificazione comunale di Protezione Civile", 06/10/2004.
DPC, 2004. Regione Lombardia: "La pianificazione di emergenza in Lombardia- Guida ai Piani di Emergenza Comunali e Provinciali", edizione 2004.
Galanti, E. 1997. DPC INFORMA – Periodico informativo del dipartimento di Protezione Civile: "Il Metodo Augustus", Giugno 1997.
Galanti, P., Curianò, I.A. & D'Angelo, L. Pianificazione e Gestione dell'Emergenza—"Struttura e gestione delle sale operative di Protezione Civile".
Galatola, E. 2005. Regione Lombardia Deliberazione n° VII / 19794 del 10/12/2004 "Linee guida per la predisposizione dell'elaborato tecnico Rischio di Incidenti Rilevanti (E.R.I.R) nei comuni con stabilimenti a Rischio di Incidenti Rilevanti", 28/01/2005.
Galatola, E. 2005. "Decreto del Presidente del Consiglio Dei Ministri 25 febbraio 2005: Linee Guida per la predisposizione del piano d'emergenza esterna di cui all'articolo 20, comma 4, del decreto legislativo 17 agosto 1999, n. 334.", 03/04/2005.
Govoni, C., Albonetti, A. 2003. Gruppo agenti chimici, cancerogeni e mutageni Regione Emilia Romagna "Il modello di valutazione del rischio da agenti chimici pericolosi per la salute ad uso delle piccole e medie imprese", Convegno Nazionale Risch 2003 del 17/10/2003.
Locatelli, C., Butera, R., Gandini, C., Minuco, G., Mazzoleni, C., Giordano, A., Candura, S.M., Varango, C., Zanuti, M., Petrolini, V., Manzo, L. 1997. "Mezzi telematici per la gestione dell'emergenza sanitaria e del primo soccorso nei luoghi di lavoro", Pavia.
Manca, D. Politecnico di Milano - Dipartimento di Chimica Materiali e Ingegneria Chimica: "Introduzione al Rischio Chimico Industriale".
Mazzini, D. 2002. Università degli Studi di Pisa Dipartimento di Costruzioni Meccaniche e Nucleari; "Dispersione atmosferica degli inquinanti aeriformi", Politecnico di Torino 06/06/2002.
PTCP della provincia di Ravenna, 2005. Adeguamento alla L.R. 20/2000—"*Valsat: Allegato 2—Stato dell'ambiente 2004—Meteoclimatologia*" adottato con delibera del consiglio provinciale n° 51 del 06/06/2005.
Provincia di Macerata, Ufficio Territoriale del Governo, 2003. "Piano Provinciale di Emergenza di Protezione Civile", Genio Civile.
Provincia di Torino, Servizio di Protezione Civile, 2003. "*La Carta dei 34 Rischi*: Parte III, Prima valutazione dei rischi sul territorio della provincia: Rischi Antropici", Torino luglio 2003.
Regione Piemonte, Assessorato Ambiente Energia Pianificazione e Gestione delle Risorse Idriche Lavori Pubblici e Tutela del Suolo—Protezione Civile, 1998. "La pianificazione comunale di Protezione Civile(PCPC)—I piani comunali di Protezione Civile e loro rapporti con alcuni strumenti di pianificazione", november 1998.
Regione Umbria, Giunta regionale, Sezione Volontariato formazione e informazione del Servizio di Protezione civile e prevenzione dai rischi, 2003. "Vademecum per l'organizzazione del Volontariato di Protezione Civile nella regione umbra", aprile 2003.
Restani, W. 2004. Arpa Lombardia, U.O. Rischi Industriali "La certificazione dei Sistemi di Gestione della sicurezza per le imprese a rischio di incidente rilevante: L.R. 19/2001", Auditorium Federchimica 21/09/2004.
Ricciuti, R. 2003. Arpa Lombardia, 7ª Conferenza Nazionale delle Agenzie Ambientali "Le verifiche ispettive sui sistemi di gestione della sicurezza come strumento di prevenzione dei rischi di incidente rilevante", Milano 2003.
Rizza, U., Mangia, C., Giostra, U., Gabucci, M.F., Di Rocco, D., Marra, G.P., Passoni, L., Schipa S. & Schipa, I. 1999. Consiglio Nazionale Delle Ricerche - Istituto per lo Studio dell'Inquinamento Atmosferico e l'Agrometeorologia "Un sistema modellistico per lo studio del trasporto e la diffusione di inquinanti nella penisola salentina", Lecce luglio 1999.
Simonetti, S. 2002. "Rischio Ambientale: Il rischio della civiltà industriale", Gennaio 2002.
Swith, 2003. "Corso di Protezione Civile e di cooperazione Civile-Militare", Foggia 27/09/2003.

Risks Challenging Publics, Scientists and Governments – Menoni (ed)
© 2010 Taylor & Francis Group, London, ISBN 978-0-415-58072-4

Rapid evaluation of ecological risk in soil with DNA biosensor

G. Bagni & M. Mascini
Department of Chemistry, University of Florence, Firenze, Italy

E. Sturchio & P. Boccia
Department of Industrial Installations and Interaction with the Environment (DIPIA), National Authority for Health and Safety at Work (ISPESL), Rome, Italy

S. Marconi
National Institute of Soil Defence (ISSDS), Rieti, Italy

ABSTRACT: An electrochemical DNA-based biosensor is proposed as a fast and easy screening method for the detection of genotoxic compounds in soil. The biosensor was assembled by immobilising double stranded *calf thymus* DNA on screen-printed electrodes. The interactions between DNA and environmental pollutants can cause variations of the electrochemical proprieties of DNA when they cause DNA damage. This kind of biosensor represents a new, easy and fast way of analysis of polluted sites, therefore it can be used as early warning device in areas with ecological risk as *in situ* measurement.

1 INTRODUCTION

Ecological risk assessment (ERA) is a process that evaluates the probability that adverse ecological effects are occurring or may occur as a result of exposure to one or more stressors (USEPA 1992). It is an investigation into the actual or potential impacts of contaminants from a hazardous waste site on plants, animals and than humans.

The most common biological tests used for ERA evaluation can be divided in two categories: tests with microorganism (i.e. *vibrio fisherii, daphnia magna*) or with sentinel organisms, like animals (i.e. *lombricus rubellus, eisenia foetida, ciprynus carpio*) or plants (i.e. *vicia faba*), and molecular tests. The first class consists in the analysis of parameters (called biomarkers) of the organisms that change in relation with the pollution. These tests are slow (analysis performed generally in one-two weeks), not specific and expensive. Instead, the second type, molecular tests are fast and hence can be used as early warning devices in areas with ecological risk, moreover they are more specific and have low costs. In this category there are biosensors, which are used in our laboratory.

A biosensor is an analytical device combining a biological recognition element, which confers selectivity, with a transducer, which provides sensitivity and converts the recognition event into a measurable electronic signal (Mascini, 2005). The biological recognition element can be an enzyme, an antibody, a nucleic acid, a microorganism, a tissue or organelle, while the transducer can be optical, electrochemical, thermal, piezoelectric or magnetic (IUPAC, 1997; Thevenot et al. 2001). The major performance characteristics of a biosensor are the minimum sample preparation, the simplicity of the apparatus, the possibility to obtain fast results and continuous readings. Moreover biosensors are cost effective, small and becoming miniaturised with new technology.

Biosensors hold great promise for the task of environmental monitoring and control (Bilitewski & Turner, 1999). The specific interaction of an immobilised biological layer with target pollutants provides the basis for analytical devices for laboratory or field use. While environmental applications of biocatalytic (enzyme) and immunosensors have greatly increased during the 1990s, little attention has been given to the development of nucleic acid probes for environmental surveillance. Advances in molecular biology and biotechnology have set the stage for exciting possibilities for DNA-based environmental biosensors (Wang & Rivas et al. 1996, 1997; Wang & Chicarro et al. 1996; Chiti et al. 2001; Lucarelli et al. 2002 (a), 2002 (b)). Such recognition layers could play a major role in future environmental analysis. Since the toxic action of numerous pollutants (e.g. carcinogens and mutagens) is related to their interaction with DNA, it is logical to exploit these events for designing new environmental biosensors.

In this work an electrochemical DNA based biosensor is proposed as a rapid screening method for

Ecological Risk Assessment in soil samples. Double stranded DNA is the molecular recognition element and graphite screen printed electrode is the transducer. DNA is immobilised at the electrode surface and its electrochemical behaviour is being studied in relation with compounds with affinity for DNA. We propose the use of this biosensor after a fast and simple analytical extraction method to mimic the bioaccessibility determined by a contaminated Italian ex industrial area, the Associated National Chemical Companies (ACNA) (Cengio, SV, Italy) site. This is a closed organic chemical industrial factory active since 1882 with the production of explosives (nytroglycerin, dynamite and trinitrotoluene), paints, nitric and sulphuric acids, phenols and amines, with serious levels of contamination of soil and surface waters, where a remediation and bonification plan started in 1999.

2 EXPERIMENT

2.1 *Materials*

Milli-Q grade water (18 MΩ) was used for the preparation of all solutions (Milli-Q water purification system, Millipore, UK). Inorganic salts for buffers preparation were obtained from Merck (Milan, Italy). *Calf thymus* double-stranded DNA type XV was purchased from Sigma (Milan, Italy). Using sterile pipettes, aliquots containing 1000 µg/mL of DNA in milli-Q water were prepared and stored in freezer (–20°C). Acetate buffer concentration was 0.25 M (pH = 4.75) with 10 mM KCl. The effect of three benzene, two naphthalene and four anthracene derivatives were analysed in this work with DNA biosensor because widely present in the ACNA site (Esposito et al. 2002; Conte et al. 2005; Avidano et al. 2005): sodium benzenesulfonate, 4-chloroaniline, 3,4-dichloroaniline, 2-naphtylamine, sodium 2-naphthalenesulfonate, sodium anthraquinone 2-sulfonate monohydrate, 2-anthracencarboxylic acid, 1,2-diaminoanthraquinone and 2-anthramine (all from Sigma-Aldrich - Milan, Italy).

2.2 *Apparatus*

Electrochemical measurements were performed with a PalmSens (Palm Instruments BV, Hoten, Netherlands) interfaced to a Compaq iPAQ Pocket PC with a PalmScan 1.3 software package (Palm Instruments BV, Hoten, Netherlands) and screen-printed electrodes (SPE) (Cagnini et al. 1995).

An autovortex mixer SA2 (Stuart Scientific, Surrey, UK) and filters 0.45 µm (Schleicher & Schuell Italia, Legnano (MI), Italy) were used respectively to homogenise the solvent-soil suspensions and to filter the extracts.

The ultrasound probe was a model VC100 Vibra Cell from Sonics and Materials (Danbury, USA).

Fluorescence measurements were performed in a quartz cuvette on a Jasco FP 750 luminescence spectrometer (Essex, UK). Slit widths were set at 10 nm for both excitation and emission wavelengths, sensitivity medium and response 1 sec. The signal level in solvent blank was subtracted and the fluorescence level was expressed as relative fluorescence intensity.

2.3 *Preparation of screen printed electrodes*

The electrodes were screen-printed in house using a DEK 248 screen-printing machine (DEK, Weymouth, UK).

The screen printing process consisted of forcing inks of different characteristics through a screen onto a surface of the substrate plate (polyester foil) with a squeegee. Typical thickness of the film was around 20 µm. The open pattern in the screen defines the pattern which will be printed on the substrate. The inks consisted of finely divided particles of different materials (silver for pseudo-reference electrode, graphite for working and auxiliary electrodes and titanium dioxide for insulating layer) in a blend of thermoplastic resins. The first layer printed consisted of the silver ink, in order to obtain the conductive tracks and the pseudo-reference electrode. The second layer was the carbon ink, to obtain both the auxiliary and working electrodes. After every step, the sheets were heated at 120°C for 10 min to achieve the polymerisation of the printed films. The insulating ink was finally used to define the working electrode surface (∅ = 3 mm). A curing period of 20 min at 70°C was applied. The sensors were produced in sheets of 20 electrodes. Each electrode printed on the polyester flexible foil is easily cut by scissors and fits a standard electrical connector. To facilitate handling, the screen-printed electrochemical cells were sticked onto a rigid polycarbonate-based support. Each electrode was used as disposable. The schemes of these three printing steps and of a screen-printed electrode sheet are reported in Figure 1.

2.4 *Biosensor principle*

Guanine peak was used as the transduction signal in systems detecting DNA interacting agents. As a result of interaction of double stranded calf thymus DNA with a pollutant agent a decrease of guanine peak (measured by square-wave voltammetry) was detected (Figure 2). DNA modification were estimated with the value of the percentage of guanine peak height change (Signal %), which is the ratio of the guanine peak height after the interaction with

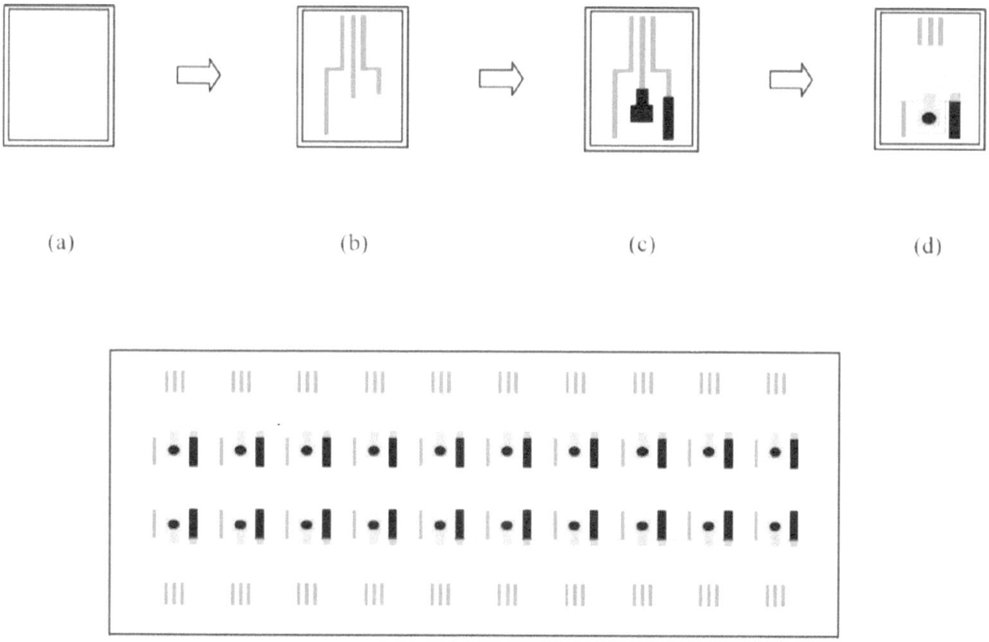

Figure 1. Scheme of screen-printing electrodes. Fabrication process of the sensor consisted of three steps including consecutive printing of the silver layer (b), carbon layer (c) and insulating layer (d) on the polyester substrate. The sensors were produced in sheets of 20 electrodes.

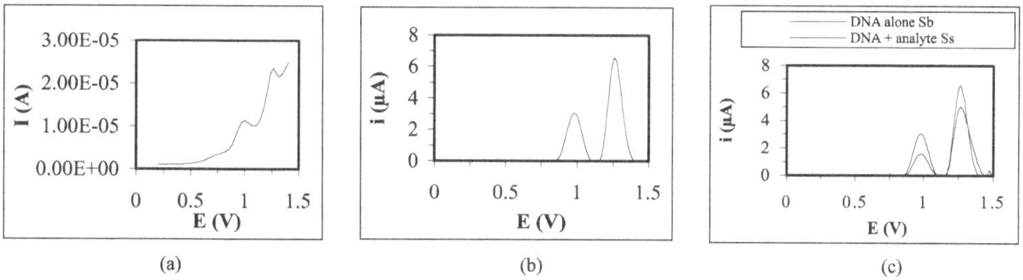

Figure 2. Redox behavior of guanine (+1.0 V) and adenine (+1.25 V) bases after a square wave voltammetric scan carried out with graphite screen printed working electrode. It can be noted the normal signal (a), the signal after a baseline correction (b) and the decrease of the DNA peaks after the interaction with a pollutant agent (c). Electrode pre-treatment: +1.6 V (vs. Ag – SPE) for 120 s and +1.8 V (vs. Ag – SPE) for 60 s in 5 mL of 0.25 M acetate buffer containing 10 mM KCl (pH = 4.75) under stirred conditions; DNA immobilisation: 50 ppm *calf thymus* dsDNA (in 0.25 M acetate buffer with 10 mM KCl) applying a potential of +0.5 V (vs. Ag – SPE) for 5 min under stirred conditions; blank or sample interaction: 10 μL of the sample solutions onto the working electrode surface for 2 min; final measurement: square wave voltammetric scan in 0.25 M acetate buffer containing 10 mM KCl from +0.2 V to +1.4 V (E_{step} = 15 mV, $E_{amplitude}$ = 40 mV, frequency = 200 Hz).

a sample (S_s), and the guanine peak height after the interaction with the buffer solution (S_b): Signal % = $(S_s/S_b)*100$.

Adenine oxidation peak can also be observed and detected, but it has been chosen to use guanine peak only because it gave more reproducible signals.

Groove binding, electrostatic interactions, hydrogen and/or van der Waals bonds and intercalation of planar condensed aromatic ring systems between adjacent base pairs are the interactions that the DNA electrochemical biosensor can detect.

The electrochemical procedure to build up the DNA biosensor included four main steps:

electrochemical conditioning of the electrode surface in order to oxidize the graphite impurities and to obtain a more hydrophilic surface to avoid DNA immobilisation, *calf thymus* double stranded DNA (dsDNA) immobilisation, interaction with the sample solution, and electrode surface interrogation.

The experiments were performed at a room temperature, according to the following steps:

1. Electrode pre-treatment: applying potential +1.6 V for 120 s and +1.8 V for 60 s; electrode in 5 mL of 0.25 M acetate buffer, containing 10 mM KCl (pH = 4.75), under stirred conditions.
2. DNA immobilisation: 50 ppm *calf thymus* dsDNA in 0.25 M acetate buffer with 10 mM KCl, applying potential +0.5 V for 5 min, under stirred conditions.
3. Blank or sample interaction: 10 μL of sample solutions onto the working electrode surface for 2 min. Soil samples have been extracted and filtered before the analysis.
4. Measurement: a square wave voltammetric (SWV) scan was carried out to evaluate the oxidation of guanine residues on the electrode surface. The height of the guanine peak (at +0.95 V *vs*. Ag screen-printed pseudo-reference electrode) was measured. SWV in 0.25 M acetate buffer, containing 10 mM KCl; parameters: scan from +0.2 V to +1.40 V, E_{step} = 15 mV, $E_{amplitude}$ = 40 mV, Frequency = 200 Hz.

The analysis of a sample took 11 minutes.

2.5 *Soil samples*

Reference soil was collected in a reference site in Florence (Italy). Soil samples were collected from the soil layers (0–30 cm) in a specific ACNA site, called hill n°5, on March 2003. The hill was made of waste from the industry accumulated during the years and was divided into four zones: zone 1 with low contamination level; zone 2 pseudo-reference; zone 3 moderate pollution level and zone 4 high ecological risk.

The soil was dried 24 h at r.t. (Berset et al. 1999), pulverised in a mortar, homogenise passing trough a metal sieve (Ø = 0.8 mm) and stored at r.t. until extraction. The extraction was performed on 0.5 g of soil, according to an optimised procedure obtained spiking a reference soil, then extract with an ultrasound probe for 2 min in 10 mL of 50 mM phosphate buffer pH 7.5, followed by an equilibration of 3 min at r.t. and a filtration.

2.6 *Statistical analysis*

A one-way analysis of variance (ANOVA) was used to compare between the different sample groups. Dunnett's test was used to determine the significance of differences between the sample groups and the control (blank assay) group. When the ANOVA indicated that significant differences existed, Fisher's "least-significant difference" (LSD) multiple comparison test was then used. Statistical analysis was performed with Excel software. Parametric tests were preceded by tests for normal distribution per site, and for homogeneity of variance between sites. A significance level of $P < 0.05$ was applied in all statistical tests.

3 RESULTS AND DISCUSSION

3.1 *Analysis of standard solutions*

The effect of three benzene, two naphthalene and four anthracene derivatives were analysed in this work with DNA biosensor because widely present in the ACNA site. The molecular structures of the compounds analysed are reported in Figure 3. Different concentrations for each compound were analysed and the inhibition of the guanine oxidation peak increased with analyte concentration (Table 1).

The results demonstrated that the compounds presented different genotoxic effect. Generally, the effect was higher with increased ring number of the molecule, and also the type of the group had relevance. For example, the amino group was more toxic than a sulfonate one at the same molecular concentration. The results showed the highest toxic effect for molecules with three rings. In fact the naphthalene and anthracene derivatives analysed are DNA intercalating agents (Chiti et al. 2001; Lucarelli et al. 2002; Snyder et al. 2002) while the benzene derivatives are not double helix intercalant but only a weak DNA interactive (i.e. as reported in Snyder et al. 2002, 4-chloroaniline is described as non genotoxic and its cytotoxicity, measured as cell viability assay, is relatively low compared with other chemicals).

This work demonstrated that naphthalensulfonate derivatives with two and three aromatic rings and aromatic amines with two and three rings gave a positive result with the biosensor in a concentration range of micro and submicro-molar. The EC_{50} value was 2.0 μmol/L for 2-anthracenecarboxylic acid and 1,2-diamineanthraquinone and 1.0 μmol/L for 2-anthramine.

3.2 *Analysis of real soil samples*

In ACNA site the environmental risk arises from pollution diffused in the ground, accumulation of waste materials and pollution of the Bormida River.

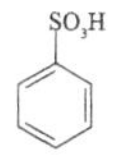

sodium benzenesulfonate
(BS)

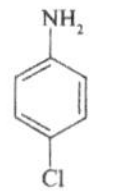

4-chloroaniline
(CA)

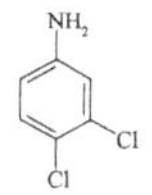

3,4-dichloroaniline
(DCA)

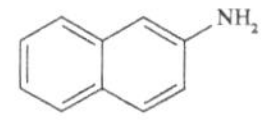

2-naphtylamine
(NA)

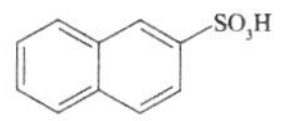

sodium 2-naphtalenesulfonate
(NS)

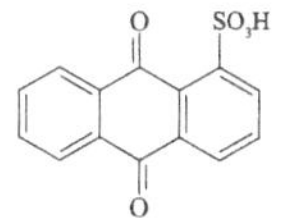

sodium anthraquinone 2-sulfonate
monohydrate
(AQS)

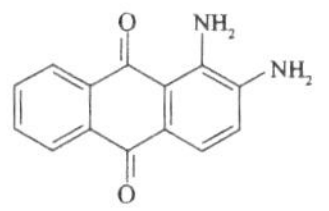

1,2-diamineanthraquinone
(DAA)

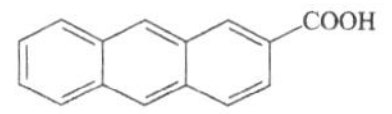

2-anthracencarboxilic acid
(AAC)

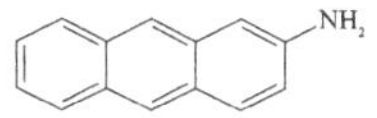

2-anthramine
(AT)

Figure 3. Structure of the molecules analysed.

Table 1. DNA biosensor results for benzene, naphthalene and anthracene derivatives standard solutions present in ACNA area. Experimental conditions as in Figure 2.

Compound	*Concentration* (*μmol/L*)	*S guanine%*	*Std. dev.* (*N = 4*)
sodium benzenesulfonate (BS)	100	75	6
	50	91	8
	25	96	6
4-chloroaniline (CA)	100	78	8
	50	92	6
	25	85	10
3,4-dichloroaniline (DCA)	100	43	8
	50	75	8
	25	84	6
	10	97	7
2-naphtylamine (NA)	50	54	7
	5	61	9
	1	85	12
sodium 2-naphtalenesulfonate (NS)	150	47	8
	100	58	9
	50	79	6
sodium anthraquinone 2-sulfonate monohydrate (AQS)	25	34	12
	10	51	7
	5	55	9
	1	90	6
2-anthracencarboxylic acid (AAC)	5	42	10
	2	61	7
	1	71	5
1,2-diamineanthraquinone (DAA)	5	59	10
	2	65	10
	1	79	6
2-anthramine (AT)	2.0	33	10
	1.0	47	6
	0.5	73	8

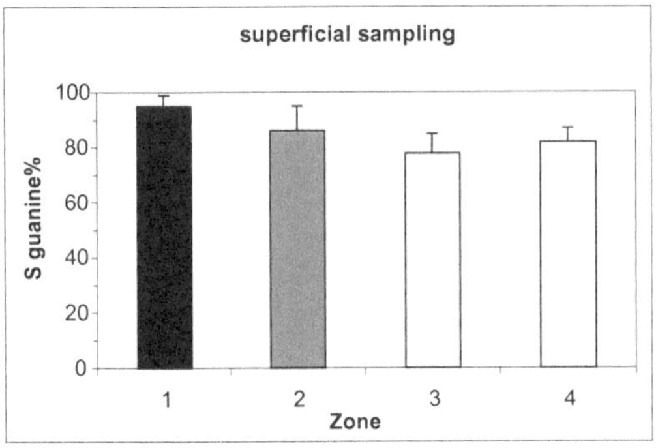

Figure 4. DNA biosensor results for superficial sampling in the four zones of hill n°5 in ACNA area. Extraction with 50 mM PBS pH 7.5, ultrasound probe 2 min. Electrochemical experimental conditions as in Figure 2.

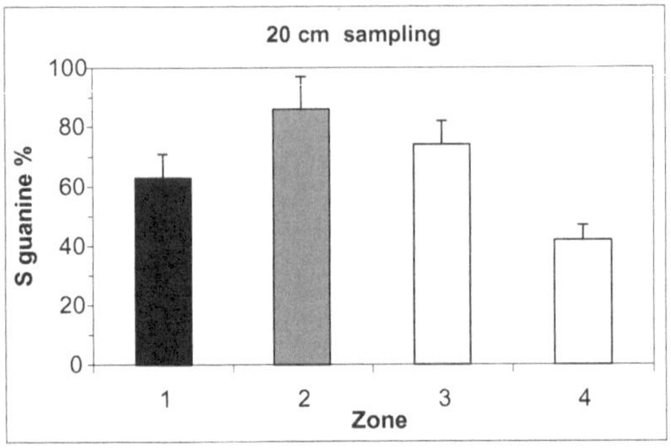

Figure 5. DNA biosensor results for 20 cm sampling in the four zones of hill n°5 in ACNA area. Extraction with 50 mM PBS pH 7.5, ultrasound probe 2 min. Electrochemical experimental conditions as in Figure 2.

ACNA samples were analysed with DNA biosensor and the different zones were compared.

The comparison of the zones with the superficial sampling (max 5 cm depth) didn't showed a significant difference between them (Figure 4). The comparison of the zones with 20 cm sampling depth was different, showing a trend of pollution as expected (Figure 5): the biosensor was able to distinguish different soil contamination sites.

A standard semi-quantitative method for aromatic ring system determination as fixed wavelength fluorescence (FF) was used for the analysis of 2-anthramine standard solutions and some ACNA samples were tested with this methodology obtaining promising comparative results. The optimal wavelength pair, regarding sensitivity was obtained by analysis of 0.5 μM AT standard solution and was 350/510 nm.

Soil samples from zone 4 were analysed with FF as AT-like compounds. The comparison of FF

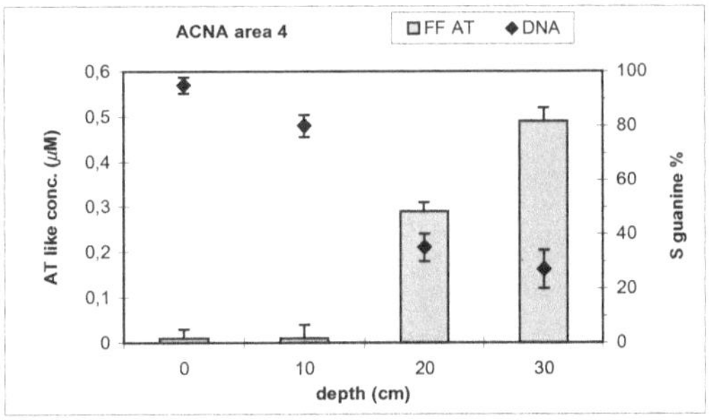

Figure 6. DNA biosensor and $FF_{350/510}$ results for different depths in zone 4 of hill n°5 in ACNA area. The samples were extracted with 50 mM PBS pH 7.5. Electrochemical experimental conditions as in Figure 2.

and biosensor results is reported in Figure 6, which shows that soil contamination increased with the sampling depth for both methods. These results confirmed the fact that this zone is the most polluted of the hill n°5. ACNA samples were analysed also with micronuclei and Comet tests performed by ISPESL/DIPIA laboratory (Rome, Italy). These are simple and sensitive techniques for analysis and measure of DNA damage in individual mammalian (and to some extent prokaryotic) cells. Seed of *Vicia faba var minor* were put in ACNA soil samples and the top roots were used for the micronuclei and Comet tests in order to analyse a vegetable mutation after 5 days of exposure. The length of the top roots was also calculated in order to establish a phyto-toxicity. After, half of the roots were prepared for the micronuclei test, the other half for the Comet test. The genotoxic effect was based on the frequency of irregular anaphases and micronucleate cells. The results showed an increase in pollution from zone 1 to 4 and within the zone, with the sampling depth for all methods (Table 2): when there was a lower DNA signal there was a lower primary roots length, a higher irregular anaphases or micronuclei frequency and a higher damage class with Comet test.

Interestingly, the results obtained analysing the ACNA soil samples with DNA biosensor were confirmed by other bioassays, hence this kind of biosensor can be very useful as a rapid screening method of analysis.

4 CONCLUSIONS

In this work, a DNA based biosensor was used for Environmental Risk Assessment in an ex-industrial area in Italy. DNA biosensor for qualitative analysis of pollutant presented a good correlation with a standard semi-quantitative method for PAH

Table 2. Comparison of biosensor results performed on ACNA samples with other bio-assays for toxicity and genotoxicity detection.

Sample code	*DNA biosensor S (%)*	*Phyto-toxicity Primary roots length (mm)*	*Genotoxicity Irregular anaphase frequency (%)*	*Genotoxicity Micro nucleate cells frequency (%)*	*Comet test Damage class*
1A	99	32.7	0.05	0.18	1
1B	87	36.7	0.06	0.20	1
1C	63	18.1	n.p.	n.p.	n.p.
2A	86	35.0	0.01	0.01	1
3A	78	25.9	0.01	0.03	0
3B	60	28.2	n.p.	n.p.	0
3C	74	22.8	n.p.	n.p.	2
4A	82	21.9	0.46	0.09	n.p.
4B	70	29.5	0.11	0.08	3
4C	42	14.5	0.25	0.12	1
4D	26	12.7	0.52	0.64	4

A: superficial sampling, B: 10 cm, C: 20 cm, D: 30 cm.
n.p.: not performed.

determination and with other common biomarkers. The advantages to use DNA biosensor is that it gives the global content of pollution and can be used for monitoring screening at contaminated sites. It has several benefits: low cost, short measurement time (10 min) and it can be easily used in field experiment when coupled with a small portable potentiostat and disposable carbon strips.

REFERENCES

Avidano, L., Gamalero, E., Cossa, G.P. & Carraro, E. 2005. Characterisation of soil health in an Italian polluted site by using microorganisms as bioindicators. *Applied Soil Ecology* 30(1): 20–33.

Berset, J.D., Ejem, M., Holzer, R. & Lisher, P. 1999. Comparison of different drying, extraction and detection techniques for the determination of priority polycyclic aromatic hydrocarbons in background contaminated soil samples. *Anal. Chim. Acta* 383: 263–275.

Bilitewski, U. & Turner, A.P.F. 2000. *Biosensors for environmental monitoring: 1–409*. Harwood Academic Publishers.

Cagnini, A., Palchetti, I., Lionti, I., Mascini, M. & Turner, A.P.F. 1995. Disposable ruthenized screen printed biosensors. *Sensors and Actuators B: Chemical* 24: 85–89.

Chiti, G., Marrazza, G. & Mascini, M. 2001. Electrochemical DNA biosensor for environmental monitoring. *Anal. Chim. Acta.* 427: 155–164.

Conte, P., Agretto, A., Spaccini, R. & Piccolo, A. 2005. Soil remediation: humic acids as natural surfactants in the washing of highly contaminated soils. *Environ. Poll.* 135: 512–522.

Esposito, A., Del Borghi, A. & Vegliò, F. 2002. Investigation of naphthalene sulphonate compounds sorption in a soil artificially contaminated using batch and column assays. *Waste Manag.* 22: 937–943.

IUPAC Compendium of Chemical Terminology 2nd Edition 1997.

Lucarelli, F., Kicela, A., Palchetti, I., Marrazza, G. & Mascini, M. 2002. Electrochemical DNA biosensor for analysis of wastewater samples. *Bioelectrochemistry* 58: 113–118.

Lucarelli, F., Palchetti, I., Marrazza, G. & Mascini, M. 2002. Electrochemical DNA biosensor as a screening tool for the detection of toxicants in water and wastewater samples. *Talanta* 56: 949–957.

Mascini, M. 2005. Editorial - *Bioelectrochemistry* 67(2): 129.

Snyder, R.D. & Arnone, M.R. 2002. Putative identification of functional interactions between DNA intercalating agents and topoisomerase II using the V79 in vitro micronucleus assay. *Mutat. Res.* 503: 21–35.

Thevenot, D.R., Toth, K., Durst, R.A. & Wilson, G.S. 2001. Electrochemical biosensors: recommended definitions and classification. *Biosensors & Bioelectronics* 16: 121–131.

USEPA 1992. *Framework for Ecological Risk Assessment.* Washington DC, Risk Assessment Forum. EPA/630/R-92/001.

Wang, J., Chicarro, M., Rivas, G., Cai, X., Dontha, N., Farias, P.A.M. & Shirashi, H. 1996. DNA biosensor for the detection of hydrazines. *Anal. Chem.* 68: 2251–2254.

Wang, J., Rivas, G., Cai, X., Palecek, E., Nielsen, P., Shiraishi, H., Dontha, N., Luo, D., Parrado, C., Chicarro, M., Farias, P.A.M., Valera, F.S., Grant, D.H., Ozsoz, M. & Flair, M.N. 1997. DNA electrochemical biosensors for environmental monitoring. A review. *Analytica. Chimica. Acta.* 347: 1–8.

Wang, J., Rivas, G., Luo, D., Cai, X., Valera, F.S. & Dontha, N. 1996. DNA-modified electrode for the detection of aromatic amines. *Analytical Chemistry* 68: 4365–4369.

Risks Challenging Publics, Scientists and Governments – Menoni (ed)
© 2010 Taylor & Francis Group, London, ISBN 978-0-415-58072-4

Bioindicators to detect environmental damage after "fall out" of pollutants

E. Sturchio & P. Boccia
Department Industrial Installations and Interaction with the Environment (DIPIA), National Authority for Health and Safety at Work (ISPESL), Rome, Italy

S. Marconi
Istituto Sperimentale per lo Studio e la Difesa del Suolo (ISSDS), Rieti, Italy

P. Ferrazza
Opera research, Italy

C. Beni, R. Aromolo & B. Felici
Agricultural Research Council ISNP, Italy

B. Ficociello & L. Casorri
Department Industrial Installations and Interaction with the Environment (DIPIA), National Authority for Health and Safety at Work (ISPESL), Rome, Italy

A. Gatteschi & C. Conti
Fondazione Biomedica Europa, Italy

ABSTRACT: Within a project jointly supported by the Italian Ministry of Health and ISPESL (National Authority for Health and Safety at Work), in order to evaluate the environmental damage after an accidental release of toxic substances in the atmosphere or/and soil, a simulation was set up with three substances: Formaldehyde, Fenilisocyanate and Brome. The aim of this study was to evaluate the effects of the contamination on soil functional qualities and plant nutrition, and to detect DNA damage with short-term genotoxicity tests performed on polluted soils and on gravitational water. The results showed that some of the parameters considered are applicable indicators for the evaluation of the soil-plant system pollution. Our study demonstrated that the comet assay is a sensitive, rapid and economic technique for the detection of DNA damage, which is ideally suited as a biomarker of genotoxicity for biomonitoring.

1 INTRODUCTION

In order to evaluate the environmental damage after an accidental release of toxic substances in the atmosphere or/and soil, a simulation was set up with three substances: Formaldehyde, Fenilisocyanate and Brome. These dangerous substances are widely used in industrial productions and, in case of accidental release, can cause pollution of soil (Formaldehyde) or/and atmosphere (Formaldehyde, Fenilisocyanate and Brome). In particular, Fenilisocyanate is a product of the decomposition of chloridazon, a pesticide commonly used in agriculture; the major risk connected with this product is the possibility of fire during storage.

The aim of this study was to evaluate the effects of the contamination on soil functional qualities and plant nutrition, and to detect DNA damage with short-term genotoxicity tests performed on polluted soils and on gravitational water.

In natural or agricultural ecosystems, the activity of soil micro-organisms, in decomposing plant and animal residues and in formation and mineralization of soil organic matter, determines the fertility of the soil. The soil-plant ecosystem may be damaged, either in the long- or short-term, by toxic agents that inhibit or stop the natural functioning of the soil. Frequently, damages occur in relation to soil chemical characteristics, plant morphology and physiology and gravitational water composition.

Also the measurement of microbial biomass appears an useful indicator of environmental stress due to contaminants, while microbial activity measurements at soil toxic concentration commonly occurring in agricultural and natural soils may not. Combining the two measurements, to give rates of activities (CO_2 emission) has been shown to be a much more subtle indicator of environmental stress, that causes a diversion of energy from biosynthesis to microbial activity in the soil microbial biomass (Jenkinson & Ladd 1981; Jenkinson & Powlson 1976).

The evaluation of the genotoxic effects was performed by studying Micronucleated cells (MC), Aberrant Anatelophases (AA) (Kilman et al. 1978; Kilman, 1956; De Simone, 1999) and DNA damage by Comet assay in root meristems of *Vicia faba* seedlings grown on soil samples polluted with the three substances.

The comet assay on plants has become a useful method for the assessment of the environmental and experimental genotoxic impact. The assay is ideal to detect DNA damage because of its high sensitivity and specificity and because it is a non-invasive technique. It may complement other test systems measuring different endpoints of genotoxicity (Menke, 2000). Comet assay not only provides an estimate of how much damage is present in cells, but what form it takes. Although it is essentially a method for measuring DNA breaks, the introduction of lesion-specific endonucleases allows detection of, for example, ultraviolet (UV)-induced pyrimidine dimers, oxidized bases, and alkylation damage (Angelis, 2000).

The evaluation of the genotoxic effects was performed by studying MC and AA in root meristems of *Vicia faba*. In previous works, the micronuclei test on this plant has been successfully used for detecting genotoxic damages in seedlings grown on a sandy soil added with different doses of composts from different starting materials. *Vicia faba* was chosen as test plant because it has large chromosomes amenable to the study of chromosome aberrations in somatic cells during mitotic division and as micronuclei in root tip cells following the mitotic division.

2 MATERIALS AND METHODS

2.1 *Experimental set up (Figure 1)*

The simulation was set up in two sets of lysimetric boxes, testing two different types of soil: sandy and clay-loamy soil. The simulation was conducted in contained environment.

The experimental set consisted of 32 boxes, half sown with spinach (*Spinacia oleracea L.*), a plant specie with leaves (Bianco & Pimpini, 1991). The 32

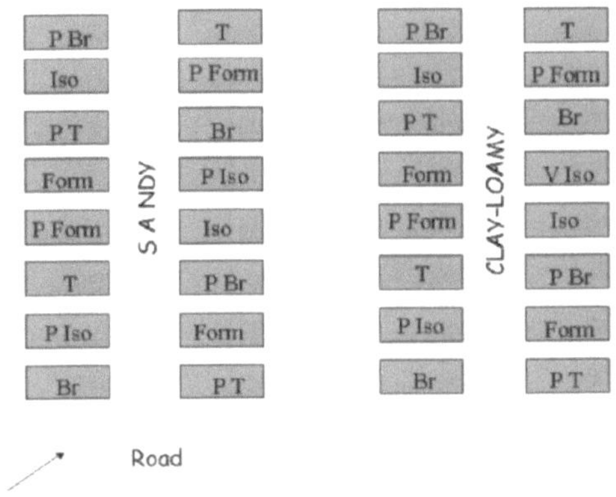

Figure 1. Experimental set up.

boxes were differently polluted with the mentioned three substances at the concentration of 10 mM (Fenilisocyanate and Brome) and 50 mM (Formaldehyde), while two boxes (only soil and planted soil) were used in terms of comparison.

2.2 *Biological, chemical and physical characterization*

Of these two soils were investigated some physical, chemical and biological parameters, as reaction, organic matter, nutrients (VV.AA, 1999) and trace elements content, carbon dioxide emission related to the microbial activity (Jenkinson, 1998). Moreover, gravitational water composition, plant nutrients uptake (Kalra, 1998), chlorophyll content and plant morphology were examineted to identify some indicators concerning the soil-plant system (VV.AA, 1994).

2.3 *Genotoxicity tests*

Soil samples (about 500 g each) were collected in plastic bags from surface soil (30 cm) and were placed in aluminium basins. Each basin containing 100 seeds/basin, treated with 120 ml of H_2O, was allowed to germinate in a climatic chamber at 20°C ± 1 for five days.

The gravitational water (about 1 litre) was collected from each polluted box; 500 gr of control sandy soil containing 100 seeds/basin were used to test the genotoxicity of this water. For this purpose each control sandy soil basin was treated with 120 ml of gravitational water.

2.3.1 *Micronuclei test*

The seedlings were taken out, the primary roots length of seedlings was measured in order to study the eventual toxic effects. The root tips were then fixed in ethyl alcohol and acetic acid 3:1 (v/v). The Feulgen method was used for staining. Micronuclei are Feulgen positive corpuscles, localised within the cell wall in the cytoplasmatic area surrounding the main nucleus. MC are formed by chromosome or chromosome fragments that are not incorporated into daughter nuclei at the time of cell division. AA are characterised by the presence of bridges and/or lagging and sticking chromosomes.

The genotoxic effects were evaluated by following the frequency of micronucleated cells and aberrant anatelophases in root meristems of *Vicia faba*. (De Marco et al. 1990; De Simone et al. 1992; Ma, 1982) The micronucleated cells and aberrant anatelophases frequency was scored from 15.000 cells (15 root tips, 1.000 cells for tip). For each experimental group, means and standard deviations were obtained by Analysis Of Variance with Schefflè test significant at 0.05 level, that has been used for analysing the significant differences among the treated groups and the positive control treated with Maleic Hydrazide 10^{-5} M.

2.3.2 *The comet assay*

The comet test was performed under alkaline unwinding/ alkaline electrophoresis (A/A) protocol by Angelis K. et al. (2000). Briefly, root tips were chopped using a razor blade, the suspension with released nuclei was filtered through a 20 μm filter to remove most of the tissue debris. Fifthy μl of the filtrate were mixed with agarose and set on a microscopic slide. Nuclei embedded in agarose, were lysed for 1 hour and a half and the slides then were placed in the electrophoresis buffer at pH>13 for 40 min, before electrophoresis.The protocol was modified in the electrophoresis applying 300 mA, 25 V for 45 minutes instead of 30V for 20 min. Slides were stained with ethidium bromide (5 μg/ml) and comets were viewed by an epifluorescence microscope and analyzed with the IAS 2000 analysis system (Delta Sistemi).

We performed an additional step to the standard protocol involving DNA digestion in the nucleoids with lesion-specific endonucleases: Endonuclease III (EndoIII) and Formamidopyrimidine glycosylase (Fpg). This comet assay reveals additional breaks at sites of base oxidation increasing comet tail intensity.

The Comet lenght is used as a parameter of DNA damage. For each experimental point, median and standard errors were calculated of at least 25 comets by ANOVA with Schefflè test significant at 0.05 level .

2.4 *Phytotoxicity*

The phytotoxicity was calculated by measuring the roots length of *Vicia faba* seedlings exposed to the polluted soil, infact the root is the first plant zone influenced by soil pollutants and its length decreases after a stress.

SOIL ANALYSIS

Boxes	pH	C. org g/kg	s. organica %
Clay-loamy test.	7,3 b	13,38	2,3
Clay-loamy test. plant	7,3 b	13,06	2,3
Clay-loamy Brome	7,1 b	13,01	2,2
Clay-loamy Br plant	7,1 b	13,16	2,3
Clay-loamy Isocyanate	7,1 b	13,02	2,2
Clay-loamy Iso. plant	7,1 b	11,92	2,1
Clay-loamy Formaldehyde	7,9 a	13,22	2,3
Clay-loamy form. plant	7,8 a	13,28	2,3
Sandy test	7,7 a	2,74	0,5
Sandy test. plant	7,7 a	2,65	0,5
Sandy Brome	7,8 a	3,08	0,5
Sandy Br.plant	7,8 a	3,01	0,5
Sandy Isocyanate	7,7 a	3,05	0,5
Sandy Iso. Plant	7,8 a	2,82	0,5
Sandy Formaldehyde	7,9 a	2,95	0,5
Sandy Form. Plant	7,8 a	2,81	0,5

Figure 2. Total organic carbon and organic matter contents. Different letters show a significant difference with $p < 0.05$ (ANOVA).

3 RESULTS AND DISCUSSION

3.1 *Soil characteristics*

There was no significant effect due to pollution on total organic carbon and organic matter contents. There was a significant increase in soil reaction in clay-loamy soils treated with formaldehyde (Figure 2).

The cumulative CO_2 production rate showed a significant decrease in all polluted clay-loamy soils, compared to the control (Figure 3). In sandy soils, the pollution with formaldehyde induced only a significant microbial activity reduction.

3.2 *Plant chemistry*

In both soils, brome release caused a significant increase in chlorophyll (Figure 4), total nitrogen and bromides concentration. The nitrates content decreased (Figure 5).

Formaldehyde contamination induced a general reduction in fluorides, chlorides, phosphates and sulphates concentration. In this treatment, the uptake of monovalent exchangeable cations, as Na^+ and K^+, diminished while bivalent cati-

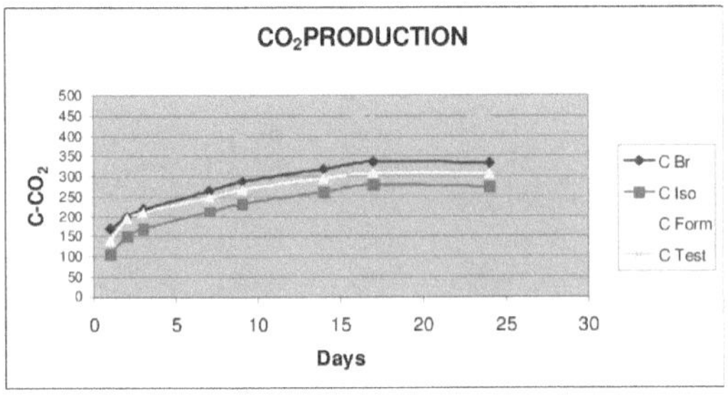

Figure 3. Cumulative CO_2 production rate.

Plant analysis	chlorophyll A mg/l	chlorophyll B mg/l	chlorophyll tot mg/l	N tot. %
Clay-loamy test	4.4 b	1,6 b	6,0 b	1,85 b
Clay-loamy Isocyanate	4,2 b	1,6 b	5,8 b	2,27 b
Clay-loamy Brome	4,9 a	1,8 a	6,7 a	2,88 a
Clay loamy Formaldehyde	4,5 b	1,5 b	6,0 b	1,83 b
Sandy test	4,2 b	1,4 b	5,6 b	1,90 b
Sandy Isocyanate	4,2 b	1,8 a	6,0 b	1,90 b
Sandy Brome	4,7 a	1,8 a	6,5 a	2,64 a
Sandy Formaldehyde	4,3 b	1,6 b	5,9 b	2,30 b

Figure 4. Plant analysis. Different letters show a significant difference with $p < 0.05$ (ANOVA).

Plant Chemistry
mg/kg

Boxes	Fluorides	Chlorides	Bromides	Nitrates	Phosphates	Sulphates
C V test	4,7 b	148,1 a	2,8 b	2,5	163,9 a	26,4 b
C V Br	25,8 a	205,2 a	45,3 a	0,0	189,9 a	48.6 a
C V Iso	27,9 a	174,0 a	1,7 b	0,8	176,9 a	22,6 b
C V form	6,8 b	94,0 b	4,0 b	2,5	85,3 b	26,7 b
S V test	21,5	132,0	2,0 b	0,1	171,8 b	26,7
S V Br	21,4	161,7	33,7 a	0,0	331,6 a	27,8
S V Iso	16,3	112,9	3,4 b	2,2	227,2 b	17,6
S V form	16,1	100,1	1,8 b	0,0	206,0 b	20,0

Anions content in gravitational water
mg/l

Boxes	Fluorides	Chlorides	Bromides	Nitrates	Phosphates	Sulphates
C test	0	129,4 a	1,4 b	48,5 a	0	253,2 a
C Br	0	102,9 a	17,7 a	33,2 a	0,1	231,7 a
C Iso	0	96,1 a	1,6 b	34,4 a	0,2	250,0 a
C form	0	6,2 b	1,1 b	1,1 b	0,2	42,6 b
C plant test	0	237,5 a	1,5 b	35,9 a	0,1	224,4 a
C plant Br	0	256,1 a	25,0 a	2,8 b	0,4	223,4 a
C plant Iso	0	267,6 a	1,4 b	14,7 a	0,9	244,2 a
C plant Form	0	15,4 b	0,4 c	0,0 c	0,4	44,3 b
S test	0,0	148,5 a	1,9 b	16,8 a	2,3	334,8 a
S Br	0,0	143,8 a	52,8 a	27,7 a	2,6	309,4 a
S Iso	0,0	54,3 b	1,7 b	2,2 b	9,8	162,5 b
S form	0,0	2,5 c	1,4 b	0,0 c	5,1	24,1 c
S plant test	0.0	277,9 a	2.0 b	0,7 a	1,4	367,2 a
S plant Br	0,2	207,5 a	39,6 a	1,2 a	1,6	222,4 a
S plant Iso	0.0	208,7 a	1,9 b	1,0 a	2,7	357,6 a
S plant form	0,0	17,0 b	0,8 b	0,0 b	2,7	39,5 b

Figure 5. Plant Chemistry. Different letters show a significant difference with $p < 0.05$ (ANOVA).

ons enhanced (for example Ca^{2+}) or remained the same (Mg^{2+}) (Figure 6). This trend was due to the higher soil adsorption activity of bivalent cations, which maintained a high available amount for the plants. The Ni and Mn content showed a significant increase in plants grown on soils polluted with formaldehyde.

3.3 *Plant morphology*

The ratio between the leaf area index and the primary root length showed a significant reduction in all treatments, compared to the control, especially in formaldehyde release and in sandy soils (Figure 7).

3.4 *Gravitational water composition (Figure 8)*

Brome administration induced an increase in bromide leaching for each type of soil and cultivation systems.

Formaldehyde pollution determined a significant decrease in major part of anions concentration. In this treatment, as regards the nutrients content, Na e K increased, due to their desorption from the soil colloidal complex (organic matter and clay minerals).

In each treatment, the total concentration of metal cations remained the same.

3.5 *The comet assay*

3.5.1 *Formaldehyde*

The results obtained by the soils polluted were significant both in sandy and clay-loamy soil (Figure 9).

The results obtained by the control soils polluted by the gravitational water were significant only in sandy soils.

Pollution with formaldehyde determins a genotoxic damage both in sandy and clay-loamy soil (Figure 10).

3.5.2 *Brome*

The results on soils polluted showed no significant genotoxic effects (data not shown).

Total metal nutrients in plants
mg/kg

Boxes	Ca	Cd	Cu	Fe	K	Mg	Mn	Ni	Na	Pb	P	Zn
C test	8087,5 a	0,27	11,00 a	3109,5	20795,0 a	2855,0	113,3 a	6,4 b	1230,5 a	10,0	2350,5	32,3
C Iso	7855,0 a	0,27	9,25 b	2230,0	21422,5 a	2522,5	84,8 b	10,0 b	1212,3 a	13,8	2281,3	24,3
C Br	6660,0 b	0,27	7,40 b	1496,0	23126,3 a	2506,8	59,5 b	8,5 b	1229,3 a	4,5	2255,8	22,5
C form	8603,5 a	0,22	12,55 a	4097,0	11505,0 b	2484,3	138,8 a	84,4 a	961,0 b	7,0	2110,0	30,3
S test	10622,8 a	0,21	12,0 b	944,3	15967,5	3496,5	84,8 b	8,8 b	3080,0 a	1,6	2796,5	37,0
S Iso	10716,8 a	0,19	16,8 b	3977,8	15667,5	3235,0	87,8 b	5,5 b	1502,5 b	2,1	2496,0	37,8
S Br	8207,5 b	0,25	14,3 b	751,0	19270,0	3477,5	71,0 b	2,1 b	1746,0 b	1,2	3655,0	51,0
S form	12177,5 a	0,23	22,0 a	1407,8	15425,0	3345,0	113,3 a	21,8 a	1196,5 c	4,5	3680,0	45,8

Figure 6. Total nutrients in plants. Different letters show a significant difference with $p < 0.05$ (ANOVA).

Plant morphology	LAI/RAD
Clay-loamy T	1,48 a
Clay-loamy Br	1,03 b
Clay-loamy form.	0,39 c
Clay-loamy Iso	1,19 b
Sandy T	1,67 a
Sandy Br	0,59 b
Sandy form.	0,15 c
Sandy Iso	0,42 b

Figure 7. Plant morphology LAI/RAD: The ratio between leaf area index and primary root length. Different letters show a significant difference with $p < 0.05$ (ANOVA).

Gravitational water

mg/l

	Ca	Cd	Cu	Fe	K	Mg	Mn	Na	Ni	P	Pb	Zn
C Br	136.7	0.0	0.0	0.0	43.2	19.1	0.0	17.9	0.0	0.0	0.0	0.1
C Iso	137.8	0.0	0.0	0.0	43.0	20.5	0.0	20.1	0.0	0.0	0.0	0.1
C T	149.9	0,0	0.2	0.0	34.2	22,6	0.0	16,9	0,0	0,1	0.0	0,3
C plant Br	174.6	0.0	0.0	0.0	33.1	35.9	0.0	46.2	0.0	0.0	0.0	0.0
C plant Iso	205.7	0.0	0.1	0.0	33.2	38.5	0.0	47.2	0.0	0.0	0.0	0.1
C plant T	167.4	0.0	0.0	0.0	33.9	32.9	0.0	45.6	0.0	0.0	0.0	0.1
S Br	170.5	0.0	0.0	0.0	60.5	42.9	0.0	54.4	0.0	0.6	0.0	0.1
S Iso	232.9	0,0	0.0	0.0	44.7	39,7	0.0	77,6	0,0	0,6	0.0	0.1
S T	162.7	0.0	0.0	0.0	45.3	43.8	0.0	45.1	0.0	0.5	0.0	0.1
S plant Br	216.6	0.0	0.1	0.0	31.8	50.3	0.0	119.9	0.0	0.3	0.0	0.2
S plant Iso	172.5	0.0	0.1	0.0	42.1	45.0	0.0	130.2	0.0	0.5	0.0	0.1
S plant T	197.2	0.0	0.1	0.0	32.5	60.7	0.0	119.8	0.0	0.3	0.0	0.1

Figure 8. Gravitational water composition.

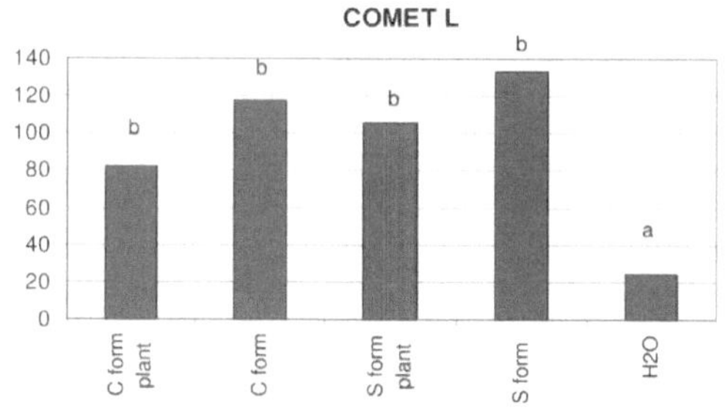

Figure 9. Comet assay on polluted soil by formaldehyde. Different letters show a significant difference with $p < 0.05$.

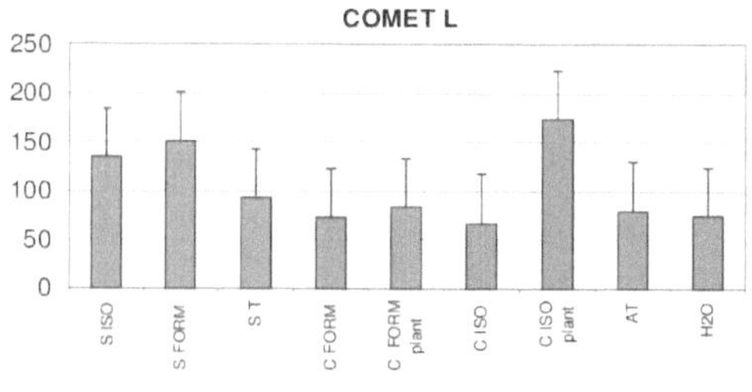

Figure 10. Comet assay on polluted soil by gravitational water.

Control soils polluted with gravitational water determined a significant genotoxic effects (Figure 11).

Such molecule probably tends to perculate determining the genotoxic effect where it accumulates.

3.5.3 *Isocyanate*

A significant genotoxic effect was found only in clay-loamy soil with plants (Figure 12) and in clay-loamy soil polluted by the gravitational water (Figure 10).

3.6 *Micronuclei test*

3.6.1 *Brome*

The results obtained on the soils polluted were significant in sandy soil with and without plants (Figures 13–14).

The results obtained on the soils polluted by the gravitational water were significant in sandy soils with and without plants (data not shown).

3.6.2 *Formaldehyde*

The results obtained on the polluted soil and on gravitational water showed no significant genotoxic effect (Figures 13–14).

3.6.3 *Isocyanate*

The results obtained on the polluted soil and on gravitational water showed no significant genotoxic effect (Figures 13–14).

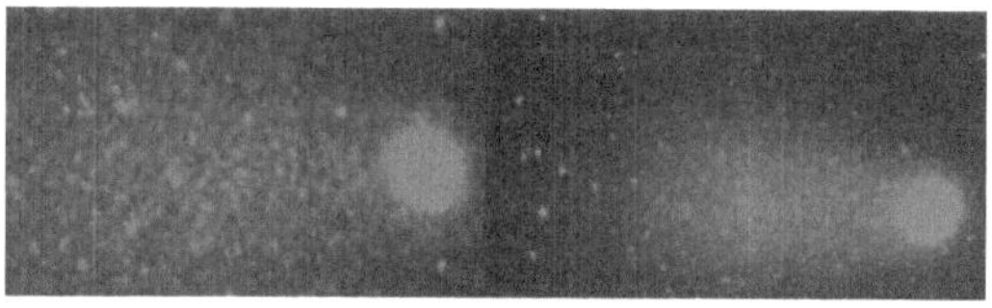

Figure 11. Comets on soils polluted by gravitational water of Brome boxes.

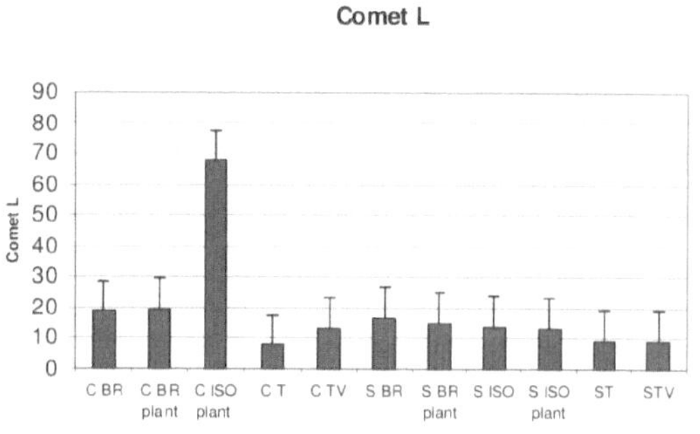

Figure 12. Comet assay on soils polluted by brome and fenylisocyanate.

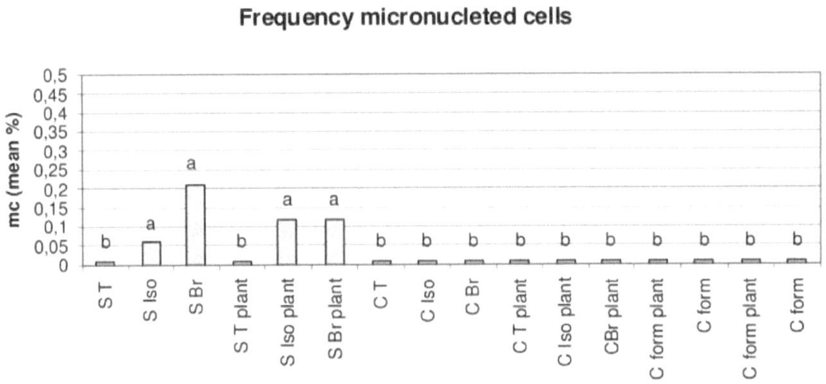

Figure 13. Micronuclei test on polluted soil by the tree substances.

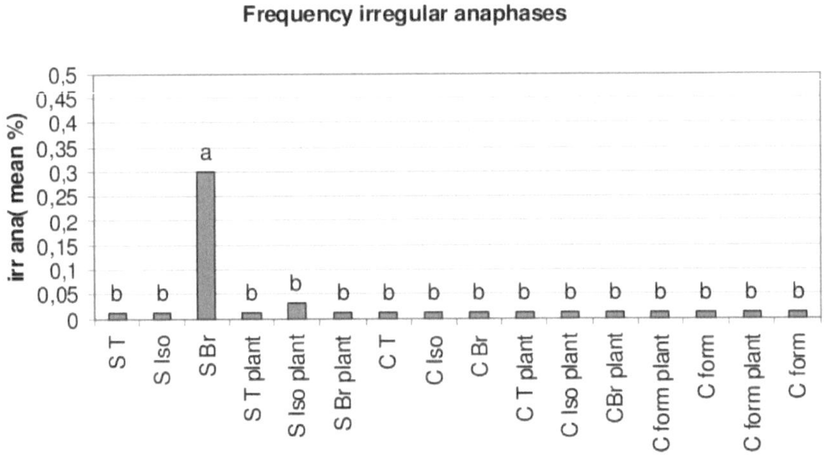

Figure 14. Frequency of irregular anaphases.

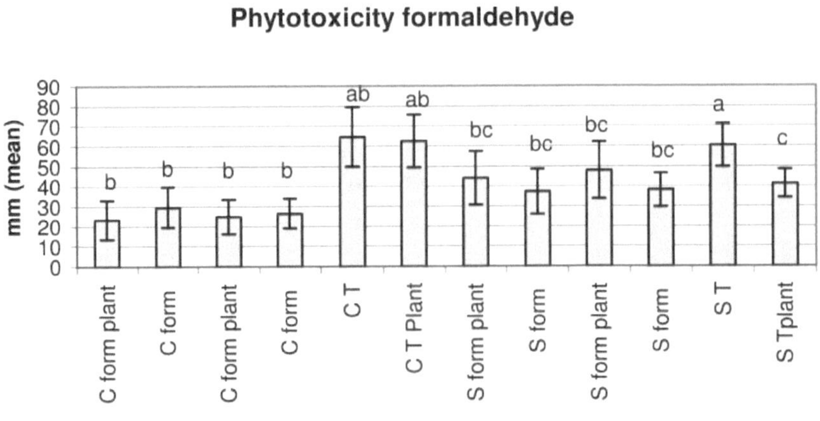

Figure 15. Phytotoxicity of soils polluted with formaldehyde.

3.7 *Phytotoxicity*

The results were significant only in clay-loamy soil polluted with formaldehyde (Figure 15).

4 CONCLUSION

The results showed that some of the parameters considered are applicable indicators for the evaluation of the soil-plant system pollution.

For example, variations on soil reaction and microbial CO_2 emission measure the effect on chemical and biological functioning of soil in relation to the release of toxic substances as formaldehyde.

Plant chemical parameters, like nitrogen, chlorophyll and ions concentration might be used as pollution indexes, if the leaf standard content of volunteer and cultivated species are well known.

Also the modifications on plant morphology seem to be a valid indicator of pollutant release.

The composition of the gravitational water collected in polluted sites might be affected by the leaching of any elements, caused by the shift of the ions adsorption equilibrium in soils.

Our study demonstrated that the comet assay is a sensitive, rapid and economic technique for the detection of DNA damage, which is ideally suited as a biomarker of genotoxicity for biomonitoring.

The comet assay seemed to be more sensitive than the micronucleus test to assess the DNA damage induced by chemicals, but the micronuclei test detects lesions which survived at least one mitotic cycle, while the comet assay identifies reparable DNA lesions or alkali-labile sites. There is a need to combine the comet assay with the use of other genotoxicity tests such as the micronuclei test, to get a more comprehensive understanding of the pollution situation of certain environment.

REFERENCES

Angelis, K.J., Dušinska, M. & Collins, A.R. 1999. Single cell gel electrophoresis: Detection of DNA damage at different levels of sensitivity. *Electrophoresis* 20: 2133–2138.

Angelis, K.J., McGouffie, M., Menke, M. & Schubert, I. 2000. Adaptation to alkylation damage in DNA measured by the comet assay. *Envir. Mol. Mutagenesis* 36: 146–150.

Bianco, V.V. & Pimpini, F. 1991. Spinacio (*Spinacia oleracea L.*): 270–355. Orticoltura.

Baskin, C.C. & Baskin, J.M. 1998. *Seeds: ecology, biogeography and evolution of dormancy and germination.* San Diego: Academic press.

De Marco, A., Boccardi, P., De Simone, C., Piccolo, A., Raglione, M., Testa, A. & Trinca, S. 1990. Induction of micronuclei in *Vicia faba* root tips treated in different soils with the herbicide alachlor. *Mutation Res.* 241: 1–6.

De Simone, C., Angelucci, R., Errichetti, M.F., Marconi, S., Rossi, M. & Selvi, S. 2003. Genotossicità: un nuovo parametro nella valutazione della qualità dei composts?. *I Quaderni di Ecomondo Rimini* 22–25 Ottobre, 29–35.

De Simone, C., Piccolo, A. & De Marco, A. 1992. Genotoxic effect induced by erbicides atrazine gliphosate in plants of *Vicia faba* grown in different soil. *The Sci. of the Total Environ.* 233: 123–124.

De Simone, C., Tomati, U., Galli, E., Owczarek, M., De Marco, A., D'Ambrosio, C., Alianello, F. & Cortellini, L. 1999. Evaluation of genotoxic activity of eleven composts from different origin. *Fresenius Environ. Bull.* 9: 683–690.

Jenkinson, D.S. & Ladd, J.N. 1981. Microbial biomass in soil. The effect of rate addition. *Journal of Soil Science, Marcel Dekker* 5: 415–471.

Jenkinson, D.S. & Powlson, D.S. 1976. The effects of biocidal treatment on metabolism in soil. *Soil Biology and Biochemistry* 8: 189–202.

Jenkinson, D.S. 1988. Determination of microbial biomass carbon and nitrogen in soil. In Advances in nitrogen cycling in agricultural systems: 368–386. Cab international.

Kalra, Y.P. 1998 Determination of cadmium,chromium, cobalt, lead and nickel in plant tissue, in Kalra, Y.P (ed.), *Reference Methods for Plant Analysis*: 193–197. Washington DC: CRC Press.

Kalra, Y.P. 1998. Determination of total nitrogen in plant tissue, in Kalra, Y.P (ed.), *Reference Methods for Plant Analysis*: 75–81. Washington DC: CRC Press.

Kilman, B.A. & Sturelid, S. 1978. Effects of caffeine on frequencies of chromosomal aberrations and sister chromatid exchanges induced by chemical mutagens in root tips of *Vicia faba*. *Hereditas* 88: 35–41.

Kilman, B.A. 1956. Factors affecting the production of chromosome aberrations by chemicals. *J. Biophys. Biochem. Cytol.* 2: 543–545.

Koppen, G. & Verschaeve, L. 1996. The alkaline comet test on plant cells: a new genotoxicity test for DNA strand breaks in *Vicia faba* root cells. *Mutation Res.* 360: 193–200.

Lewis Sax, J.L. 1984. *Dangerous properties of industrial materials, 7th edition.* Boston Mass.: CRC Press.

Ma, T.H. 1982. *Vicia* cytogenetic tests for environmental mutagens. A report of the US Environmental Protection Agency Gene-Tox Program. *Mutation Res.* 99: 257–271.

Manahan, S.E. 1994. *Environmental Chemistry*. Lewis Publishers.

Menke, M., Angelis, K.J. & Schubert, I. 2000. Detection of Specific DNA Lesions by a Combination of Comet Assay and FISH in Plants, *Environmental and Molecular Mutagenesis* 35: 132–138.

VV.AA. 1994. *Metodi Ufficiali di Analisi Chimica del Suolo*, Ministero delle Risorse Agricole, Alimentari e Forestali.

VV.AA. 1999. *Determination of total organic carbon (TOC) in sludge*, European standard cen.

Risks Challenging Publics, Scientists and Governments – Menoni (ed)
© 2010 Taylor & Francis Group, London, ISBN 978-0-415-58072-4

The soil contamination with cadmium after applications of sewage sludge

K. Hyblerova
Department of Applied and Landscape Ecology, Mendel University of Agriculture and Forestry in Brno, Brno, Czech Republic

F. Bozek
Department of Civil Protection, University of Defence, Brno, Czech Republic

J. Hlusek & T. Losak
Department of Agrochemistry and Plant Nutrition, Mendel University of Agriculture and Forestry in Brno, Czech Republic

ABSTRACT: A fundamental regarding issue city sewage plants is to find an efficient, safe and low-cost method of sewage sludge disposal. One option is to use the sludge to fertilise and reclaim agricultural land. The target of this article is to identify the effect of stabilised, hygienic and safe sewage sludge applications on biomass growth of poplars and on changes in the level of cadmium in the soil. Soil analyses showed that the cadmium content did not change considerably after sewage sludge use. Detected cadmium amounts are markedly below the limit estimated for application on agricultural soil (1 mg kg^{-1} of soil). Increasing doses have a positive effect on biomass growth of poplars. Considering these facts there is a good reason to use stabilised and hygienically safe sewage sludge for fertilisation in plantations of energy tree species. In this way the sludge is effectively used in the process of biological recycling.

1 INTRODUCTION

In keeping with principles of sustainable development it is necessary to reduce the consumption of natural resources, the amount of produced waste and air pollution. An important type of waste, both in terms of the amount and of raw material and energy potential, is sludge from sewage treatment plants. It is therefore highly relevant and pragmatic to reduce the amount of stockpiled sludge. An effective solution, besides the energy aspect, is to use sludge in agriculture as organic fertiliser, for soil treatment and for compost production.

2 THEORY

Sludge from sewage treatment plants is a source of basic nutrients (N, P, K, Ca etc.), including trace elements, of biodegradable organic matter, and it has a high microbiological activity. On the other hand, however, it contains heavy metal ions, particularly Cd^{2+}, Cu^{2+}, Pb^{2+}, Ni^{2+}, Cr^{3+} etc., toxic organic substances, i.e. linear alkylbenzenesulphonans from detergents, micro-organisms (viruses, bacteria, lower fungi, protozoa) and embryos of lower animals, all this limiting the possible use of sludge in agriculture. Consistent chemical analysis, stabilisation and disinfection of the sludge are therefore necessary (Bozek et al. 2003).

Despite the above measures, re-valuation of sewage sludge in the agro-system is often associated with fears of soil contamination, namely with heavy metals. Integration of these elements into plant metabolism generates the danger of a negative effect on plant growth and development and potential danger of the entry of hazardous ions into the food chain (Baxter et al. 1983; Häni et al. 1995). In this sense, cadmium atoms, which are characterised by high toxicity (ORNL 2006) and ecotoxicity (An, 2004), play a specific role as they cause relevant functional and morphological changes in the plant organism (Azmat et al. 2005). The EU Directive specifies 10 mg kg^{-1} of dry matter as the limit concentration (EEC 1986).

National legal regulations in this sense are stricter specifying 5 mg kg^{-1} of dry matter as a limit concentration of cadmium, if used in agriculture (MoE CR, 2001).

According to results of comprehensive investigations into the quality of sludge carried out by CISTA (Central Institute for Supervising and Testing in Agriculture) in the Czech Republic, the required national limit in the past decade was not only satisfied, but permanent improvement was monitored (CISTA 20005). The level of cadmium in dry matter gradually decreased from the average 4.6 mg kg^{-1} in 1994 to 3.0 mg kg^{-1} in 2003 (Balik et al. 2004).

Cadmium finds its way into the environment either naturally, or through anthropogenic activities. Natural sources include dust, forest fires, vegetation, volcanic activity, sea aerosols and groundwater. Anthropogenic activities are e.g. metallurgy, combustion of fossil fuel and oil, industrial production of metals and compounds, electro-chemical treatment of metals etc. (Nriagu & Pacyna, 1988; Benes & Benesova, 1993). Average contents of cadmium in soils of the Czech Republic range between 0.13 and 0.52 mg kg^{-1}. According to the Decree (MoE CR. 1994) the maximal admissible content of cadmium in light soils is 0.4 mg kg^{-1} of soil and in other soils it is 1.0 mg kg^{-1} of soil, whereas most important for uptake by the plant is the water-soluble and exchangeable form of the metal (Zhao & Saigusa, 2007; Richter & Hlusek, 2003). To reduce the intake of cadmium by plants it is recommended, in the first place, to modify the pH value to neutral, when it becomes considerably immobilized (Basta et al. 2001; Bolan et al. 2003; Bozek et al. 2003). Humus substances in the soil can also partly bind heavy metals (Kurek et al. 1996).

Well known is the positive effect of sulphur in the mechanism of metabolic adaptation of plants to heavy metals. According to Paulsen (2001) thiols, for instance, have a high affinity to some heavy metals (Cd, Hg, Cu) and withhold them in roots or stems. Important is also the role of glutathione as precursor for the production of phyto-chelatins, which can detoxify some heavy metals, particularly cadmium, but also zinc and copper, and have a positive effect on their transport in the plant (De Kok et al. 2003). The other authors (Yurekli & Kucukbay, 2003) noticed the induced growth in the content of phyto-chelatins in plants with increasing cadmium levels in the soil.

There are many papers dealing with the use of different kinds of plants with higher content of sulphur substances for reducing concentration of heavy metals in contaminated soils. (Brown, 1995; Salt et al. 1995; Schnoor et al. 1995). Thus the potential risk for the plants and the heavy metal migration to the environment can be reduced under these conditions.

The objective of the present study was to prove if it is possible to use sewage sludge as a fertiliser in plantations of energy tree species, poplars, at the same time keeping below-limit concentrations of heavy metals in the soil, with a focus on the presence of cadmium.

3 MATERIAL AND METHODS

The field trial was established in 2003 on a plot in the cadastre of Bystrice above Pernstejn, part Domaninek, ca 45 km northwest of Brno, altitude 560 m. The soil of the locality can be characterised as light, in terms of the soil type as acid cambisol. Table 1 shows the agro-chemical properties of the soil before the establishment of the trial.

The taking of samples was carried out in three locations of the observed area for every variant and the measured values were averaged arithmetically. The pH value was monitored with a potentiometer. The available nutrients in the soil (P, K, Ca, Mg) were extracted with a Mehlich III solution and specification proper was conducted using optical emission spectrometry with induction-bound plasma (ICP–OES).

The content of available nutrients, with the exception of phosphorus, was good, which means that fertilisation is necessary to achieve the standard requirement for plants. Soil reaction was between slightly acid and neutral. The level of mineral nitrogen in the soil (N_{min}) was 8.42 mg kg^{-1}, corresponding to ca 3.8 g m^{-2}. Based on the methods of CISTA we can evaluate this level as low (CISTA. 2005).

An experimental plot was established in the plantation of two-year-old poplar trees in three variants, i.e. control variant (A), sludge application once a year (B) and sludge application twice a year (C). The experimental chart demonstrated in Table 2 and the number of experimental trees was drawn by means of the G POWER programme.

Table 1. Agro-chemical characteristics of the soil in [mg kg^{-1}] (Mehlich III, 11. 04. 2003).

pH/KCl	P	K	Ca	Mg	Cd	Hg	Pb
6.52 slightly acid	122 high	181 good	2040 good	150 good	0.149 below-limit	0.0170 below-limit	10.80 below-limit

Table 2. Chart of experiment conducted in 2003.

Variant	Sludge applications [number year^{-1}]	Dose of fresh stabilised sludge [t area^{-1}]	Area [m^2]	Number of trees
A	–	–	97.55	104
B	1	1.293	84.32	101
C	2	1.458 + 1.540	91.56	102
Total for 2003		4.291	273.43	307

The dose of applied fresh sludge respected the chemical composition of the sludge and ranged between 0.50 and 1.00 kg of dry matter per m^{-2} with a content of ca 20 g nitrogen per m^{-2}. In variants B and C the sludge was applied on 9 May 2003 and in variant C again on 5 August 2003 using hose applicators in the form of furrow podmoku by the trees.

Figure 1 shows the plantation of 3-year-old poplars. The poplars were planted out in 2001, in double rows, 0.7 × 0.7 m spacing and 2.5 m between the rows.

The applied sewage sludge was analysed twice a year. The results are summarised in Tab. 3, which also shows the limit concentrations of the monitored contaminants according to the Directive (EEC 1986). From these data it is evident that none of the foreign elements exceeded the concentration limit.

Figure 1. View of plantation of 3-year-old poplar trees (29 July 2003).

4 FINDINGS AND DISCUSSIONS

Cadmium is a component of the bedrock and primary minerals and that is why all soils have their own characteristic background. Nevertheless, anthropogenic activities may considerably increase the level of the background. The parts available to plants are affected by changes in soil properties, namely pH, content of organic matter and chemical properties of the element (Sichorova et al. 2004; Basta et al. 2001).

On the basis of data summarised in Table 4 we see that the content of cadmium in the soil between spring and autumn 2003 did not undergo relevant changes. After sludge application the expected level of cadmium in the soil in relation to the level in the applied sludge did not increase, moreover, compared to the control variant, the concentration decreased. In spite of the slight increase in the cadmium concentration in the soil containing a higher amount of sludge, no important differences were found between the variants. The cadmium content ranged between 0.126 and 0.137 mg kg^{-1}; these values are in correlation with Fecenko and Lozek (2000), whose data say that the average cadmium content in the soil ranged between 0.100 and 2.500 mg kg^{-1}, while Cermak and Provaznik (2004) stated 0.240 mg kg^{-1}. Jasiewicz and Antonkiewicz (2004) presented analogical data stating that the average levels ranged between 0.200 and 0.500 mg kg^{-1}. However it is

Table 3. Chemical analysis of the dry matter of sewage sludge.

Date	pH	Na [%]	K [%]	Ca [%]	Mg [%]	P [%]	N [%]	Dry matter [%]
4/2/2003	7.4	0.15	0.34	1.43	0.38	1.84	3.26	3.84
9/7/2003	6.9	0.12	0.45	2.72	0.37	1.74	2.82	5.99
	Cu [mg kg^{-1}]	Zn [mg kg^{-1}]	Cd [mg kg^{-1}]	Ni [mg kg^{-1}]	Pb [mg kg^{-1}]	Cr [mg kg^{-1}]	Hg [mg kg^{-1}]	As [mg kg^{-1}]
4/2/2003	205	1018	2.30	44.0	72.3	33.0	3.18	3.05
9/7/2003	144	980	2.09	35.4	41.3	33.9	2.69	8.13
Limit	*500*	*2500*	*5*	*100*	*200*	*200*	*4*	*30*

Table 4. Average content of cadmium in the soil after sludge application in [mg kg^{-1}] in 2003.

Element	Variant	Content in soil before sludge application	Dose in sludge after first application	Dose in sludge after second application	Theoretical content in soil in autumn	Actual content in soil in autumn
	1	0.149	–	–	*0.14900*	0.126
Cd	2	0.149	0.00471	–	*0.15371*	0.129
	3	0.149	0.00471	0.00494	*0.15865*	0.137

important, that in no variant and no taken sample even the strictest standard required for light soils, i.e. 0.4 mg kg^{-1}, was not exceeded.

It has been mentioned that the uptake of foreign elements by plants is a function not only of their concentration in the soil but chiefly of their acceptability by plants (Zhao & Saigusa 2007; Sichorova et al. 2004). The mobility of cadmium in the soil increased considerably after a decrease in the soil reaction when pH < 5.5 (Basta et al. 2001; Bolan et al. 2003; Richter & Hlusek, 2003). It is true that in the course of the experimental period a slight reduction in pH was observed in all three variants down to the autumn values of 5.37–5.61, nevertheless the drop in pH was not reflected in a higher concentration of cadmium in the soil. The increased cadmium concentration was neither detected in the soil in the vicinity of the observed plantation and its concentration corresponded to the average cadmium concentration in the plantation soil with 95% level of statistical reliability. On this basis we can conclude that good conditions were created in the soil for its immobilization relatively. However it is necessary to emphasise that the cadmium concentration was determined neither in groundwater nor in the different depths of unsaturated zones.

Table 5 compares the growth characteristics of poplars in spring and autumn 2003. The data show that an increasing dose of sludge increases the average half-year growth in height of the poplars, i.e. by 6.49% after one application of sludge and by 15.30% after two applications compared to the unfertilised control variant. In terms of increased stem girth the differences were more marked; related to the applied sludge doses it was 12.66% compared to 31.77% in the control variant.

The biomass increase corresponds with the results of Repshas et al. (1999) who proved that 4.5 - 13.5 kg m^{-2} of applied sludge had a positive effect on tree growth. The conclusions of Tripepi et al. (2000) were similar; he noted that the height of the plants increased by as much as 40% after the application of 13.5 kg m^{-2} of sludge. According to other authors the application of 5 tonnes of sludge dry matter supplied the soil with as much as 192 kg N, 40 kg P, 9 kg K, 30 kg Mg, 263 kg Ca and 2400 kg of organic matter, creating favourable conditions for plant growth and development (Hanc et al. 2004).

Table 5. Changes in growth of poplar biomass during 2003.

	Average poplar height [m]		Change in height	
Variant	Spring 2003	Autumn 2003	[m]	[%]
1	2.00	3.18	1.88	59.00
2	2.26	3.74	1.48	65.49
3	1.79	3.12	1.33	74.30
	Average stem girth [m]		Change in girth	
1	0.0580	0.0823	0.0295	55.87
2	0.0591	0.0996	0.0405	68.53
3	0.0437	0.0820	0.0383	87.64

5 CONCLUSION

Basing on the results it appears that applications of sewage sludge in plantations of energy tree species are effective, the precondition being consistent chemical analysis, stabilisation, hygienic safety of the sludge, and guaranteeing the soil pH ≥ 6. Increasing doses of sludge have a positive effect on the growth of the biomass of poplars, which can be a valuable source of energy. At the same time the cadmium level in the soil, in relation to the original value (control variant), actually decreases and is markedly lower than the adequate limit concentrations. In order to prove the sufficient cadmium immobilization it will be interesting to evaluate the application of sludge with a higher content of cadmium in relation to the resulting occurrence of this element in the soil, biomass, groundwater, and in the different depths of unsaturated zone. Thus the evidence would be given for the absence of cadmium migration to the environment.

The present results are an integral part of the research project MSM 432100001 of the Faculty of Agronomy of Mendel University of Agriculture and Forestry in Brno. This study was supported by

Research plan No. MSM6215648905 "Biological and technological aspects of sustainability of controlled ecosystems and their adaptability to climate change", which is funded by the Ministry of Education, Youth and Sports of the Czech Republic.

REFERENCES

An, Y.J. 2004. Soil Ecotoxicity Assessment Using Cadmium Sensitive Plants. *Environmental Pollution* 127(1): 21–26.

Azmat, R., Zill-e-Huma, Hayat, A., Khanum, T. & Talat, R. 2005. The Inhibition of Bean Plant Metabolism by Cadmium Metal I. Effect of Cd Metal on Physiological Process of Bean Plant and *Rhizobium* Species. *Pakistan Journal of Biological Sciences* 8(3): 401–404.

Balik, J. et al. 2004. The Sources of Risk Elements in the Environment. In *Proceedings of the 10th International Conference on Reasonable Use of Fertilizers:* 22–29. Prague: Czech University of Life Sciences.

Basta, N.T., Gradwohl, R., Snethen, K.L. & Schroder, J.L. 2001. Chemical Immobilization of Lead, Zinc, and Cadmium in Smelter-Contaminated Soils Using Biosolids and Rock Phosphate. *Environ. Qual.* 30(4): 1222–1230.

Baxter, J.C., Aguilar, M. & Brown, K. 1983. Heavy Metals and Persistent Organics at a Sewage Sludge Disposal Site. *J. Environ. Qual.* 12: 311–316.

Benes, S. & Benesova, J. 1993. Balance of Risk Elements in the Environment. *Plant Production* 39(10): 941–958.

Bolan, N.S., Adriano, D.C., Mani, P. & Duraisamy, A. 2003. Immobilization and Phytoavailability of Cadmium in Variable Charge Soils. II. Effect of lime addition. *Plant and Soil* 251: 187–198.

Bozek, F., Urban, R. & Zemanek, Z. 2003. *Recycling.* [Monography.]. 1st Ed. Pustimer: MoraviaTisk. 238 p. ISBN 80–238–9919–8.

Brown, K.S. 1995. The Green Clean: The Emerging Field of Phytoremediation Takes Root. *BioScience* 45(9): 579.

Central Institute for Supervising and Testing in Agriculture (CISTA). 2005. *Agricultural Soil Testing.* [on line]. [2005–03–15]. URL: <http://database.zeus.cz/bokrs/index_ oapvr.php?id = azp>

Cermak, P. & Provaznik, K. 2004. Risk Elements and Hazardous Substances in the Soils of the Czech Republic. In *Proceedings of the 10th International Conference on Reasonable Use of Fertilizers:* 42–47. Prague: Czech University of Life Sciences.

De Kok, L.J. et al. 2003. Sulphur in Plant Physiology. *Nawozy i nawozenie*, (2), 55-80.

European Economic Community (EEC). 1986. *Council Directive 86/278/EEC on the Protection of the Environment, and in Particular of the Soil, when Sewage Sludge is Used in Agriculture.* Brussels: EEC.

Fecenko, J. & Lozek, O. 2000. *The Nutrition and Fertilisation of Field Crops.* 1st Ed. Nitra: Slovak University of Agriculture and Duslo Sala. 442 p. ISBN 80-7137-777-5.

Hanc, A., Tlustos, P. & Balik, J. 2004. The Influence of Sewage Sludge Application on the Yield of Oat Biomass Grown on Three Different Soils. In *Proceedings of the 10th International Conference on Reasonable Use of Fertilizers:* 138–141. Prague: Czech University of Life Sciences.

Häni, H., Siegenthaler, A. & Candinas, T. 1995. Soil Effects Due to Sewage Sludge Application in Agriculture. *Nutrient Cycling in Agroecosystems* 43(1–3): 149–156.

Jasiewicz, C. & Antonkiewicz, J. 2004. Contents of Heavy Metals (Cd, Cu, Ni, Pb and Zn) in Soils of Poland. In *Proceedings of the 10th International Conference on Reasonable Use of Fertilizers:* 85–90. Prague: Czech University of Life Sciences.

Kurek, E., Kaczorowska, R., Nadulskan I., Ochal, M., Puacz, E. & Patkowska, E. 1996. Retention of Cd by Soil Constituents under Different Environmental Conditions. *Chemosphere* 33(2): 277–284.

Nriagu, J.O. & Pacyna, J.M. 1988. Quantitative Assessment of World-Wide Contamination of Air, Water and Soils by Trace Metals. *Nature* 333: 134–139.

Oak Ridge National Laboratory (ORNL). 2006. *The Risk Assessment Information System.* [on line]. [2006–10–09]. <http://rais.ornl.gov/tox/profiles/cadmium.shtml>

Paulsen, H.M. 2001. *Sulphur Day.* [Personal Advice], FAL Braunschweig, SRN.

Repshas, E.A., Gradyatskas, A.A., Kubertavichene, L.V., Riepsas, E.A. & Gradeckas, A.A. 1999. Effects of Wastewater Sludge Fertilizers on the Growth of Biomass Plantations on Mineral Soils. *Lesovedenie* 2: 35–41.

Richter, R. & Hlusek, J. 2003. *The Soil Fertility.* 1st Ed. Prague: Institute of the Agricultural and Food Information. 44p. ISBN 80-7271-130-X.

Salt, D.E., Blaylock, M., Kumar, N.P.B.A., Dushenkov, V., Ensley, B.D., Chet, I. & Raskin, I. 1995. Phytoremediation: A Novel Strategy for the Removal of Toxic Metals from the Environment Using Plants, *Bio/Technology* 13: 468–474.

Schnoor, J.L., Licht, L.A., McCutcheon, S.C., Wolfe, N.L. & Carreira, L.H. 1995. Phytoremediation of Organic and Nutrient Contaminants. *Environmental and Science Technology* 29(7): 318–323.

Sichorova, K., Tlustos, P. & Szakova, J. 2004. Changes of Cd a Pb Contents in Plants Growing at Polluted area. In *Proceedings of the 10th International Conference on Reasonable Use of Fertilizers:* 168–171. Prague: Czech University of Life Sciences.

The Ministry of Environment of the Czech Republic (MoE CR). 1994. *Decree No. 13/1994 Coll. on Protection of Agricultural Land Fund.* Prague: MoE CR.

The Ministry of Environment of the Czech Republic (MoE CR). 2001. *Decree No. 382/2001 Coll. on the Conditions for Using Treated Sludge on Agricultural Land updated by amendment 504/2004 Coll.* Prague: MoE CR.

Tripepi, R.R., Zhang, X.G. & Campell, A.B. 1996. Use of Raw and Composted Paper Sludge as a Soil Additive or Mulch for Cottonwood Plants. *Compost-Science-and-Utilization* 4(2): 26–36.

Yurekli, F. & Kucukbay, Z. 2003. Synthesis of Phytochelatins in Helianthus Annuus is Enhanced by Cadmium Nitrate, *Acta-Botanica-Croatica* 62(1): 21–25.

Zhao, X. & Saigusa, M. 2007. Fractionation and Solubility of Cadmium in Paddy Soils Amended with Porous Hydrated Calcium Silicate. *Journal of Environmental Sciences* 19(3): 343–347.

Embedding social and economic perspectives into risk assessment and management

Risks Challenging Publics, Scientists and Governments – Menoni (ed)
© 2010 Taylor & Francis Group, London, ISBN 978-0-415-58072-4

Perceptions of risk

D. Zaruk
Risk Management Consultant, Risk Perception Management, VUB, Belgium

ABSTRACT: The text that follows is based on my experience in the field of chemical risk management, developed within industry activities, in European Union stakeholder forums, regulatory debates and UN conferences, dialogues with environment and health NGOs the work implied also developing models for scientists to communicate risks to non-specialists. This analysis of my practical observations will hopefully be of value to those more academically active in the field of risk communication, perhaps providing incentive for further study.

1 INTRODUCTION

In December, 2004, I was serving as the reporter to the Risk Communications session of the Dutch Presidency to the European Union's conference on children's health issues (SCALE).[1] Several days before the conference, we were requested to remove the word 'risk' from the session title and from the text we had prepared for the event. Several environment and health NGOs had protested that risk was a major issue of consideration on the subject of children's health assessment and the Dutch government found it prudent to avoid a public conflict. This initiated a discussion as to why the word 'risk' would have proven to have been so offensive (as well as the somewhat comic situation of how we could speak about risk communications issues without using the word 'risk').

At the three-day conference, I observed that scientists were speaking about risks to children's health in a manner which was alien to the policy-makers present, and the industry perspective on risks and benefits came across as completely outside of the perfect world expectations that the NGOs were expressing. Thus four stakeholder groups from government, science, industry and NGOs came together to discuss the issue of children's health, with each having a different perception of what a risk is and its normative value. It came as no surprise that the debate on precaution was even less satisfactory due to the linguistic oceans dividing the participants on what precaution was meant to be about.

Defining what a risk is, in general, is fairly easy: an environmental health risk is the level of uncertainty of an effect arising from exposure to some form of hazardous product or activity. Perceiving what a risk means to someone—its normative value—and how this is interpreted in risk management decisions, is more difficult.

In this paper, I plan to outline how the four main stakeholder groups actively involved in environment and health debate (policy-makers, scientists, industry and NGO representatives) perceive risk, and how this perception affects the implementation of some of the regulatory tools used in risk issue management (namely, precaution, risk assessments, regulatory impact assessments and stakeholder dialogue). Although the emphasis in the European Union today is to encourage a wide dialogue to build consensus, when stakeholders are not speaking the same language, a consensus position could barely be anything more than weak compromises. For this paper, I will focus on the domain of chemicals risk issue management within the environmental health debate.

2 THE SCIENTIST'S PERCEPTION OF RISK

Toxicologists, following Paracelsus, consider environmental health risks as relative to the dose. "Everything is poison. Just the dose makes the difference between poison and non-poison."

The scientific endeavour is one of discovery—pushing the frontiers and testing the hypotheses. Discovery often involves a leap in the dark. There is an inherent openness to risk-taking, whether it is in the lab, in outer space or inside the human body. Trial and error can lead to accidents, and accidents are often the source of many great scientific discoveries, as well as demands for human creativity and ingenuity to solve problems as they arise.

[1] the conference website at: http://www2.vrom.nl/pagina.html?id=9102

Today, the scientist is active in the regulatory process by furnishing the necessary data in the form of a risk assessment. The role of the scientist is to provide the information necessary for the policy-maker to adequately take a decision. After the BSE crisis, the European Commission has insisted on the separation of the risk assessment from risk management. The role of the scientist in a risk assessment is to only answer the questions posed to them—to give advice within certain parameters.[2] They are not to be invited to comment on the political decisions that would ensue, or to voice an opinion on whether precaution would be necessary due to their advice.

Some scientists might find these restrictions rather debilitating, especially if policy-makers are more concerned about non-scientific influences in the decision-making process (as in the GMO issue in Europe), but recalling the BSE scandal, the British government's inability to lead or take difficult political decisions coincided with an acute ability to blame the scientists for giving them poor advice, hurting the image of science.[3] The Edelman Trust Barometer measured a low-point in 2001 when, in the UK, the government-appointed scientists were ranked third from the bottom in the UK in response to the question: Whom do you trust?[4]

So what does a risk assessment entail, if it is not to tell us whether we should proceed with precaution regarding a substance, product or process? A risk assessment should provide the decision-makers with estimates of thresholds of harm to human health. For example, if a prolonged exposure to a chemical substance raises a certain degree of risk of cancer, increasing the exposure by x would increase the risk of cancers per million by *y*. Toxicologists might be requested to provide possible means of preventing or reducing the exposure levels.

The policy-maker (the effective risk manager) would be expected to take the information from the risk assessments and determine whether the appropriate level of risk would be acceptable. In our example, the risk manager would have to determine whether it would be desirable for an increase or decrease in exposure which might result in, say, an increase or decrease of two cancers per million. The policy-maker would have to determine whether this would be an acceptable risk given the economic and social benefits which might run parallel to such an exposure. The toxicologist is not expected to provide any further information outside of the assessment. This at times can be frustrating for scientists who feel perhaps that the information is clearer than the policy-makers are making it, or that other stakeholders have clouded the information so that the risk assessment is not being interpreted or used properly.

There has been some momentum recently to get scientific information better communicated to decision-makers, to have the wider public better able to understand scientific and technological information and to have scientists take better care in how they communicate their information to non-scientists.[5] But is it a question of better science communication leading automatically to better decision-making? Do the other stakeholders, namely policy-makers, really want the scientists' information better communicated?

3 THE POLICY-MAKER'S PERCEPTION OF RISK

The policy-maker (decision-maker, cabinet minister, Commissioner, public official, high-level civil servant) is the ultimate risk manager. The main tool remaining at the policy-maker's disposal is the regulatory process (today fiscal, social and moral leadership tools face too many legislative and popular obstacles), thus risk management debates in the political arena are inevitably regulatory debates.

A good part of the regulatory debate is based around the regulatory impact assessment (RIA). Science is one element of a RIA which is to be assessed in light of other elements (social, political, economic, environmental and moral).[6]

What the regulatory risk manager has to do is gather the scientific evidence provided by the risk assessment, and determine, with the economic, social and moral impacts, what would be an acceptable risk. Back to the previous example, the risk manager must decide if two cancers per million is worth accepting a lower exposure regulation vis-à-vis employment and business advantages.

---

[2] See the European Food Safety Authority (EFSA) raison d'être at http://www.efsa.eu.int/science/catindex_en.html

[3] Gummer, John UK Minister of Agriculture at the height of the BSE crisis, felt compelled to blame his experts for the advice he received. See http://www.guardian.co.uk/bse/article/0,2763,387701,00.html for an analysis of the 2000 Phillips Report on the UK BSE crisis.

[4] http://www.edelman.com/image/insights/content/Edelman_ Trust_Barometer-2005_final_final.pdf

[5] See in particular the UK's Science and Society Programme, the EU's Science and Society in Europe, the Science Media Centre and www.greenfacts.org

[6] The European Policy Centre released a working paper entitled *Enhancing the role of science in the decision-making of the European Union* (Bruce Ballantine, EPC Working Paper # 17, Brussels: The European Policy Centre, March 2005). The paper argues that science must play a greater role in the EU decision-making process, particularly in the RIAs.

A complication for the regulatory risk manager is that environmental health risks are not the only risks in the political arena. Economic risks (financial, employment, trade issues) and, in most Western democracies, political risks (lobbyists, political action committees, international approbation) loom heavily in the impact assessment considerations, especially near election times. So what a scientist may determine as an acceptable risk from the scientific data, will not be so clear-cut for the policy-maker. Precaution is meant to be a political tool used when the scientific information is unclear, but more often than not, it is used when the political risks are too high. A good case in point is the European Union's political manoeuvring over GMOs. The scientific community is fairly united on the science of most genetically modified organisms (as seen in the 2004 FAO SOFA report[7]), but many European politicians find themselves arguing that scientific uncertainty is not the main reason for precaution on this issue.

What are the political risks of following through with regulations based on a scientific assessment of an acceptable risk? In an economically mature atmosphere, the survivors of those two cancer victims may take public action on what they determined to have been a preventable and unnecessary exposure. In an atmosphere of a more developing country, 20 more cancers per million may turn out to be a less-challenging acceptable risk from the regulatory impact assessment position. Of course these different global standards of acceptable risk lead to different multinational investment strategies, which must be monitored through industry's voluntary initiatives, international agreements and NGO watchdog surveillance.

Take as an example the recent decisions (June, 2005) in the European Parliament regarding the continued ban on phthalates in children's toys. Phthalates is a general term for a wide range of chemicals used, among other things, as a softening plasticizer for PVC used in the manufacturing of toys. Phthalates have been under sustained attack by the NGOs (particularly Greenpeace[8]) for more than ten years as a tactic in its global campaign to remove a market for PVC, which in itself is part of Greenpeace's continuing effort, for more than 20 years, to deny chlorine a sustainable market. Faced with the relentless pressure of the NGOs, in July, 2005, the European Parliament chose to ignore the European Union's Joint Research Centre's 2003 commissioned scientific assessments which concluded that there was negligible health risks to children chewing on toys containing phthalates[9], and voted to ban phthalates in children's toys. The scientific assessments were ignored on the basis that the political pressure was more intensive and that the toy manufacturing industry had, in any case, effectively been largely off-shored to China. The political risks far out-weighed the actual environmental health risks in what is a text-book case of political cowardice.

There have been many calls from better regulation circles for a greater use of science in the decision-making process.[10] While it is logical to conclude that a greater, clearer communication of the scientific information will lead to more rational public policies, there is also a problem that policy-makers themselves may not always welcome clearer, incontestable scientific information. While sounding like a solution to a long-standing environmental health risk issue, a clear pronouncement coming from a risk assessment, whether it is over phthalates in children's toys or BT-10 maize for human consumption, does not necessarily make the political risk assessment any easier. In other words, a scientific community in disagreement can be more politically expedient, allowing policy-makers to sit on the fence and point the 'vacillation-blame-game' in the direction of others. For many years, climate change was a matter of scientific dispute, allowing politicians to avoid making hard decisions on energy-management issues. On climate issues, now is the time for political courage and leadership (or perhaps one more scientific study!).

This ability for policy-makers to rely on scientific uncertainty is accentuated by the *'need for further funding'* impulse scientists habitually write into their conclusions. Almost every scientific report closes with the statement of a need for further studies. From a scientific perspective, there is always a need for further studies (and funding), but politically, this can be used by those who do not agree with the general conclusions to question the level of certainty of the scientists who had written the report.

William Leiss, in 2001, spoke of the trend to de-scientize the policy-making process in a shift from the technical risk management paradigm, based on the policy-maker's trust in scientific evidence, to the socially-culturally embedded risk management paradigm.[11] Amplifying dissenting scientific

[7] http://www.fao.org/es/esa/en/pubs_sofa.html

[8] http://www.greenpeace.org/international/news/toxic-toy-story

[9] http://ecb.jrc.it/DOCUMENTS/Existing-Chemicals/RISK_ASSESSMENT/SUMMARY/dinpsum046.pdf

[10] Reference to the EPC report in note 6. See also the US National Research Council's "Strengthening Science at the US Environmental Protection Agency", Washington DC: National Academy Press, 2000.

[11] Chapter 5 of William Leiss' **In the Chamber of Risks**, with Michael D Mehta, Montreal and Kingston: McGill-Queens University Press, 2001.

views and calling into question the motives of the decision-makers to represent the real public will, NGOs, social, cultural and community groups have de-legitimized the role of scientific information in policy-making.

Having de-scientized the process, NGOs in the chemical risk issue debates are threatening to place policy-makers into a regulatory straight-jacket with three coordinated activities.

1. They have challenged the Paracelsus Principle, *the dose is the poison*, with the NGO principle, *the timing is the poison*, focussing on children's health issues, particularly new-borns, who are particularly vulnerable to any low-dose chemical exposure traces detected in indoor air environments.
2. Couple that with the NGO ability to call into question the distinction of potential hazard and risk with the precision of biomonitoring tools to detect chemical contamination of humans even prior to birth, and there are the means to show that even the lowest dosage of chemicals might just require immediate removal from markets.
3. Finally, whatever issues regulators are able to resist from the pressure of the NGOs' campaigns, the Aarhus Convention will ensure that the international courts will be tied up in cases for the next generation.

There seems to be very little left for the regulator to effectively regulate on.

How will the regulators be able to manoeuvre around the obstacles set upon them by these limitations? A decision-vacuum has been created, with stakeholder dialogue becoming the recent trend in the means to shift the decision-making process to a 'consensus-oriented policy evolution process'. Regulation today is becoming a process rather than a decision.[12] Into this decision-making void on environmental health issues, the lobbyists from industry and NGOs have taken stronger positions as quasi-regulatory bodies.

4 INDUSTRY'S PERCEPTION OF RISK

Of all the stakeholders in the environment and health debate, industry representatives view risk in the 'least negative' manner. There is a recognition, in industry circles, not only of the harms and uncertainties risks may pose, but also the benefits and opportunities that they produce. The World Business Council for Sustainable Development declared that, in the business world, risks imply opportunities.[13]

That risks produce opportunities is not so odd in the world of financial risks ('the greater the risk, the higher the potential gain' is a maxim any venture capitalist lives by). What business draws from this is that risks should not be feared, and even environment and health risks provide opportunities for innovation and new product development.

Behind every risk is a benefit. The most dangerous and potentially harmful products also provide enormous benefits. Chlorine has saved billions of lives through its disinfecting and purifying properties yet it provides enormous risks to environmental health.[14] Industry is concerned with the proper management of risks in order to provide maximum benefits. Managing risks implies limiting the exposure, controlling the potential risks and thus making the benefits more evident (and indeed, more profitable).

Managing exposure is never 100% and this is a failure in the industry communication model to have not had this understood and accepted. Risks always linger, and today, there will always be someone to exploit that potential risk. The debate over REACH presently going on in the European Union is an example of how other stakeholders have intervened to try to improve the exposure management. REACH stands for Registration, Evaluation and Authorization of Chemicals and aims to have all chemicals in use in Europe above a certain tonnage assessed for their health and environmental effects. The debates in Brussels have become legendary, mainly because the two main adversaries have different ideas of what REACH is for. NGOs see REACH as a means to substitute all harmful chemicals with safer ones. Industry sees REACH as a means to ensure a safer downstream use of chemicals. After six long years, regulators just want the whole process to be over.

One milestone in the REACH debate was the acceptance of the European Commission of the difference between risks and hazards. Risk, of course, equals hazard plus exposure.[15] Prior to the October 2003 revision of the REACH dossier, the European Commission had not made that distinction and many intermediate chemicals were due

[12] REACH is perhaps the best example of how a decision to introduce new regulations into the European chemicals management policy became, as of September, 2005, a six year process (and counting).

[13] See the WBCSD's paper entitled "Running the Risk", March, 2004 at http://www.wbcsd.org/web/publications/running-the-risk.pdf

[14] See the International Programme on Chemical Safety, Environmental Health Criteria 216, Disinfectants and Disinfectant By-products, Geneva: World Health Organization, 2000, Chapter 8.

[15] See the Cefic brochure entitled: "Risk and hazard: How they differ" at http://www.cefic.org/Files/Publications/Risk%20&%20Hazard2.pdf

for registration even though, through the process of chemical reactions, they would never have been exposed to the general population.

One interesting element is that REACH shifts the burden of proof. The present European chemicals management system relies on national bodies to prove that a substance is a health risk. Under REACH, industry, in making its registration, will have to prove that a substance is safe. Proving that something may be a risk is easier than proving that something is safe (and how safe is safe?). What industry might be able to prove is that a certain substance is the safest of all available alternatives. This is the essence of product stewardship.

Product stewardship is a corporate concept coming into wide use today. Besides the widespread understanding of protecting a product's proper use downstream, the concept also implies continuous improvement and, if necessary in certain applications, substitution for other, more sustainable, products. Developing better products to replace more obsolete (less beneficial) ones is at the core of the product stewardship business model, justifying research and development and driving innovation. If an existing product has too many risks, the business model pushes creativity and innovation to develop its substitution or enhancement. Business leaders feel that this can be done without regulation—other stakeholders feel that there needs to be a regulatory push to encourage alternatives.

CFCs is a good case in point. While it was widely accepted that CFCs were depleting the ozone layer and causing risks to human health, the development of alternatives in cooling systems was led by industry. Was it the attraction to the markets for safer products or the non-binding regulations imposed by governments that served as the impetus to develop replacements? Continuous improvement and substitution where necessary allowed for opportunities to arise from the environmental health risks of a product that had become obsolete. Any company that continues to market a product that is becoming obsolete does not have a sustainable business model.

For several years at Solvay, a Belgian chemicals and pharmaceuticals multinational, I was responsible for promoting the group's strategy of innovation. My job, essentially, was to reward the risk-takers who were providing the greatest benefits to the company and introducing innovations that would lead to new products and markets. A successful business today is one that can innovate and add value; innovation is risk taking. Thus we can find risk at the core of industry—it is not hidden or avoided, but courted and utilized.

Industry has brought great innovations to society, from the developments in personal transportation, chemicals, pharmaceuticals, communications and food safety. Industry has also profited from the developments it has brought to market. The consumer, in paying the price for the product, does not feel gratitude for the great strides in development and innovation, but rather, he feels that he owns the new product. Having paid for it and contributed to the company's profits, there is no feeling that industry has reaped great benefits on society. Microsoft is perhaps the most publicly reviled company in corporate history, clouding the great contributions it has made for societies. When these products are determined to also contain certain risks to human health, it is not merely a manner of risk-benefit trade-off—the consumer feels that he has paid for a product that should have zero risks. Industry is unable to deliver great benefits at zero risks, and thus there is great ease to paint industry in a negative light.

Such has been the game of the NGOs.

5 NGO PERCEPTIONS OF RISK

Many parts of this article have shown how the NGOs have been able to influence or affect the other stakeholders' risk management approaches, and have literally handcuffed regulators on certain issues. NGOs have played the risk card to their benefit, highlighting the uncertainties in the body of knowledge, implying that innovations are merely profit-oriented and not adding real value to society, but rather, jeopardizing human health.

For an NGO activist, a perfect world is one without any risks. This perfect world scenario is a goal NGOs work towards with each campaign, with progress measured in the lessening of each risk.[16] The present world is seen as a victim of man's aggression, and the subsequent human health problems are the consequences.

For NGO activists, the word 'risk' is indeed a four-letter word, one rarely found in any NGO documents or speeches. Recall how this paper began, with the Dutch government being pressured to remove the word 'risk' from the title of an event and the summary document. When talking about chemical risk issues, why don't NGOs use the term "risk"?

A risk implies benefits. Any risk a person takes, whether it is crossing the street or taking medicine, is done to attain a preconceived benefit. A harm or a danger is undesirable and does not imply any

[16]On energy issues, for example, NGOs have been calling for zero-carbon-emission energy production (wind and solar). WWF and Greenpeace will not accept other carbon-free alternatives like hydroelectricity or nuclear energy and assume that improving the wind and solar technology will meet all of mankind's energy needs.

benefits. Chlorine, for example, has been portrayed by NGOs as a danger (potential source of dioxins, an endocrine disrupter). If chlorine were considered as a risk, it would be allowed to be weighed against its benefits, which, as mentioned earlier, the WHO has acknowledged to be much higher than any potential risks.

Risk also implies that it can be managed. If a chemical company is working with an intermediate in a controlled environment, and in a chemical reaction, this product no longer exists in the new product, one would conclude that the risk is managed and there would be no need for this risk to be regulated.

Public fear of unquantifiable unknowns, particularly of elements potentially inside one's body (chemicals, pharmaceuticals, GMOs) is an emotive force that NGOs have been able to exploit in their fundraising campaigns. It is a commonly cited joke in Brussels that GMO stands for Greenpeace Membership Opportunity. The more people are afraid of something (and the more detached they are from recognizing the benefits), the more willing they are to contribute to organizations committed to stopping the potential cause of that fear.

The bottom line is the bottom line. People will give money to the major NGOs (Greenpeace, Friends of the Earth and WWF) to have them stop chemicals in their blood (no matter how small the traces of exposure may be) or GMOs in the food chain, but they will not give them money to remove other potentially serious environmental threats like the car or the cell phone, which the public perceives as providing enormous benefits and freedoms. None of the three big NGO organisations are running campaigns against cars or cell phones.

One of the founders of Greenpeace, Patrick Moore, made a scathing attack on the organization in his article, *Environmentalism for the 21st Century*[17], where he showed how Greenpeace had lost its way and had become a political organ rather than one committed to improving the environment.

With the introduction of the concept of sustainable development, NGOs had arrived at a crossroads in the mid-nineties: to continue on the path of confrontational activism or move in the direction of consensus and working together to improve the environment. In other words, to continue to create fears about risks or to generate means to lessen the real risks people face.

This sustainable development crossroads created a schism in the NGO world between the idealists who postulate a perfect world without risks and the pragmatists who see their activities as promoting a better world and environmental progress. The idealists opposed conciliation and dialogue with other stakeholders and radicalized even more to include opposition to globalization. Pragmatists began to work with industry, engage in stakeholder dialogue and accept corporate funding.[18] For Patrick Moore, there is no doubt that Greenpeace chose the wrong path. Other NGOs, like WWF, are continually battling internally between the two factions of idealism and pragmatism.

NGOs are meant to be 'Watchdog' organizations. Watchdogs are always welcome and easy to justify keeping. If my dog barks at the moon 99 times out of a hundred, I will tolerate this for the one time that it scares off a prowler. An NGO can be right just once (for example, on BSE) and the general public can forgive it for the other times when their campaigns proved to be unfounded. Environmental NGOs enjoy a deep public trust[19] and have been successful in portraying themselves as the noble defenders against the corrupt business interests. Contrast this with industry, which has to make just one mistake in a hundred projects to lose all public trust it had built up. Moral crusaders on a selfless mission to save the world are very difficult targets for profit-oriented capitalists to engage in substantial debate—the credibility deficit industry suffers is just too debilitating and facts just cannot fit into this picture. The European Union Transparency Initiative[20] has called on all sides to be transparent and accountable, but all of the attention has been on industry lobbyists (NGO activists are, for some reason, not often considered as lobbyists).

But the problem is that the NGOs are not dogs. They know that they enjoy this credibility surplus with the public and they have developed a tendency to go 'fishing with the facts'. Strong NGO declarations are based on clever phraseologies like: *this might happen; potential problems; it is not certain; there could be a risk; or science has not proven this to be safe.*[21] NGOs are able to seamlessly make the

[17] http://www.greenspirit.com/21st_century.cfm

[18] The landmark deal was WWF's global partnership with cement giant, Lafarge, which committed to work with the NGO to lower its greenhouse gas emissions. Shell has regularly sought out partnerships with environmental organizations as a centrepiece of it corporate social responsibility programme.

[19] NGOs are consistently at the top of the Edelman Trust Barometers. the sixth report, February, 2005 at http://www.edelman.com/image/insights/content/Edelman_Trust_Barometer-2005_final_final.pdf

[20] European Commission Vice-President, Siim Kallas launched the Transparency Initiative in March, 2005. http://europa.eu.int/comm/commission_barroso/kallas/transparency_en.htm

[21] A simple scan, on any given day, of the news titles on WWF's Detox website show this risk elevated wordplay in action. http://www.panda.org/campaign/detox/news_publications/newsroom.cfm

move from possible to potential to probable, often without having to reference any credible scientific studies.

And when science cannot prove with certainty that an undesirable effect is not impossible, NGOs then call for precaution to be applied.

6 PRECAUTION: UNCERTAINTY MANAGEMENT OR RISK MANAGEMENT?

It is not surprising, given the differences in the stakeholders' perceptions of the concept of risk, that they would also disagree on precaution, when it should be applied and how.[22] But is precaution a means of managing risks?

The use of precaution has evolved from its inception at the 1992 Earth Summit in Rio. Principle 15 of the 1992 Rio declaration states: *Where there are threats of serious or irreversible damage, lack of full scientific certainty shall not be used as a reason for postponing cost-effective measures to prevent environmental degradation*. The EU, in their 2000 Communication on the Precautionary Principle states that the precautionary principle should be applied *where preliminary objective scientific evaluation, indicates that there are reasonable grounds for concern that the potentially dangerous effects on the environment, human, animal or plant health may be inconsistent with the high level of protection chosen for the Community.*[23] The document went further to say that the precautionary principle is relevant to risk management.

But the EU seems to be talking about uncertainty management rather than risk management. What is the difference here?

- Not knowing is not a reason for doing nothing—in other words, if there is uncertainty, then one must not be inactive. This is risk management.
- Not knowing is a reason for doing something—in other words, if there is uncertainty, one must do something. This is uncertainty management.

Precaution has moved from risk management in Rio—where uncertainty was not a reason to not act (that precaution should not be avoided because we do not know)—to uncertainty management in Brussels, where uncertainty was a reason to act (that precaution should be applied because we do not know).[24] This is not simply a clever splitting of double negatives. In the business world it may seem clearer: risk management would imply that I should not avoid investing if I do not know the outcome; uncertainty management would imply that I should avoid investing if I do not know the outcome—which implies no investment.

These two approaches lead to different applications of precaution based on different perceptions of risk. If you are a scientist or an industry representative, and you believe that risks are inherent in all activities, and that risk-taking can lead to benefits, then precaution is a question of managing the risks. Chlorine is a product with enormous benefits but its hazards must be managed properly (not being certain is not a reason for not doing something, thus benefits would imply managing the risk). If, like NGO representatives, you believe that all risks should be avoided in order to protect the environment and human health, then precaution is an uncertainty management question. The environment and health effects of chlorine are not completely certain, thus we must act to limit this uncertainty.

Where the benefits are not patently obvious, policy-makers are closer to the NGO position given the political consequences of losing control of the risks.[25]

Thus how precaution is used is dependent on how one perceives risk. If one perceives risks negatively, then precaution is about uncertainty management—to limit any exposure to risks. If one sees opportunities from risk taking, then precaution is about how to manage those risks to avoid any possible negative aspects.

[22] The European Commission and the European Environment Agency have introduced what they call the Precautionary Principle. I will avoid using the term 'principle' here because I find this application to be nonsensical. A principle is a rational articulation of a normative concept (justice, equality)—of values that are appreciated by all individuals. Precaution is a natural tendency (nobody wants to do harm to themselves). When you try to articulate natural tendencies within normative confines, it should not be surprising that strange things happen. If I were to structure a principle of love to justify my emotions for my wife, I could expect to see these emotions being affected, and would probably end up with marital problems.

[23] Communication from the Commission on the precautionary principle, Brussels: 2000, page 2.

[24] It should be noted that in those eight years, environmental priorities for precaution had shifted. Rio was very concerned about the possibility of climate change in 1992 and the dangers of political inaction. The European Commission was dealing with BSE and chemicals management in 2000 and the need for political action.

[25] Precaution, politically speaking, is also a convenient trade management tool, but that is for another discussion.

7 CONCLUSION: IS THERE ANY CHANCE FOR STAKEHOLDER DIALOGUE?

It might be assumed that I am beginning to paint a negative picture towards pursuing stakeholder dialogue, one of the grand scenarios foreseen in developing the tools of sustainable development. The stakeholders have different perceptions of risks, their normative value and what needs to be done to limit them. There is a lack of trust between many stakeholders (particularly between NGOs and Industry). On precaution, there is a disagreement on whether it is a risk management tool or an uncertainty management tool.

In short, given these differences, could stakeholder dialogue on environment and health risk issues be fruitful or worthwhile? Or should we just give up?

The concept of dialogue is often misunderstood. It is not a nicer, more polite form of debating. It is not a means to reach out to adversaries to look for common ground (that is negotiating). Dialogue is, first and foremost, listening to the other stakeholders and trying to understand their positions. The first step in dialogue is understanding, understanding where you differ and why. If we can better understand how other stakeholders perceive and value risks, then those involved in environment and health debates may be in a better position to listen.

I hope, to some degree, this paper will serve to help make this dialogue more fruitful.

REFERENCES

Ballantine, B. 2005. EPC Working Paper number 17, Brussels: The European Policy Centre, March 2005.

Leiss, W. 2001. *In the Chamber of Risks: Understanding Risk Controversies.* Montreal: McGill-Queen's University Press.

The International Programme on Chemical Safety, 2000. Environmental Health Criteria 216, Disinfectants and Disinfectant By-products, Geneva: World Health Organization.

US National Research Council's, 2000. *Strengthening Science at the US Environmental Protection Agency.* Washington DC: National Academy Press.

Risks Challenging Publics, Scientists and Governments – Menoni (ed)
 ISBN 978-0-415-58072-4

Perceived ways to adapt to the potential effects of the climate change in the northernmost Europe

M. Rusanen & T. Rusanen
University of Kuopio, Finland

ABSTRACT: It has been estimated that the potential effect of global warming in Europe will be first visible in the northernmost areas. That is why the experiences and opinions on extreme weather conditions in these areas may be relevant, and potentially applicable to other regions that would be hit in later stages. This paper explores the strategies "lay professionals" have to adapt to the global warming in the Barents Sea Region, in the northernmost Norway, Sweden, Finland and Russia. The aim is to get to know the lay strategies of adaptation to climate change. This information is probably useful in looking for strategies to adapt permanently changing weather conditions and to complement the "academic" expert knowledge on potential risks and effects of Climate Change. The aim of the study is to clarify, what is the perceived role of the potential effects of climate change in person's professional life and how people are ready to adapt to/mitigate the risks of Global Climate Change.

1 BACKGROUND

It is essential that the research on climate change is multidisciplinary and topic-specific. Only by means of various disciplines there is the possibility to get to know the multidimensional phenomenon "climate change" (Kuusisto et al. 1996, p. 253). This is important to understand the causes, mechanisms and impacts of climate change as well as mitigation and adaptation methods (Finland's… 2001, 160). In the Northernmost countries of Europe, it is even possible that the countries might economically benefit from the climate change. However, the indirect changes may make the overall consequences of climate change strongly negative (Kinnunen, 1992). Climate change may have as well symbolic and cultural meanings. Although warmer climate might be appreciated in the North, losing touch to snow and ice means losing contact to the essential part of historic tradition (Heinonen, 1990, p. 138).

2 MATERIAL AND METHODS

The material for analysis consisted in interviews with "lay professionals" in the four countries (Norway, Sweden, Finland & Russia), living in various geographical settings. Interviews' responses were analyzed qualitatively to reflect the perceived effects of Climate Change.

The target groups for the interviews are professionals in indigenous or otherwise nature oriented professions: reindeer herders, fishermen, professionals in forestry and agriculture and representatives in tourism related professions. Interviews with professionals, based on earlier comparable risk assessment and vulnerability identification has taken place e.g. in Arvidsjaur, Piteå and Torneå river region in Sweden, Rovaniemi, Sodankylä and Inari in Finland and Vadsö and Alta in Norway.

The interviews were conducted using the focus group method, each group was made of 4 people in average. Altogether 27 interviews were conducted. Participants in each focus group were selected to represent the same profession. The purpose of the focus group research is to get to know people's attitudes, feelings, beliefs, experiences and reactions by discussing on the topic based on a relatively flexible interview structure. These attitudes may be partly independent on group, partly revealed via the interaction among group members.

The paper will present the results of the interviews, showing in particular, which aspects of life will gain benefits and which will suffer drawbacks and which are the factors relevant to this development, both in relation to nature and society, according to the involved people.

In addition to the focus group interviews, a mailed survey was conducted among lay people in four countries, 200 persons in each. This survey is used in this paper as a reference material. The mailed survey study were conducted in the following selected areas: Inari, Rovaniemi, Kemi and Kittila in Finland; Övertorneå, Haparanda, Jokkmokk and Boden in Sweden; and Tana,

Karasjok, Alta and Vadsö in Norway. In Russia, there were conducted 100 face-to-face interviews in Murmansk and 100 in Apatity were carried out. The response rate was 59%.

The results are reported especially on the aspects of a) sustainability of local communities, b) agriculture and farming, c) forestry and forest professions, d) reindeer herding and indigenous professions, e) fishery and f) tourism and other "modern" means of living. Specifically, social and cultural aspects are taken into consideration and a model of ways to mitigate the effects of climate change will be created.

3 RESULTS

The relevant spontaneous response of lay experts in interviews refers to the effects to each one's own profession, although they are very well able to comment each other's professions as well. Especially aspects such as tourism and sustainability of communities are often commented. In general, lay experts are much more concerned on climate change than lay people in general, however they do not perceive climate change as a very serious threat but as an issue that need to be adapted and mitigated.

When first taking into consideration the professional's views on their professions, significant changes are predicted to happen due to the climate change.

Most changes are expected in agriculture and tourism, while in forestry and reindeer herding change are predicted a smaller. On the other hand, professionals think that for agriculture and tourism, it will be easier to adapt to changes, for forestry and especially reindeer herding it might be more difficult since they are considered professions rather vulnerable to external changes.

Farmers see changes to their profession as positive: in agriculture the positive effects identified include change in farming zones, crop increase and new staple crops. The possible presence of new pests is identified as a negative drawback. The profession will gain on prolonged heat sum, crops are expected to increase significantly, new staple crops are expected which make a significant difference in the northernmost agricultural setting. These effects surpass the risk of increase in number of expected pests. It might be possible that in the future agriculture may be more feasible in the north, although specialization, e.g. organic production or special products like herbs are needed. Agriculture may remain subsidized profession or produce only for local needs or very specialized products, since free competition within the EU setting against farmers in more southern European regions seems unlikely. Some farmers experienced the wish that climate change would lead to changes in the pan-EU directives perceived as ill-matched to the agricultural practices in the northernmost frozen edges of Europe.

In forestry, the potential increase in timber yields is seen as a positive effect, as the estimated presence of more "southern" tree species. Experts perceive the latter change to happen in much later future than the lay people's estimates do. The negative effects are the weeds, pests and shift of natural habitats, the natural balance is perceived possibly not being able to recover. It is seen that increased timber will improve the economical feasibility of forestry, but in the northernmost setting, the quality of the timber is the key issue, not the amount. Slow-growing, dense timber is excellent material for small-scale construction (cottages, sauna's) and furniture. In the long run, the benefits might be surpassed by increased number of pests and decreasing quality of the timber. Natural balance in the forest may suffer and it may take very long time for it to recover.

Reindeer herders see their profession to be challenged by climate change, and they deem those changes to be chose in time irritably. Nature is changing, becoming milder and this affects the profession. In reindeer herding the most essential perceived issues are connected with pastures and shifts in annual herding "rhythms". Less snow is perceived as a benefit for grazing, but it may also lead the to soil freeze, which would make more difficult for the reindeer to find its feed. Mosquitoes may lose the "optimum", the sufficient number of them is needed as "cowboys", to irritate the reindeer pushing them uplands winter pastures. The circadian rhythm in reindeer herding might easily suffer as well.

Tourism is seen to be the profession that might most benefit from the climate change. Summers warm enough for tourism will be more probable although increased wind and rain might decline the positive effect. Most benefits are expected for the winter tourism. Although winters might be shorter and milder in objective terms than in present times, they will remain very much "the real thing" opposite to the conditions in winter resorts in more southern settings. Tourism might become the growth factor in Lapland's economy in the future partly thanks to climate change.

Professionals expect the effects of warming climate for communities to be positive and significant. It helps and makes more possible to keep small northern villages and farms inhabited in conditions where presently the population density is very small, around one inhabitant per square kilometer opposite to the EU-15 population density 120/square kilometer. Increased possibilities to maintain living might prevent migration and unemployment and bring more service-sector employment in the local settings.

Outside the independent aspects affecting professions, lay experts pay attention to increased water demand due to draught. Nature and professions might need increased amount of water due to increased heat. Rainfall might not be increased that much so draught might result. Especially in Finland experts pay attention to difficulties in road transport. "Sharper" seasons, shortening period of ground frost, alternating rainfalls and droughts might result in serious difficulties in transport. In Norway there is a concern on terrestrial-aquaculture interaction. Sea might be a factor whose conditions might be extremely difficult to predict.

The potential influx of the more southern wildlife species are seen as a benefit, for hunting and fishing supposing that they will not replace the current flora and fauna. The potential losses of existing northern natural habitats are perceived as a loss in general, and particularly in agriculture and forestry.

A positive effect would be reduced energy consumption: warmer climate is positively correlated to saving energy for heating. This aspect covers both housing and shelters for professional use. Currently winter temperatures may reach 40C for weeks so heating is a major economic burden for these people.

Finances and insurance are perceived to suffer losses due to the costs in adaptation processes. According to respondents, insurances suffer losses because extreme weather conditions will become more frequent. Many respondents have experiences on receiving compensation from insurance companies and they expect insurance premiums to rise in case damages and losses become more frequent.

All these professionals have experiences on earlier extreme weather conditions. E.g. Reindeer herders speak most often about changes in snow cover and number of mosquitoes, in agriculture the temperature is a very significant issue. They all have met weather conditions, which scientific experts predict to be results of temperature changes in the coming decades. On the other hand, these experiences have been mainly negative. e.g. tourism might have lost the whole season and farmers the whole crop due to the weather conditions. When weather becomes mode difficult to predict, it will only be a negative aspect.

Results obtained for lay experts can be compared to these observedwith the mail survey and direct interviewing with lay people. Health effects due to climate change are barely recognized in interviews. Extreme weather events cause more health concern among scientific experts than lay experts. The health aspects of climate change (e.g. UV, heat, allergy) are largely ignored among the lay experts. Lay people, "men in the street", are much more positive in their response to the effects of climate change than the lay professionals. Lay people appreciate the climate change as a source of "positive effects", much more than the lay professionals, including decreasing period of snow cover, warmer climate, longer summer periods and smaller energy consumption. Attitudes to climate change are almost evenly distributed: 29% of lay people think climate change is a positive issue, 30% it is neutral, and 41% think it has predominantly negative consequences. Finns take climate change most positive together with Russians. Swedes and especially Norwegians are more critical.

In Finland lay people think climate change will bring benefits on "way of life" and "standard of living" issues, sustainability of communities will improve as well as it will promote possibilities for tourism. In Norway fishery-related issues will benefit, in Sweden traditional lifestyle will gain benefit, as forestry and reindeer herding will be promoted. In Russia people think climate change will improve standard of living by reducing costs for heating and bringing better crops. Energy consumption is a main benefit, warmer climate demands less energy for heating leading into ecological and economic gains.

On the other hand, in Finland and in Sweden people think transport will become more difficult. Road conditions may be the facts when the period of ground frost is shorter roads being more vulnerable in wet seasons in spring. Surprisingly many interviewed Norwegians think sustainability of communities to be at risk even if one of the major means of living in the Northernmost areas, fishery will be promoted. This result is difficult to interpret. In Russia there is more open-ended response available since the face-to-face interviews were conducted there opposite to mailed surveys. There are so much more acute issues to worry, why to bother on climate change on very distant future, is common response. Anyhow there was presented some concern on sustainability of reindeer herding and estimated problems with transport in case the permafrost will melt.

4 CONCLUSION

Interviewed lay people considered the potential effects of Climate Change on life in northernmost areas as positive, "making life easier and more feasible". They generally considered negative effects take place in a "very distant future". This is the Nordic View. In Russia people expect similarly positive effects, but there are other environmental and economic concerns which outweigh the positive impact of climate change.

The lay/stakeholder perceptions presented in pilot testing and preliminary results are in many aspects aligned with the expert views identified on the base of the literature and academic knowledge of risk perception and climate change. This is particularly the case when compared with the data in the Nordic Countries (see the Finish Research Program on Climate Change, SILMU) and to certain extent data from the Arctic region in general (Arctic Climate Impact assessment - ACIA - project). The ongoing year 2004 fieldwork, however, reveals some interesting differences arising between "the expert view" and "the lay/stakeholder view". These similarities and differences will be analyzed and taken into consideration in the further analysis including dissemination of results in academic articles. The view of lay professionals is quite in line with the scientific experts' view. Lay professionals have detailed view on practical implementation of the effects of climate change to their own profession. The lay people seem to have the most positive attitude on the effects of climate change. This may lead to some positive adaptation capabilities to climate change.

REFERENCES

Finland's Third National Communication under the United Nations Framework Convention on Climate Change. 2001 Karisto Oy: Hämeenlinna.

Heinonen, J. 1990. Finland – 60th Parallel Syndrome. *In Man's Future in Arctic Areas. Proceedings of the 13th Polar Libraries Colloquy*. Arctic Centre Publications 1: Rovaniemi.

Kinnunen, J. 1992. *Ilmastonmuutoksen taloudelliset vaikutukset. Elinkeinoelämän tutkimuslaitos*. Helsinki: ETLA.

Kuusisto, E., Kauppi, L. & Heikinheimo, P. 1996. *Yliopistopaino*. Helsinki: Helsinki University Press. www.acia.uaf.edu.

Risks Challenging Publics, Scientists and Governments – Menoni (ed)
© 2010 Taylor & Francis Group, London, ISBN 978-0-415-58072-4

Integrated risk analysis: How to consider coping capacity?

C. Frischknecht & J.-J. Wagner
University of Geneva, Earth Sciences, CERG, Geneva, Switzerland

F. Romerio
University of Geneva, CUEPE, Switzerland

ABSTRACT: In this study, in an integrated risk index that has been developed, risk reduction measures and potential human coping capacities are incorporated. Distinguishing coping capacities from global vulnerability aims at showing existing positive assets and therefore, encourage concerned stakeholders to pursue their effort in risk reduction. Such an approach implies the development of two sets of indicators: related to generic vulnerability and to coping capacities. The case study in Guatemala shows that departmental scale of investigation is already suitable for planners and administrators dealing with risk management inside a country. As data are defined in various units, the integrated risk index will be obtained through normalization and aggregation procedures. This study concludes that developing an integrated risk index at departmental level is a first step, it demonstrates that one should not only act on vulnerability but also consider coping capacities, which will be a challenge for stakeholders dealing with risk management, from emergency to risk reduction.

1 INTRODUCTION

Disasters that affect people stem from the realization of risk. This results from interactions between natural phenomena and human societies, which exist in a specific area and for a given period of time. The risk is dynamic. By virtue of its components, it varies according to time and space.

Usually risk analysis aims at being quantitative and is calculated in terms of potential physical losses (number of victims, economic losses), and vulnerability is restrictively defined as a degree or percentage of potential losses. However, in order to implement risk management in a sustainable development frame, this requires the consideration of factors that influence human vulnerability, such as socio-economic, political, cultural and ideological environment (Wilches-Chaux, 1993; Cardona, 2001).

On the other hand, looking only at the vulnerability side can induce victimization and neglecting the individuals and communities own capacities (Anderson, 1994) that make them competent to be resilient to disasters. Moreover identifying and acting on coping capacities will help building resilience to withstand the effects of natural and human-induced hazards (ISDR, 2004).

Indeed, resilience to disasters can be seen as composed of two aspects, coping and adaptive capacity. Coping capacity is the short term component. It depends on all the available strengths and resources existing on site when risk is realized. Adaptive strategy, primarily defined in the field of climate risk, concerns the ability or capacity of a system to adjust in order to cope better with external pressures (Adger et al. 2004) and can be seen as the long term component.

The purpose of this present study is to realize integrated risk analysis, considering the various contributors, i.e. hazard, vulnerability, coping capacity and mitigation. As this analysis considers various hazards, vulnerability and coping factors are those that are susceptible of influencing consequences, and they are not hazard dependent.

Coping, as defined by Blaikie et al. 1994, is a series of strategies that are used in order to maintain or to respond to a certain level of needs, such as food, health, shelters in an unusual and adverse situation, such as a disaster. However, in order to be able to put such strategies in place, there is also a need to know the available strengths and resources. Therefore in the contrary of the ISDR[1] definition (2004), we will regard coping capacity not as the manner of acting but as the existing strengths and resources that exist at an institutional or individual level that can help reducing the effects of disaster.

Mitigation as defined by ISDR (2004) regards structural and non structural measures that can

[1] International Strategy for Disaster Reduction (ISDR)

limit the negative consequences. Structural measures can be building retrofitting, dike construction and non-structural measures can be land-use planning, building code, hazard and risk mapping as well as monitoring.

Recent studies integrate the dimension of coping capacities in their vulnerability assessment (Cardona, 2001; Cardona et al. 2004; Hossain, 2001). However, contrary to these authors, we will separate vulnerability, coping capacities and mitigation aspects. Defining what would be a factor contributing to vulnerability or to coping capacity is related to the approach followed. Indeed it depends whether vulnerability and capacities are considered as two ends of a spectrum or represent two distinct characteristics (Cannon, 2003). In the former case, indicators will be seen either characterizing high vulnerability or high capacity. In the latter case, it requires two sets of indicators. In this study, we follow the second option.

This approach aims at being constructive, showing that some positive points already exist and to encourage concerned authorities to pursue their efforts.

The pilot study is in Guatemala, a country facing various kinds of natural hazards. The investigation will be carried out a departmental scale. As opposed to the global scale that allows only a comparison between countries (e.g. UNDP[2], 2004), the scale selected is already suitable for planners and administrators dealing with risk management inside a country. It allows identifying departments being more at risk in the case of a major event and therefore to highlight where to act to finally increase resilience in front of natural events. This analysis follows an earlier study (Frischknecht et al. 2003) on this topic where some weaknesses existed regarding the choice of parameters (redundancy) and classification type.

This work on risk indicator development is in accord with the Hyogo framework for action to build resilience of nations and communities to disaster, as one of its priorities for action is the identification, assessment and monitoring of risk (ISDR, 2005).

2 METHODOLOGY FOR INDEX CONSTRUCTION

In this work, we define risk as a function of hazard, element at risk, vulnerability, coping capacity and mitigation measures. However, due to the fact that the effect of coping capacities and risk reduction on disaster impacts is still not quantitatively measured, we are going to employ and combine various indicators.

For each category, i.e. hazard, vulnerability, coping and mitigation, suitable indicators have been chosen based on available information. They were then passed through the following steps to finally obtain a risk index (Figure 1).

1. correlation analysis
2. normalization
3. aggregation
4. ranking
5. combination

Before normalization, a correlation analysis was carried out between indicators chosen as contributing to vulnerability and those for coping capacities in order to determine whether the selected parameters are correlated, the hypothesis being that if factors had a value close to 1, they should be considered as redundant.

Then as the chosen indicators are heterogeneous in terms of unit (monetary unit, percentage, proportion of people, etc.), a normalization procedure was adopted. Various techniques exist (for a review see Saisana & Tarantola, 2002). In this case each

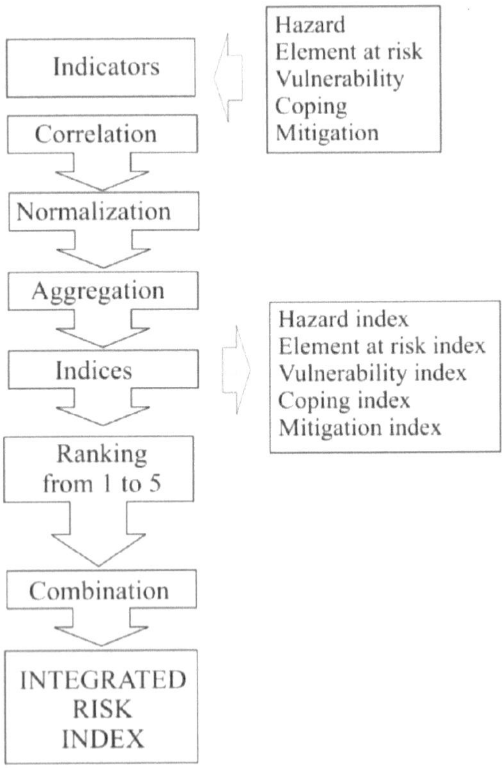

Figure 1. Procedure for the development of an integrated risk index.

[2] United Nations Development Programme (UNDP)

indicator has been normalized using the z-score method, i.e.:

$$I = \frac{actual\,value - \bar{x}}{\sigma}$$

I = indicator, actual value = value under consideration, $\bar{x}$ = mean value, σ = standard deviation

Indices, or composite indicators, are elaborated simply by averaging all the concerned indicators. They are considered to be equally weighted. Before risk calculation, each index will be ranked from 1 to 5. This attribution has been achieved on the basis of quantiles classification. This method has been chosen as it allows to characterize the department position. Finally risk will be calculated by applying a specific formula as described later.

2.1 *Hazard index*

Guatemala is affected by a number of hazards; the most "common" being earthquakes, volcanic eruptions, landslides, floods and forest fires.

Hazard is defined by ISDR (2004) as a physical event that can potentially trigger loss of life, injury, damages to properties. Hazard major components are frequency of occurrence, size, duration and extension. Usual hazard analysis is based on probabilistic or deterministic basis, the former tends to define areas with a specific parameter and a specific return period and a specific probability of occurrence, whereas the latter is based on a specific event of a certain size.

Such analyses are based on catalogs of past events, whatever their size and without considering their impact. The objective of these analyses aims at giving the maximum level of hazard that can affect an area and therefore susceptible of triggering a large-scale disaster depending on the vulnerability.

In this study, as we are dealing with multi-hazards and are looking for indicators, there is no need to use probabilistic hazard analyses. The objective is to obtain information that can be used as a proxy for hazard-proneness at a departmental level.

At this scale, the available database for Guatemala is the catalog of events put together by the LA RED[3] network (Gellert, 1999). This network has developed the database with the aim of making an inventory of all the phenomena that affect a country, whatever its size, by monitoring them as closely as possible. This database mainly contains the type of phenomenon, the time, day, month, year of occurrence, the number of people dead, injured, unaccounted for, affected, as well as housing damages and geographic information.

Events concerning earthquakes, landslides, floods, wildfires and volcanic eruptions have therefore been extracted. It is then possible to deduce an average annual frequency of a damaging event in a department, in so far as a phenomenon has occurred. The observation period is 11 years, between 1988 and 1998. In this time period no major event, such as the 1976 earthquake, occurred and the Mitch event was not included (Gellert, 1999). Consequently this temporal frame only accounts for medium or small events that are not triggering impacts across department boundaries.

In terms of energy involved, small and medium events cannot be compared to large-scale events. However, in terms of accumulative impacts over the same time period, they tend to be comparable (Lavell, 1999). Moreover as small and medium events often occur, i.e. annually, they continuously erode the livelihood and development of the exposed population (Lavell, 1999), who might therefore be more vulnerable to a large-scale, but who, on the other side, might also develop strategies to cope.

Event frequency was calculated by extracting the number of events for each department, scaled by the department size and divided by the number of years of observation. The values were then added together and normalized to obtain the hazard index (Table 1).

2.2 *Vulnerability index*

Vulnerability is multidimensional, with for example physical, socio-economical, environmental components (Wilches-Chaux, 1993; Cardona, 2001); and these factors are not independent and they interact to result in a global vulnerability.

In detailed analysis, vulnerability has to be analyzed related to each considered hazard. At departmental scale, we can only consider generic vulnerability, i.e. factors that determine vulnerability to various hazards (Adger et al. 2004). Therefore, we emphasize those factors (Table 2) that increase the potentiality of people to be affected, such as health status, physical exposure, socio-economic and education level as well as environmental degradation. Most of the available data were for the period 1994–1995.

Regarding the socio-economic dimension, the composition of the population plays an important role. Indeed Guatemala is characterized by a high percentage of native people. In the 1994 census, they represented around 42% of the global population. Moreover, as in other places around the world, native people in Guatemala are out of main political, economical and socio-cultural currents

[3] Red de estudios sociales en Prevención de desastres en América Latina

Table 1. Estimation of hazard index.

	Volcanic eruptions [N/km²/an]	Earthquakes [N/km²/an]	Floods [N/km²/an]	Fires [N/km²/an]	Landslides [N/km²/an]	Total	Hazard index
Alta Verapaz	0.0000	0.0000	0.0002	0.0001	0.0002	0.0005	–0.3252
Baja Verapaz	0.0000	0.0000	0.0002	0.0001	0.0003	0.0006	–0.2882
Chimaltenango	0.0000	0.0001	0.0001	0.0000	0.0001	0.0003	–0.4092
Chiquimula	0.0000	0.0000	0.0002	0.0001	0.0002	0.0005	–0.3453
El Progreso	0.0000	0.0000	0.0000	0.0000	0.0004	0.0005	–0.3168
Escuintla	0.0003	0.0001	0.0011	0.0001	0.0002	0.0018	0.2799
Guatemala	0.0001	0.0003	0.0024	0.0011	0.0063	0.0102	4.1917
Huehuetenango	0.0000	0.0000	0.0000	0.0001	0.0001	0.0002	–0.4730
Izabal	0.0000	0.0000	0.0002	0.0000	0.0000	0.0002	–0.4652
Jalapa	0.0000	0.0000	0.0000	0.0000	0.0001	0.0001	–0.4972
Jutiapa	0.0000	0.0000	0.0002	0.0001	0.0001	0.0003	–0.4012
Peten	0.0000	0.0000	0.0000	0.0001	0.0000	0.0001	–0.5174
Quezaltenango	0.0003	0.0001	0.0013	0.0000	0.0009	0.0027	0.6759
Quiche	0.0000	0.0000	0.0000	0.0000	0.0000	0.0001	–0.5032
Retalhuleu	0.0000	0.0000	0.0004	0.0000	0.0000	0.0006	–0.2855
Sacatepequez	0.0000	0.0000	0.0014	0.0002	0.0008	0.0023	0.5318
San Marcos	0.0000	0.0000	0.0006	0.0001	0.0003	0.0010	–0.1128
Santa Rosa	0.0000	0.0001	0.0002	0.0000	0.0002	0.0005	–0.3156
Solola	0.0000	0.0002	0.0001	0.0001	0.0012	0.0015	0.1582
Suchitepequez	0.0000	0.0000	0.0001	0.0000	0.0001	0.0002	–0.4577
Totonicapan	0.0000	0.0001	0.0009	0.0003	0.0006	0.0018	0.2777
Zacapa	0.0000	0.0000	0.0002	0.0000	0.0001	0.0003	–0.4016

Table 2. Vulnerability factors.

Theme	Sub-theme	Indicators	Date	Source
Health	drinking water	percent of household without access to water	1994–1995	PNUD, 1998:200
	malnutrition	percent of children below 5 years with moderate to insufficient weight	1994–1995	PNUD, 1998:200
	mortality	percent of people with no hope to live above 40 years	1994–1995	PNUD, 1998:200
Physical	concentration	density hab/km²	1994	Elias et al., 1997:253
	urban population	mean annual urban growth in percent 1994–2002		Calculation for this work
Socio-economic	rural population	percent of rural population	1994	INE, 2001, 1996 (PNUD, 1998:219)
	native population	percent of native population	1994	INE, 2001, 1996 (PNUD, 1998:219)
	gender	percent of women being native and living in rural conditions	1994	Calculation for this work
	elderly people	percent of people over 65 years	2002	PNUD 2004:268
	poverty	percent of people below the general poverty line in percent	1999	PNUD, 2001:A31
Education	analphabetism	percent of illiterate adult	1994	PNUD, 1998:200
Environmental	soil degradation	over exploitation of soil in percent of land	1992	PNUD, 1998:224

(BDRL[4], 2001) and are usually the most affected by major natural events, as demonstrated by the 1976 earthquake (Blaikie et al. 1994).

Regarding the gender issue in the context of disasters of natural origin, women who are susceptible to be highly vulnerable are those who are native, single or widowed, suffering of malnutrition (Enarson, 2000) and having low income. Unfortunately, in Guatemala, native women living in rural areas correspond

[4] Bureau of Democracy, human rights and labor (BDRL)

Table 3. Correlation analysis among vulnerability factors.

	Density	Urban_ growth	Rural popu- lation	Native popu- lation	Women (native_ rural)	Aged popu- lation	General_ poverty	Death before 40	Without water access	Children malnu- trition	Adult analpha- betism	Soil overex- ploited
Density	1.000											
Urban_ growth	−0.002	1.000										
Rural population	−0.752	−0.404	1.000									
Native population	−0.012	−0.524	0.221	1.000								
Women (native_rural)	−0.117	−0.647	0.438	0.959	1.000							
Aged population	0.141	0.387	−0.244	−0.678	−0.691	1.000						
General_ poverty	−0.536	−0.519	0.809	0.607	0.722	−0.441	1.000					
Death before 40	−0.099	−0.726	0.311	0.645	0.676	−0.529	0.515	1.000				
Without water access	−0.390	0.051	0.422	0.087	0.187	−0.482	0.299	−0.190	1.000			
Children malnutrition	−0.191	−0.411	0.256	0.610	0.562	−0.289	0.552	0.565	−0.361	1.000		
Adult analphabetism	−0.536	−0.397	0.727	0.696	0.807	−0.633	0.807	0.472	0.472	0.316	1.000	
Soil overexploited	0.299	−0.248	−0.016	−0.173	−0.115	0.459	−0.061	−0.207	−0.231	−0.069	−0.241	1.000

to these classes (Trujillo et al. 2000). In order to take this fact into consideration, an indicator regarding native and rural women has been added.

Poverty is not synonymous with vulnerability, but the general poverty line, as defined for Guatemala as the cost of food that satisfies the minimum caloric requirement plus the cost for living, education and health items (PNUD, 2001), can be considered as a proxy for indicating that portion of population that has severely limited resources and precarious survival conditions, which can be worsened by external factors (Trujillo et al. 2000) such a natural phenomenon.

Soil over-exploitation can be used as an indicator of zones that are potentially vulnerable in case of a natural hazard occurrence. Indeed overexploitation also means deforestation. As a consequence soils are less susceptible to absorb water during heavy rainfall, therefore increasing the potentiality of flooding or landslides as natural barriers disappear. In this context, flooding and landslide can be considered as socio-natural hazards (Lavell, 1999), as they are strongly influenced by human activity.

In order to evaluate the relationship between selected indicators, a correlation analysis was carried out (Table 3).

Depending on sampling (Lamotte, 1962), a limit of significance of the correlation coefficient can be defined, below which correlation is not significant, consequently the parameter can be retained. In the case of this study, the sample is made up of 22 departments, therefore the limit is set to 0.54 with a security of P = 99%. As seen in table 3, some combinations (e.g. rural population-general poverty line) exhibit a correlation coefficient higher than 0.54. However, none of these shows a correlation coefficient close to 1 or greater than 0.54 for all the possible combinations together. As a result, we assumed that all the selected parameters bring useful information for evaluating vulnerability for each department.

The vulnerability index results from the normalization of each parameter, their addition and average as shown in table 4.

2.3 *Coping capacity index*

At an individual scale, coping capacity will rely on resources for livelihood, healthcare structures, information access, and physical displacement possibility. At an institutional level, coping capacity will rely on information transmission, access to affected people, response capacity and financial resources.

Consequently, the chosen indicators (Table 5) are health care capacity, road length per 100 km^2, radios per 1000 inhabitants, mean income (US$), active population (% of population), fiscal autonomy indicator, departmental coordination center for natural disaster reduction (CODRED[5]) which depend on CONRED[6], the national coordination for natural

[5] Coordinadora deparamentales para la reducción de desastres (CODRED)

[6] Coordinadora nacional para la reducción de desastres (CONRED)

Table 4. Normalized indicators for vulnerability index construction.

Department name	Density	Urban growth	Rural population	Native population	Women (native _rural)	Pop>65 years old	General poverty	Death before 40 years	Without water	Children malnutrition	Adult analpha-betism	Soil over-Exploited	Aggre-gation	Vulne-rability index
Alta Verapaz	–0.50	–0.24	0.82	1.47	1.69	–1.53	0.88	–0.48	3.10	–0.87	2.03	–0.42	5.95	0.50
Baja Verapaz	–0.57	–0.11	0.57	0.39	0.51	0.04	0.62	–0.66	0.47	–0.77	0.78	0.87	2.13	0.18
Chimaltenango	0.03	0.51	–0.81	1.12	0.55	–0.53	–0.12	0.06	–0.56	1.68	–0.32	0.81	2.42	0.20
Chiquimula	–0.31	1.95	0.26	–0.40	–0.39	0.61	–0.59	–0.75	0.17	–0.85	0.56	–0.39	–0.13	–0.01
El Progreso	–0.54	0.22	0.13	–1.29	–1.16	1.90	–0.29	–0.66	–0.41	–0.85	–0.79	1.42	–2.32	–0.19
Escuintla	–0.36	0.08	–0.49	–1.17	–1.07	0.19	–1.35	0.41	–1.00	–0.13	–0.83	–0.92	–6.65	–0.55
Guatemala	3.88	0.38	–2.63	–0.98	–1.08	0.47	–2.63	–1.11	–0.70	–1.99	–2.14	1.43	–7.08	–0.59
Huehuetenango	–0.37	–0.71	0.89	0.68	0.90	–0.96	0.96	–0.21	–0.12	1.26	0.94	0.37	3.61	0.30
Izabal	–0.70	0.04	0.57	–0.63	–0.52	–0.53	–0.44	–0.48	1.20	–0.85	–0.32	–0.94	–3.59	–0.30
Jalapa	–0.32	1.53	0.13	–0.18	–0.18	–0.10	0.67	–0.48	0.32	–0.49	0.32	–1.97	–0.76	–0.06
Jutiapa	–0.32	0.42	0.57	–1.20	–1.06	1.04	0.20	–0.66	0.03	–0.49	–0.28	1.51	–0.24	–0.02
Peten	–0.82	–0.06	0.13	–0.53	–0.48	–1.96	–0.05	0.68	1.93	–0.91	0.05	–1.34	–3.36	–0.28
Quetzaltenango	0.58	0.32	–0.68	0.55	0.17	0.33	0.03	0.68	–0.41	0.85	–0.65	–0.39	1.38	0.11
Quiche	–0.56	–1.42	0.89	1.28	1.53	–1.53	1.13	1.84	–0.12	1.09	1.80	–0.12	5.82	0.49
Retalhuleu	–0.29	0.46	0.07	–0.31	–0.29	0.47	–0.14	–0.03	–0.85	0.85	–0.54	–0.97	–1.57	–0.13
Sacatepequez	1.30	0.61	–2.63	–0.02	–0.74	0.47	–1.45	–0.93	–1.00	–0.13	–1.48	–0.27	–6.26	–0.52
San Marcos	0.09	–0.68	1.01	–0.02	0.23	0.04	1.44	–0.30	0.47	1.40	0.01	0.65	4.33	0.36
Santa Rosa	–0.39	0.15	0.32	–1.26	–1.14	1.19	0.10	–0.39	–0.27	–0.49	–0.57	1.09	–1.65	–0.14
Solola	0.31	–0.60	–0.24	1.63	1.21	–0.96	0.88	1.75	–0.70	0.85	1.20	–1.22	4.11	0.34
Suchitepequez	–0.17	0.09	–0.05	0.46	0.32	0.19	–0.34	0.24	–0.70	0.85	–0.01	–0.35	0.51	0.04
Totonicapan	0.57	–3.17	1.14	1.63	2.07	–0.39	1.38	2.65	–0.85	0.85	0.77	0.94	7.58	0.63
Zacapa	–0.53	0.26	0.01	–1.23	–1.10	1.61	–0.89	–1.19	0.03	–0.85	–0.54	0.21	–4.21	–0.3

Table 5. Coping capacity indicators.

Theme	Sub-theme	Indicators	Date	Source
Infrastructure	Health care	Public hospital rooms per 1000 inhabitants	1996	PNUD, 1999:240
		Number of public health centers per 10000 (type B) and 20000 inhabitants (type A)	1996	PNUD, 1999:240
		Number of public health post per 2000 inhabitants	1996	PNUD, 1999:240
		Number of NGOs dedicated to health care per 10000 inhabitants	1997	PHR, 1998
		Number of private hospitals per 10000 inhabitants	1997	PHR, 1998
		Number of private clinics per 10000 inhabitants	1997	PHR, 1998
	Road infrastructure	Road length per 100 km^2	2001	PNUD 2003:257
Information		Number of radio per 1000 inhabitants	1999	En Cifras, 1999:236
Economy	Financial resources	Fiscal autonomy index	1999	PNUD, 2001:A60
	Active population	Percentage of active population over 10 years old	1999	PNUD 2002:360
	Income	GDP per capita U$/year	1996	PNUD, 1998:211
Response	Awareness and response	Number of IRCR centers per sq km^2	2005	Cruz Roja web, 2005
		Departmental center CODRED per sq km^2	2005	CONRED web, 2005

Table 6. Correlation analysis of the indicators for coping capacities.

	Redcross/ km^2	CODRED/ km^2	Road density	Nb radio/ 1000 hab	Active population	Fiscal autonomy	Care capacity	GDP
Redcross/km^2	1.000							
CODRED/km^2	–0.093	1.000						
Road density	–0.002	0.577	1.000					
Nb radio/1000 hab	0.311	–0.271	–0.155	1.000				
Active population	0.011	0.372	0.228	0.479	1.000			
Fiscal autonomy	0.205	–0.001	–0.046	0.654	0.611	1.000		
Care capacity	–0.116	0.039	–0.091	0.438	0.411	0.301	1.000	
GDP	0.093	0.056	0.193	0.500	0.783	0.654	0.496	1.000

disaster reduction, and existence of national Red Cross and Red Crescent society centers per department. Both institutions act before, during and after disasters. They aim at developing awareness and capacity of response in the event of disasters, CODRED acting at an institutional level and Red Cross at a community and individual level.

Correlation analysis (Table 6) of these indicators shows few combinations having a correlation coefficient higher than 0.54. And none of the selected parameters show a redundancy.

Comparison of maps established on the vulnerability and coping capacities indices (Figures 2–3) show that these departments characterized by a very low vulnerability index present a coping capacity index level varying from very high to moderate (e.g. Guatemala, Zacapa, Escuintla). However, the relationship is not totally inversed, i.e. department having high coping capacities do not necessary owns very low vulnerability. Indeed the correlation coefficient is only equal to –0.62.

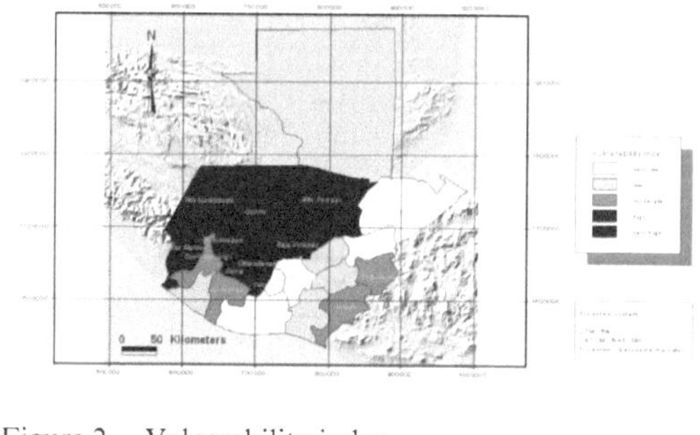

Figure 2. Vulnerability index.

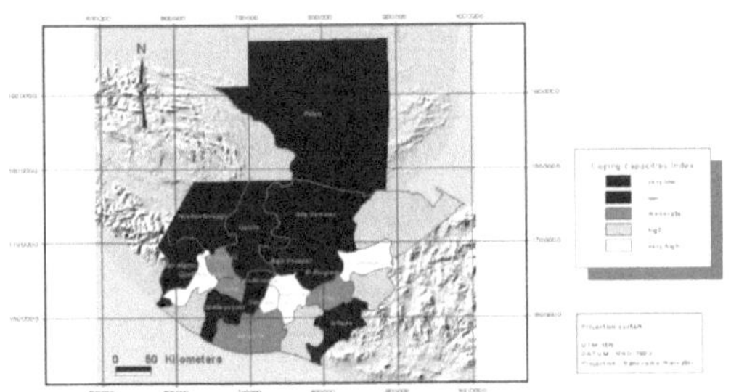

Figure 3. Coping capacities index.

The construction of the index is based on the normalization of each parameter, a summation and averaging (Table 7).

2.4 *Mitigation index*

Mitigation measures are elements that can reduce risk. Some are direct, especially structural measures, as they are acting on the level of hazard, whereas some are indirect, such as land-use planning, building codes, hazard monitoring and assessment.

Available information, at a departmental level, concerns hazard monitoring and mapping. Consequently, the information that has been used (Table 8) is the number of monitoring stations such as seismic, hydrometric and meteorological stations, the existence of volcanic observatories and the existence of hazard maps in or at departmental scale, i.e. landslide, flood and seismic maps.

Such information does not give any insight into the state of the stations, information and transmission networks, the quality of maps, the quality of the network in relation to the hazard concerned. Moreover no information was available neither regarding structural measures such as dike for flood protection, protective infrastructures for landslide nor regarding the application of building codes.

The mitigation index is composed of the combination of two sub-indicators, one combining the information regarding monitoring and the second considering mapping (Table 8).

3 RISK INDEX

The construction of a risk index can be achieved based on several mathematical approaches. Some authors (e.g. D'Ercole & Pigeon, 2000) are simply adding the different sub-indexes to obtain the risk index, but as pointed out by Cardona (2003), risk results from the convolution of hazard and vulnerability and they cannot exist on their own. Consequently one will construct the risk index following the common way of expressing risk, when considering vulnerability as an integrative part, since the initial work of Fournier d'Albe (1979).

Risk = Hazard * Value * Vulnerability

Value is also called the element at risk (UNDRO, 1991).

Table 7. Normalized indicators for coping capacities index.

Department name	Red-cross/ km^2	CODRED/ km^2	Road length/ 100 km^2	Nb radio/ 1000 hab	Active popu-lation	Fiscal auto-nomy	Care capacity	GDP	Aggre-gation	Coping index
Alta Verapaz	−0.449	−0.787	−0.842	−1.193	−0.436	−0.268	−0.583	−0.49575	−5.055	−0.63189
Baja Verapaz	−0.036	−0.326	−0.609	−0.544	−0.935	−0.774	0.354	−0.12243	−2.993	−0.37413
Chimaltenango	−0.681	0.091	1.145	−0.999	−0.496	0.322	−0.339	−0.39248	−1.349	−0.16861
Chiquimula	0.167	−0.099	0.248	0.290	0.791	−0.015	−0.067	−0.03071	1.283	0.160376
El Progreso	−0.681	0.125	−0.811	0.000	−0.358	−0.606	0.754	0.152703	−1.425	−0.17813
Escuintla	−0.222	−0.533	−0.352	0.037	0.445	0.828	−0.440	−0.34116	−0.578	−0.0722
Guatemala	0.267	0.012	1.114	1.636	3.246	2.767	1.015	4.373507	14.430	1.803806
Huehuetenango	−0.681	−0.742	−0.638	−1.212	−1.346	0.069	−0.783	−0.47109	−5.805	−0.72566
Izabal	0.211	−0.797	−1.445	1.767	−0.002	2.177	0.334	−0.1092	2.136	0.266994
Jalapa	0.296	0.045	0.620	−0.094	−0.439	−0.353	−0.093	−0.15033	−0.169	−0.0211
Jutiapa	−0.681	−0.347	−0.205	−0.503	−0.196	−0.943	1.396	−0.24511	−1.724	−0.21553
Peten	−0.569	−0.983	−1.600	0.000	0.111	−0.184	−0.237	−0.17002	−3.632	−0.45396
Quetzaltenango	3.452	0.107	0.756	−0.105	−0.434	0.491	−0.371	−0.2783	3.618	0.452268
Quiche	0.040	−0.777	−0.915	−0.510	−0.868	−1.196	−0.610	−0.46392	−5.300	−0.66251
Retalhuleu	1.491	0.166	−0.868	0.044	−0.398	0.575	−0.083	0.187742	1.114	0.139301
Sacatepequez	−0.681	3.792	0.815	0.000	1.839	0.238	0.216	−0.01098	6.208	0.77594
San Marcos	−0.681	−0.452	0.292	−0.785	−0.660	−1.027	−0.340	−0.50397	−4.158	−0.5197
Santa Rosa	1.365	−0.285	−0.140	0.243	0.388	−0.859	0.162	−0.10956	0.766	0.095691
Solola	−0.681	1.074	1.253	0.000	−0.498	−0.100	−0.213	−0.30396	0.532	0.066459
Suchitepequez	0.122	−0.150	−0.007	0.000	−0.127	−0.437	−0.360	−0.32108	−1.280	−0.15999
Totonicapan	−0.681	1.074	2.552	0.000	−0.424	−1.196	−0.499	−0.38265	0.443	0.055402
Zacapa	−0.681	−0.210	−0.395	1.928	0.796	0.491	0.791	0.188744	2.908	0.363473

Table 8. Indicators and their normalization for mitigation index.

Department name	Meteoro-logical station	Hydro-metric station	Seismic station	Volcano observatory	Sum	Monitoring indicator	Landslide mapping	Flood mapping	Seismic mapping	Sum	Mapping index	Aggregation	Mitigation index
Alta Verapaz	3	2	0	0	5	−0.05233	1	1	0	2	−0.30896	−0.36129	−0.18064
Baja Verapaz	2	2	0	0	4	−0.43609	1	1	0	2	−0.30896	−0.74505	−0.37253
Chimaltenango	3	3	0	0	6	0.331432	1	1	0	2	−0.30896	0.022475	0.011237
Chiquimula	2	3	0	0	5	−0.05233	1	1	0	2	−0.30896	−0.36129	−0.18064
El Progreso	1	2	0	0	3	−0.81986	1	1	0	2	−0.30896	−1.12882	−0.56441
Escuintla	4	7	0	0	11	2.25025	1	1	0	2	−0.30896	1.941293	0.970646
Guatemala	1	3	2	0	6	0.331432	1	1	1	3	3.089572	3.421004	1.710502
Huehuetenango	6	5	0	0	11	2.25025	1	1	0	2	−0.30896	1.941293	0.970646
Izabal	2	3	0	0	5	−0.05233	1	1	0	2	−0.30896	−0.36129	−0.18064
Jalapa	2	0	0	0	2	−1.20362	1	1	0	2	−0.30896	−1.51258	−0.75629
Jutiapa	2	3	0	0	5	−0.05233	1	1	0	2	−0.30896	−0.36129	−0.18064
Peten	1	5	0	0	6	0.331432	1	1	0	2	−0.30896	0.022475	0.011237
Quetzaltenango	2	1	0	1	4	−0.43609	1	1	1	3	3.089572	2.653477	1.326738
Quiche	4	3	0	0	7	0.715196	1	1	0	2	−0.30896	0.406239	0.203119
Retalhuleu	1	3	0	0	4	−0.43609	1	1	0	2	−0.30896	−0.74505	−0.37253
Sacatepequez	1	2	1	1	5	−0.05233	1	1	0	2	−0.30896	−0.36129	−0.18064
San Marcos	2	6	1	0	9	1.482723	1	1	0	2	−0.30896	1.173766	0.586883
Santa Rosa	1	2	1	0	4	−0.43609	1	1	0	2	−0.30896	−0.74505	−0.37253
Solola	3	0	0	0	3	−0.81986	1	1	0	2	−0.30896	−1.12882	−0.56441
Suchitepequez	0	4	0	0	4	−0.43609	1	1	0	2	−0.30896	−0.74505	−0.37253
Totonicapan	0	0	0	0	0	−1.97115	1	1	0	2	−0.30896	−2.28011	−1.14005
Zacapa	3	1	0	0	4	−0.43609	1	1	0	2	−0.30896	−0.74505	−0.37253

However, as we are dealing with indices, the formula is rewritten as:

$I_R = I_H * I_E * I_V$

Wit I_H = hazard index, I_E = element at risk index, i.e. the population and I_V = vulnerability index.

The second step is to integrate coping capacity into the formula. Some authors integrate the inverse of coping capacities in their vulnerability assessment, so that they consider only vulnerabilities (Cardona et al. 2004). In the present study, transposing Wisner idea (2000, 2001) into practice, we distinguish between vulnerability, mitigation measures, and coping capacities components. The question now is how to incorporate these elements into the risk formula.

In the present case, mitigation index only contains information that comes under either hazard evaluation (cartography) or its monitoring (stations, observatories). These elements do not reduce hazard level. But they are elements that can indirectly reduce vulnerability as monitoring can be used for early warning and mapping as awareness of individuals and institutions dealing with land-use planning and construction. Coping factors are directly related to population and can influence on vulnerability.

Therefore, the risk considering both population's capacity potential to cope in adversity as well as the existence of risk mitigation measures R_{cm} is expressed as:

$I_{Rcm} = I_H * I_E * (I_V - I_C - I_M)$

Wit I_H = hazard index, I_E = the population, I_V = vulnerability index, I_C = coping capacity index, I_M = mitigation index

3.1 *Results*

As stated above, each index is the mean value of combined indicators, but they cannot be used directly in the defined risk formula. Therefore it has been decided to attribute a grade from 1 to 5, value 1 being attributed to minimum influence on risk and value 5 to the highest. This attribution has been achieved on the basis of quantiles classification. This method is based on the distribution of features and not on values. The optimum number of categories has been determined by Huntsberger's number (quoted by the Chadule group, 1986: 31) according to the formula $K = 1 + 3.3 \log_{10} N$, N being the number of statistical units. In this case, N is equal to 22. Then, the optimal number is 5 categories. For each index, the departments have been classified by quintiles and then a value from 1 to 5 has been attributed depending on the classification of the department (Table 9).

Table 9. Integrated risk index.

	H	E	V	C	M	H*E*V	H*E*(V–C–M)
Department name	Event	Element	Vuln	Coping	Mitigation	Standard risk	Integrated risk
Alta Verapaz	3	5	5	1	3	75	15
Baja Verapaz	3	1	4	2	2	12	0
Chimaltenango	2	4	4	2	4	32	–16
Chiquimula	2	2	3	4	3	12	–16
El Progreso	3	1	2	2	1	6	–3
Escuintla	5	4	1	3	5	20	–140
Guatemala	5	5	1	5	5	25	–225
Huehuetenango	1	5	4	1	5	20	–10
Izabal	1	3	1	4	3	3	–18
Jalapa	1	2	2	3	1	4	–4
Jutiapa	2	3	3	2	3	18	–12
Peten	1	2	2	1	4	4	–6
Quetzaltenango	5	4	3	5	5	60	–140
Quiche	1	4	5	1	4	20	0
Retalhuleu	4	1	2	4	2	8	–16
Sacatepequez	5	1	1	5	3	5	–35
San Marcos	4	5	5	1	4	100	0
Santa Rosa	3	2	2	4	2	12	–24
Solola	4	2	4	3	1	32	0
Suchitepequez	2	3	3	2	2	18	–6
Totonicapan	4	3	5	3	1	60	12
Zacapa	2	1	1	5	2	2	–12

Risk has then been qualitatively evaluated using the defined formula (Table 9).

A mapped representation of the integrated risk index is presented on figure 4. This map is also based on quintiles classification. This allows the evaluation of the relative position of departments one from the other. It can be seen that the departments such as Quiche, Baja Verapaz, Totonicapan are evolving into a higher level of risk, whereas others, such as Guatemala, Escuintla, Sacatepequez, Quetzaltenango departments are located in a lower level.

It would be interesting to compare the standard risk approach, i.e. considering only vulnerability, to the integrated risk one. Looking at both maps (Figures 4 & 5) shows that some departments such as Quiche, Solola, Baja Verapaz have their relative position reaching a higher level of risk, with the reverse for others (e.g. Escuintla, Guatemala departments). This is due to the fact that the former departments have reduced potential coping capacities compared to others.

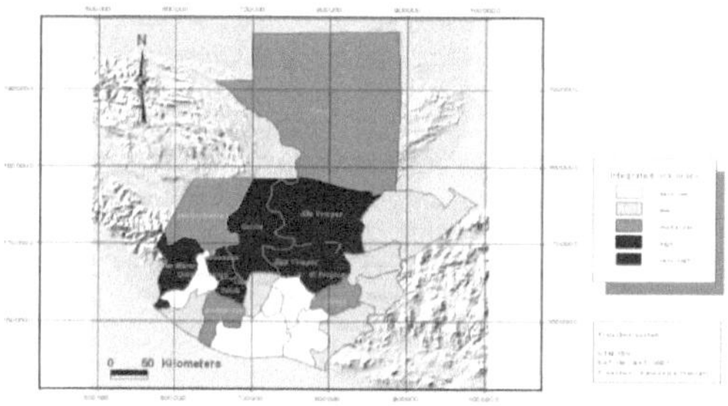

Figure 4. Integrated risk index map.

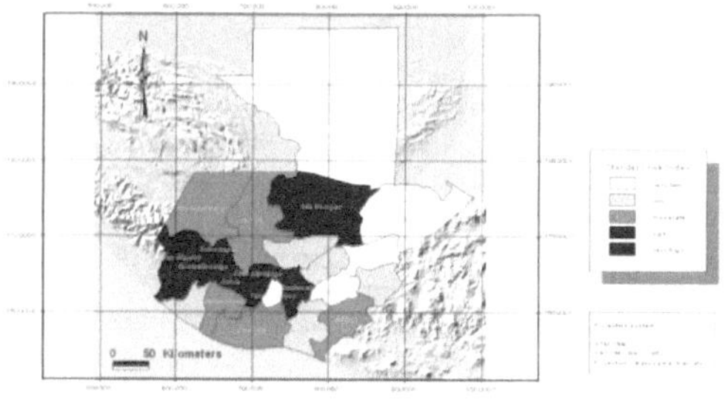

Figure 5. Standard risk index map.

4 DISCUSSION AND CONCLUSION

This investigation is an attempt to develop an integrated risk index that considers not only the negative aspect such as vulnerability factors but also positive factors on which to continue to build, i.e. coping capacities and risk reduction measures.

In the elaboration of a risk index, one question that arises is how to validate the results. It depends on what it represents. In quantitative analysis, if the model is to obtain the potentiality of fatalities, it is possible to compare between the modeled risk and the realized risk or outcome risk (Sarewitz et al. 2003). Usually outcome risk is based on killed and/or affected people. These proxies are then normalized with the population or the land area (Yohe & Tol, 2002; UNDP, 2004; Adger et al. 2004). However, in a qualitative model, it is more difficult as the objective is to highlight sensitive areas and it is based on generic indicators and not specifically related to hazards. Even so we carried out a correlation analysis between modeled risk and realized risk using the following proxies:

- killed and affected as per cent of departmental population
- killed as per cent of departmental population
- killed over affected ratio

This information was extracted from the Desinventar database for the period 1988–1998. Before carrying correlation analysis each considered parameter was expressed in terms of rank. Results show weak to no relationship between the outcome risk proxies and the integrated risk index (Table 10).

Two main arguments can be given:

- Killed and affected people is not the only outcome risk, one could also consider economic losses, loss of land, houses, livestock, etc.
- For the period considered, only "small" events occur and the number of fatalities seems to be

Table 10. Correlation analysis between outcome risk ranks and integrated risk index rank.

	Integrated risk index rank	Killed_affected rank	Killed_rank	K/aff_rank
Integrated risk index rank	1			
Killed_affected rank	−0.294622062	1		
Killed_rank	−0.189898798	0.12565399	1	
K/aff_rank	0.290150153	−0.3892877	0.520638	1

directly related to the physical vulnerability, in terms of location or buildings. For example in the case of the Guatemala department, the maximum death toll is related to landslides, a phenomenon that occurs at specific places within this department.

Moreover regarding coping capacity and mitigation measures, it is almost meaningless to achieve a correlation analysis on a risk proxy such as number of deaths. These parameters are supposed to reduce losses. But how to evaluate this "lack" of losses is still a problem to be solved. Indeed such analysis should be carried out over a certain period of time, starting from a reference year and also having precise data to characterize coping capacities and mitigation measures in order to be able to develop sophisticated risk scenarios and also to have detailed information on the outcome risk of past events. At the present time, no such information is available.

Future work should build upon the study presented here by improving certain points. First of all, this attempt was based on accessible information, of uneven quality and various years.

The second point is that as stated previously, risk is a dynamic process, as societies are evolving. To monitor the evolution of the situation, there is then a necessity to have continuity in time for the selected parameters that is not usually the case, parameters being calculated for one year or two and then suppressed. In this study, some parameters were available for one specific year and others for another year. This demonstrates the difficulty in dealing with statistical data that are not developed for the purpose of risk assessment.

Finally designing an integrated risk index, to be useful and used, should involve all stakeholders, from decision-makers to communities, and in the case of Guatemala, also indigenous communities. There is now an increasing reconnaissance (e.g. Ellemor, 2004) that native people do have developed coping capacities through their history and considering their view will certainly enhance the process of risk evaluation and risk reduction.

Despite these elements, this study illustrates that taking into account the capacities of people and existing mitigation measures already allows one to highlight the strengths and weaknesses of the areas analyzed. The departmental level already allows us to show that at the level of risk management differential measures must be applied and aspects strengthened, according to the departments involved, and acting on coping capacities and implementing mitigation measures are a challenge for disaster risk management.

REFERENCES

Adger, W.N., Brooks, N., Bentham, G., Agnew, M. & Eriksen, S. 2004. *New indicators of vulnerability and adaptive capacity.* Technical Report 7, Tyndall Centre for Climate Change Research.

Anderson, M.B. 1994. Le concept de vulnerabilité: au-delà des groupes vulnérables. *Revue internationale de la Croix-Rouge, juillet-août*, 808: 360–365.

Blaikie, P., Cannon, T., Davis, I. & Wisner, B. 1994. *At risk, Natural hazards, people's vulnerability, and disasters*. London: Routledge.

BDRL – Bureau of Democracy, human rights and Labor-, Guatemala: *Country report on human rights practices, 2000*, http://www.state.gov/g/drl/rls/hrrpt/2000/wha/775.htm, 24.12.2002, (2001), 29 p.

Cannon, T. 2003. *Vulnerability Analysis, Livelihoods and Disasters Components and variables of vulnerability: modelling and analysis for disaster risk management.* IADB/IDEA Program on Indicators for Disaster Risk Management, http://idea.unalmzl.edu.co, Manizales: Universidad Nacional de Colombia.

Cardona, O. 2003. *The Need of Rethinking the Concepts of Vulnerability and Risk from a Holistic perspective: a Necessary Review and Criticism for Effective Risk Management, in Mapping Vulnerability: Disasters, Development and People*, Bankoff G., Frerks G, Hilhorst D. (ed). London: Earthscan Publishers.

Cardona, O. 2001. *Estimación holística del riesgo sísmico utilizando sistemas dinámicos complejos*, tesis doctoral, Universitat politècnica de Catalunya, Escola tècnica superior d'enginyers de Camins, Canals I Ports.

Cardona, O.D., Hurtado, G.J.E., Moreno, R.A.M., Chardon, A.C. & Cardona, A.G.J. October 2004. *Results of the system of indicators' application on twelve countries of the Americas*, Inter-American Development Bank Information and Indicators Program for Disaster Risk Management, Execution of Component II (Indicators for Disaster Risk Management) OPERATION ATN/JF-7907-RG PHASE 2. Manizales National University of Colombia, November, http://idea.unalmzl.edu.co.

D'Ercole, R. & Pigeon, P. 2000. *L'expertise internationale des risques dits naturels: intérêt géographique, in La géographie des risques dits "naturels" entre géographie fondamentale et géographie appliquée, d'Ercole, R. and* Pigeon, P. (coord). *Cahiers savoisiens de géographie* 1: 29–36.

Elías, S., Gellert, G., Pape, E. & Reyes, E. 1997. *Evaluación de la sostentabilidad en Guatemala.* Facultad Latinoamericana de Ciencias Sociales (FLACSO), Guatemala Cit.

Ellemor, H. 2004. *Reconsidering emergency management and indigenous communities in Australia*, Environmental Hazards, article in press, doi:10.1016/j.hazards.2004.08.001.

Enarson, E. 2000. *Gender and Natural Disasters*, Working Paper 1, Infocus Programme on crisis response and reconstruction. Geneva: ILO.

Fournier d'Albe, E.M. 1979. Objectives of volcanic monitoring and prediction. *Jl geol. Soc. Lond.* ed. *The Geological Society*, 136: 321–326.

Frischknecht, C., Romerio, F., Rossiaud, J., Wagner, J.-J., Bertrand, S., Laporte, J.-D. & Harding, T.-W. 2003. Integrated "Natural" Risk management: the basic assessment; an essay, DRPI – IIASA 3rd International Conference on Integrated Disaster Risk Management, 3–5th July 2003, Kyoto, Japan.

Gellert, G. 1999. *Algunas lecturas de riesgo y vulnerabilidad en Guatemala, utilizando la herramienta DesInventar*, FLACSO.

Global InfoGroup, 1999. En Cifras: Guatemala, Perfil Integral des País, Volumen I, N°1.1, Edición 1999.

Groupe Chadule, 1986. Initiation aux pratiques statistiques en géographie, Collection Géographie, Masson, Paris.

Hossain, S.M.N. 2001. *Assessing Human Vulnerability due to environmental change: concepts and assessment methodologies*, master thesis, Division of Land and Water Resources, Department of Civil and environmental engineering, Royal Institute of Technology, Stockholm.

ISDR, Hyogo Framework for Action 2005–2015: building the resilience of nations and communities to disasters.

ISDR, *Living with Risk. 2004. A global review of disaster reduction initiatives*. Geneva: United Nations.

Lamotte, M. 1962. *Initiation aux méthodes statistiques en biologie*, deuxième édition, Mason & Cie (eds), Paris.

Lavell, A. 1999. *Natural and technological disasters: Capacity building and Human resource development for disaster management: Concept paper*, prepared under contract to ERD-UNDP Geneva.

Partnerships for health reform (PHR), 1998. Informe sobre iniciativas especiales N°10, Cuentas nacionales de Salud: Guatemala, *Septiembre de 1998*.

PNUD, 1998. *Guatemala: los contrastes del desarrollo humano*. Edición 1998.

PNUD, 2000. Guatemala: La Fuerza incluyente del Desarrollo Humano, Edición 1999.

PNUD, 2001. Guatemala: El Financiamiento del Desarrollo Humano, Edición 2001.

PNUD, 2003. *Guatemala: una agenda para el desarrollo humano*, informe nacional de desarrollo humano, 2003.

PNUD, 2004. *Desarrollo Humano y Ruralidad*, Compendio estadísico 2004.

Saisana, M. & Tarantola, S. 2002. State-of-the-art Report on Current Methodologies and Practices for Composite Indicator Development, Joint Research Centre European Commission.

Sarewitz, D., Pielke, R. Jr & Keykhah, M. 2003. Vulnerability and Risk: some thoughts from a political and policy perspective, *Risk Analysis* 23(4): 805–810.

Trujillo, M., Ordonez, A. & Hernandez, R. 2000. *Risk-Mapping and Local Capacities: Lessons from Mexico and Central America*, Oxfam WorkingPaper, http://publications.oxfam.org.uk /oxfam/display.asp?isb = 0855984201&%20, as seen 06.09.2005.

UNDP, 2004. *Global report. Reducing Disaster Risk: A Challenge for Development*.

UNDRO, 1991. *Mitigating Natural Disasters, Phenomena, Effects and Options*. A manual for Policy Makers and Planners, New-York.

Wilches-Chaux, G. 1993. *La vulnerabilidad global in Los Desastres no son naturales*, Andrew Maskrey (ed), La RED: 11–44.

Wisner, B. 2000. *From 'Acts of God' to 'Water Wars': the urgent analytical and policy role of political ecology in mitigating losses from flood: a view of South Africa from Central America*, 24p., http://www.anglia.ac.uk/geography/radix/resources_papers.htm

Wisner, B. 2001. *Notes on Social Vulnerability: Categories, Situations, Capabilities, and Circumstances*, http://www.anglia.ac.uk/geography/radix/resources_papers.htm.

Yohe, G. & Tol, R.S.J. 2002. Indicators for social and economic coping capacity – moving toward a working definition of adaptive capacity. *Global environmental change* 12: 25–40.

Risks Challenging Publics, Scientists and Governments – Menoni (ed)
© Crown Copyright 2010/HSL, ISBN 978-0-415-58072-4

Use of a national population database to aid major hazard modelling and risk estimation

H. Balmforth & H. McManus
Health and Safety Laboratory, Harpur Hill, Buxton, UK

A. Fowler
Health and Safety Executive, Bootle, Merseyside, UK

ABSTRACT: The UK's Health and Safety Executive (HSE) has recently developed a new tool for identifying and estimating population levels, which is being used for major hazard modelling and risk estimation. The National Population Database (NPD) tool provides population data from a local to a national level for England, Scotland and Wales. It is a Geographical Information System (GIS) based tool, which uses a number of different source data sets to produce detailed estimates of a variety of populations. The data sets incorporated in the NPD include tables from the UK census, national mapping on a range of scales, address and transport network data and commercial and business data sets. This paper will discuss the NPD tool methodology and present a selection of case studies to illustrate how this tool could aid the risk assessment process from risk policy and development to regulatory and enforcement duties.

1 INTRODUCTION

Britain's Health and Safety Executive (HSE) undertakes major accident hazard modelling in relation to all hazardous sites and pipelines in the UK, which potentially present a hazard to the surrounding population under EU and domestic legislation (HSE 1989). As the body responsible for the regulation of almost all the risks to health and safety arising from work activity in Britain, HSE ensures compliance with health and safety legislation for nuclear installations, mines, factories, farms, hospitals and schools, offshore gas and oil installations, the safety of the gas grid, the movement of dangerous goods and substances and railway safety, plus also many other aspects of the protection both of workers and the public.

The Health and Safety Laboratory (HSL), which is an agency of the HSE, provides scientific support and research for these duties. One area of expertise relates to risk assessment and the development of methodologies, techniques and policies in this area. Risk assessment research at HSL includes the development of Quantified Risk Assessment (QRA) methodologies and tools and the integration of these tools with Geographical Information Systems (GIS) and databases. A GIS is a set of software tools which can be used to analyse, manipulate, store and display data with a spatial component. It particulary lends itself to integration with QRA tools and population estimations, which are of particular interest to HSE.

As part of the major hazard modelling and risk estimation process HSE and HSL produce risk assessments to provide estimates of the potential risk. A risk assessment is an estimation of the level of risk and the evaluation of the significance of that risk in order to inform decisions about risk management. It is a recognised integral part of successful health and safety management and is implicitly and explicitly included in much of the UK Health and Safety legislation (HSW 1974, MHSWR 1999). Risk estimates require a knowledge of the numbers and locations of people that may be exposed to potential hazards. This has resulted in the HSE developing a National Population Database (NPD) for use in its risk estimation work.

The NPD provides a representation of patterns of potential population occupation; it can provide estimates of where people are likely to be at any time when an accident could occur. This includes estimates of concentrations of people that may build up at certain times of the day in places that may be relatively empty at other times, for example a shopping centre. The NPD is a new development that represents a substantial improvement over the previous data used by HSE to estimate spatial distributions of populations, which was largely highly generalised and of poor quality.

This paper will briefly discuss the rational behind the need for the NPD, the development of the tool and some of the details of the populations available from it. A range of case studies of how the NPD is being used by HSL for HSE will then be presented.

2 THE HSE'S NATIONAL POPULATION DATABASE

The NPD is composed of a number of different source datasets, which are combined and manipulated to produce the population estimates. The NPD uses ESRI ArcGIS software as a frontend to the underlying data layers. This allows a user to manipulation, analyse and interrogate the data within the GIS environment. The NPD tool extends a methodology previously developed for HSE (HSE 1998, HSE 2002) to produce a sophisticated tool for deriving population data, from a local to a national level for England, Scotland and Wales.

The NPD methodology is described in detail in the HSE research report number 297 (HSE, 2005). In summary it involves identifying and locating key features and then applying population multipliers to provide the detailed locations and breakdowns of the population. The populations that are available within the NPD are shown in Table 1 below.

Table 1. Populations available from the NPD tool.

Population layer	Category within population layer	Differentiation
Residential		Usual/night time Day time/ term time Day time/ non term time
Transport	Roads (motorways, A roads)	Average daily flow, Peak flow, Max capacity
	Railway stations	Location only
	Ports	Location only
	Airports	Location only
Sensitive and Communal Establishments	Schools	Daytime
	Boarding Schools	Night time Max capacity
	Care Homes	Max capacity
	Hospitals	Max capacity
	Prisons	Max capacity
Workplace	Workplace Populations	Total Workplace population
Retail	Retail Population	Core retails centre, Town centre, Retail park
Leisure Facilities	Stadia	Max capacity
	Camping Grounds	(location only) Camp sites Caravan sites
	Public Attractions	(location only) Aquarium, Historic House, Motor Racing Circuit, Racecourse, Theme Park, Wildlife Centre, Zoo

Also listed is the differentiation of the populations, for example daytime night time splits etc.

HSE requires this breakdown in populations to aid its risk assessment and risk estimation work. For example, it is important to know the locations and numbers of people that may be more sensitive to harm, such as children, the elderly or hospital patients, hence the inclusion of the sensitive layer. Populations that may be hard to evacuate are also of interest, such as those in prisons, in town centres or retail areas, or in large stadia for example. These are available from the communal and leisure layers. HSE's scenario planning work also involves examining worst-case scenarios, which requires estimates of the maximum populations in an area. This is often the night time residential population, which is available from the differentiation in the residential layer.

The populations from the NPD are available on a 100 m grid and at the individual building level for some layers. The resolution is determined by the resolution of the source data sets and also on the requirements of the user. Populations that are available down to individual building resolution are the residential, the sensitive and the communal establishment populations, which are of particular concern for HSE. A user is able to select the resolution that is most suitable for a particular study, which is usually determined by the area of concern or the hazard footprint. A variety of different datasets are used within the NPD to produce the population details and differentiation. These include UK Ordnance Survey addressing, infrastructure and mapping data, commercial business data sets and UK government data, which are used to locate many of the key features. UK census and government data is then used to populate the locations. The population multipliers provide the differentiation between the population at different times of day and between different sensitivities.

3 NATIONAL POPULATION DATABASE CASE STUDIES

HSL are exploring using the NPD tool for HSE in a number of areas, including risk assessment, land use planning, major hazard accident modelling and risk estimation. Some of this work has previously been described in Balmforth et al. (2005). This paper presents some further case studies, which cover the other areas in which the NPD is starting to be utilised. The following examples are fictional, but are based on the types of studies that the tool could be used for.

3.1 *Case study 1: populations and nuclear sites*

The Nuclear Safety Directorate (NSD) of HSE licenses, on behalf of HSE, and regulates nuclear sites in the UK. Part of NSD's role is to monitor development around such sites, and HSL is using the NPD for NSD to provide detailed estimates of the population around them. This is primarily for use with planning and development issues around the site, but can also be extended for emergency planning use.

Figure 1 shows the population around a nuclear plant, residential populations are shown in red, the road population (time-dependent vehicle occupancy) in green, and sensitive populations (hospitals and schools) in yellow. An illustrative buffer zone of 5 km around the site is shown in blue.

For this work the NPD has also been integrated with another GIS application, which allows all the planning, consultation and emergency zones to be viewed around any nuclear plant. This application keeps track of the increases in population due to new developments around the site. By combining the two tools detailed monitoring of the population density and distribution within certain areas of interest around nuclear sites is possible.

3.2 *Case study 2: development of industrial areas – land use planning issues*

Increasing development within cities of industrial, brown field or redevelopment sites has put increasing pressure on the land use of these areas and on the land use planning system. The NPD can be used to examine land use planning issues, for example, with respect to the development of areas around major city. The NPD can be used to model the density and distribution of the current population in these areas and examine how this relates to the expected distribution of the population if certain developments go ahead. These populations can also be compared to the population density and distribution for this same area from up to 40 years ago, which is available from combining the NPD with historical mapping data. This allows a picture of the evolution of the population in certain areas to be established. The population information from the different time periods can be combined with risk assessment data for the major hazard sites in this area during each period.

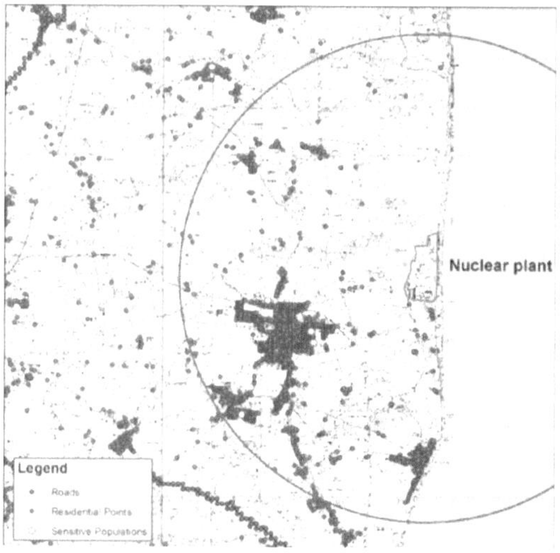

Figure 1. Populations around a UK nuclear plant.

3.3 *Case study 3: major hazard modelling*

HSE is required to regulate the risk to the population from major hazard sites through its duties under the EU SEVESO legislation. The NPD can be used to inform this process. An example is illustrated in Figure 3; here two different harm criteria have been used to model the effect of a toxic plume that could potentially escape from a major hazard site as a result of an accident.

Several scenarios are shown, which are weather dependant and represent different worst cases for different populations. The full green and purple plumes illustrate the scenario with the wind from the southwest. In this case the night time residential population is at a maximum. The dashed plumes illustrate the scenario when the wind is blowing from the west or the northwest.

In this case the plumes could potentially extend over the playing fields of either of two local schools, thus potentially exposing a large sensitive population. The school nearer the top of the picture has a population of 993, while the school further down in the figure has a population of 379.

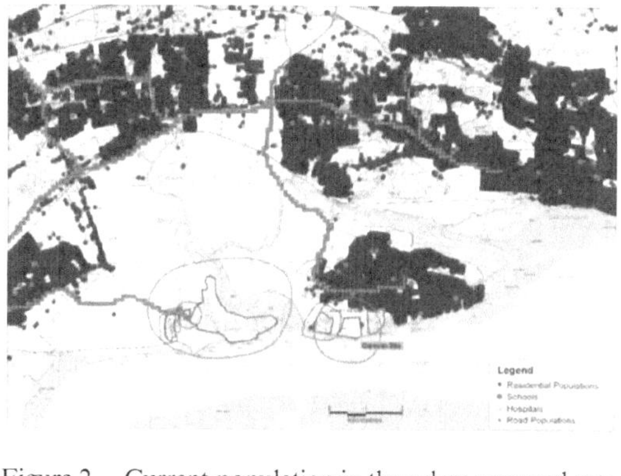

Figure 2. Current population in the urban area and consultation distances from the major hazard sites in the area.

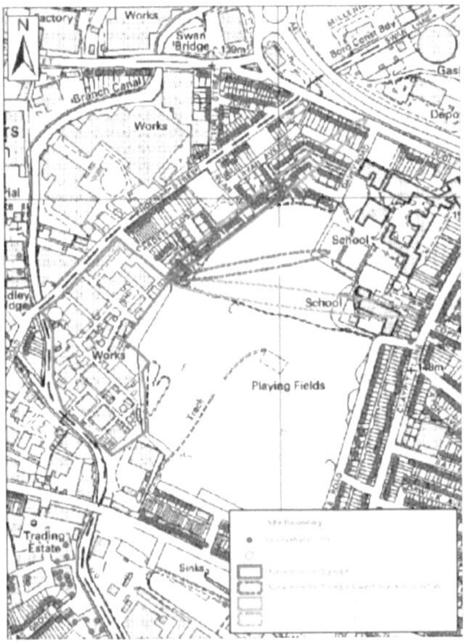

Figure 3. Populations that may be potentially at risk from a release of a hazardous substance from a major hazard site (shown in pink) are illustrated. The toxic plumes have varying extents, which are dependent on the harm criteria used. Different wind directions show that different populations of varying sensitivity may potentially be exposed.

3.4 *Case study 4: estimation of spatial risk*

Spatial risk measures the potential risk of harm to a group of people. Such risk values are often given in terms of an FN curve, which represents the frequency, F, of N or more people being harmed. Studies of this sort require detailed estimates of the numbers and locations of people, often over large areas. Figure 4 illustrates a fictitious example from the NPD over a larger area, for spatial risk work. The area is a 6 km by 6 km square around a major hazard site. The NPD can be used to illustrate the variation of the population in this area and to highlight concentrations. The residential and road transport populations are shown in this case. These data can then be combined with consequence modelling to examine the contribution that certain populations have on the risk value around major hazard sites. This type of information can be used to inform risk assessment policy development.

4 SUMMARY AND FUTURE WORK

The NPD tool is being used in a number of varied applications by HSL, as illustrated by the case studies in this paper. This is helping to provide support for HSE's regulatory duties, including land use planning and major hazard modelling and also risk assessment and estimation methodology and policy development. While the NPD should be approached as a representation of patterns of potential population occupation, rather than a definitive measure, it represents a major step forward for HSE in this area.

Further development of the NPD is planned, to take into account new datasets that become available and new areas of interest and use. It is envisaged that the data from the NPD will be of use to a wide range of other UK government departments and also for local authorities and the emergency services. Further exploitation in these areas is planned.

Figure 4. Population concentrations for use in spatial risk calculations.

REFERENCES

Balmforth, H., McManus, H. & Fowler, A. 2005. *Assessment and management of 'at risk' populations using a novel GIS based UK population database tool.* Safety and Security Engineering, Editors C. Brebbia, T. Bucciarelli, F. Garzia & M. Guarascio. WITpress.

HSE, A. 2005. National Population Data Base for Major Accident Hazard Modelling. *Health and Safety Executive Research Report Series* 297.

HSE, 2002. The derivation and use of population data for major accident hazard modelling. *Health and Safety Executive Research Report Series* 410.

HSE, 1998. Spatially referenced population data for land use planning advice. *Health and Safety Executive Research Report Series* 189.

HSE, 1989. *Risk criteria for land use planning in the vicinity of major industrial hazards.* London: HMSO.

HSW, 1998. Health and Safety at Work etc. Act, 1974. In J. Hendy, M. Ford (eds), *Redgrave's Health and Safety*, third ed. 1998. Butterworths. ISBN 0 406 89969 X.

MHSWR, 1999. Management of Health and Safety at Work Regulations, *Approved Code of Practice and Guidance*. HSC 2000, HSE L21, ISBN 0 7176 2488 9.

Risks Challenging Publics, Scientists and Governments – Menoni (ed)
© 2010 Taylor & Francis Group, London, ISBN 978-0-415-58072-4

Interdisciplinary quality-of-life parameters as a universal risk measure

U. Köhler
Dresden University of Technology, Clinic and outpatient's clinic of Dermatology, Dresden, Germany

D. Proske
University of Life Science, Institute of Mountain Risk Engineering, Vienna, Austria

M. Curbach
Dresden University of Technology, Institute of Concrete Structures, Dresden, Germany

ABSTRACT: The paper focuses on methods to represent the risk of loss of human life. Several risk measures, such as F-N-diagrams, probability of loss of life and days of lost life expectancy are introduced. All of the mentioned risk measures represent different types of loss. In addition, quality of life parameters can be understood as a risk measure. In contrast to the former mentioned risk measures the proposed one is able to cover the entire field of risk to humans. If this is true then all safety actions, such as the strengthening of bridges, can be compared to assess their efficiency. This would permit the competition of safety actions over the entire society and should yield to an improvement not only of human's safety, but also to an increase of the quality of life inside a society.

1 INTRODUCTION

One major aim of human activity is based on building and cultivating social structures. The efficiency of such structures forms the basis for the exponential growing of human society, which has been going on for two centuries now.

Despite their vastness, eye to eye contact is still today the main criterion for successful social structures. The enormous economic strains which have been used to widen the range of communicative means within the last years are proof of this development. The beginning of personal contact, regardless which means of communication is used is marked by a formalized gesture or an 'empty phrase', which is known as greeting. The form of welcome, in almost all cultures on earth also includes the question concerning the well-being of the counterpart. The answer to the question requires an evaluation of the personal life situation. Basis for such qualitative evaluation processes are usually quantitative parameters, which enable us to compare actual values and target values. The question for evaluating the personal life situation can be answered by a quality-of-life parameter.

2 QUALITY-OF-LIFE PARAMETERS

The building and developing of social structures is closely connected with the term quality-of-life. Although the famous social critics of the 19th century did not know the term quality-of-life yet, they already regarded the improvement of the circumstances of living as the main motivating force of human development. One of the first definitions of quality-of-life is therefore taken from the area of social charity science:

"Measure of the congruence between the conditions of a certain objective life standard and the subjective evaluation of the thereby marked group of population."

It was the economist Cecil Pigou who coined the term "Quality-of-life" at the beginning of the 20th century and who brought the term into the academic discussion as a target figure for social actions and as measure of individual well-being. The development and usage of the term quality-of-life has since then not only touched economy and social science, but it has also been introduced to many other areas, e.g. medicine. In 1947, the World Health Organization termed 'Health' as a condition of absolute physical, mental, and social well-being. Over fifty years ago, this definition widened aims and criteria for actions of physicians. Not only somatic aspects of health and illness, but also psychological and social aspects, the patient's well-being and his capacity to act are part of the physician's duty.

In the field of medicine new target figures were introduced via quality-of-life parameters, which among physicians are also known as quality-of-life measure instruments; the aim of this was to make quality-of-life measurable and therefore testable. Quality-of-life parameters are used today in preventive medical check-ups, in therapy research, quality security and health economy. Over 20,000 scientific publications deal with the topic of medical life quality in the German speaking countries (Bullinger, 1996), whereas probably 2,000 new publications are added per year (Frei, 2003). These scientific strains led to the development of over 800 life quality parameters in the field of medicine (Ahrens & Leininger, 2003; Frei, 2003; Spilker et al. 1990). Exemplary, some medical quality-of-life parameters are mentioned in Table 1. The trodden path of metrical representation of the life quality required at first a definition of the term "quality-of-life". In the following, a definition from the field of medicine is quoted:

"By health-related quality-of-life is meant a psychological construct, which describes the physical, psychological, mental, social and functional aspects of the well-being and the function capacity of the patients from their view." (Bullinger, 1996).

As the short description illustrates, the term quality-of-life goes far beyond pure medical questions, so as to other areas other definitions were formed:

"Quality-of-life is the individual perception of the personal living situation in the context of the respective culture and the respective value system in relation to personal aims, expectations, judgment scales and interests." (WHOQOL-Group)

"Quality-of-life is the result of an individual, multi-dimensional evaluation process of interaction between an individual and its environment. As evaluation criteria can be used social norms as well as individual value judgments and affective factors." (Frei, 2003)

Numerous other definitions are to be found in Proske 2004.

According to the respective definition, quality-of-life depends on a number of introductory figures, which are partly hard to seize numerically. In Table 2 possible introductory figures are collected tabular. Based on the high multi-dimensionality, the insecurity in choosing and evaluating introductory figures and the very specific question in many fields, special quality-of-life parameters were developed in a wide range, which limit the number of introductory figures and which seize the quality-of-life in a certain situation, comparable to the development in medicine. The quality-of-life parameters differ tremendously in the assembly. Even the assembly of various quality-of-life parameters for identical questions reveals difficulties in the fixing of decisive introductory figures in the functional connection. Table 3 gives proof of this statement by the example of the number of introductory figures for various quality-of-life parameters for psychiatric patients.

In some of the publications, the possibility of a metrical description of quality-of-life is categorically excluded (Küchler & Schreiber, 1989). There, a rather coarse fixing of quality-of-life is considered to be possible, as is schematically represented in Figure 1.

The topic quality-of-life and the problems connected to this topic are fortunately not only subject of discussion in academic journals, such as the "Journal of Social Indicators" or at academic conferences, such as the "International Society for Quality-of-life Studies", a conference held in

Table 1. Health-related quality-of-life parameters.

Illness-comprehensive Quality-of-life parameters	illness-specific quality-of-life parameters
• Nottingham Health Profile	• Quality of Life Index – Cardia Version III (QLI)
• Sickness Impact Profile	• Seattle Angina Questionnaire (SAQ)
• SF-36 (SF-12)	• Angina Pectoris Quality of Life Questionnaire (APQLQ)
• WHOQoL	• Minnesota Living with Heart Failure Questionnaire
• EuroQol	• Asthma Quality of Life Questionnaire (AQLQ)
• McMaster Health Index Questionnaire	• Fragebogen zur Lebensqualität bei Asthma (FLA)
• MIMIC-Index	• Fragebogen für Asthmapatienten (FAP)
• Visick-Skala	• Asthma Questionnaire (AQ20/AQ30)
• Karnofsky-Index	• Osteoporosis Quality of Life Questionnaire (OQLQ)
• Activities-of-Daily-Living Index	• Quality of Life Questionnaire for Osteoporosis (OPTQol)
• Health-Status-Index	• Osteoporosis Assessment Questionnaire (OPAQ)
• Index-of-Well-being	• QOL Questionnaire of the European Foundation for Osteoporosis (QualEFFO)
• Rosser-Matrix	• Juvenile Arthritis QOL-Questionnaire (JAQQ)
• Rosser & Kind Index	• Schmerzempfindlichkeitsskala (SES)
• Quality of Well Being Scale	• Pain Disability Index (PDI)

Table 2. Possible introduction variables for quality-of-life parameters (Hudler & Richter 2002).

Predominant objective variable	Predominant subjective variable	Societal variables
– living conditions – family – social conditions – participation in social life – life standard – income – health – education and work	– life contentment – happiness – carefreeness – subjective class membership – optimism pessimism about future developments – judgment of the personal living conditions	– social conflicts – trust in other people – security, freedom, justice – social integrity

Table 3. Quality-of-life measure instruments for psychiatric patients.

Quality-of-life measure instruments	number of parameters
Social Interview Schedule (SIS)	48
Community Adjustment Form (CAF)	140
Satisfaction of Life Domain Scale (SLDS)	15
Oregon Quality of Life Questionnaire (OQoLQ)	246
Quality of Life Interview (QoLI)	143
Client Quality of Life Interview (CQLI)	65
California Well-Being Project Client Interview (CWBPCI)	304
Quality of Life Questionnaire (QoLQ)	63
Lancashire Quality of Life Profile (LQoLP)	100
Quality of Life Index for Mental Health (QLI-MH)	113
Berliner Lebensqualitätsprofil (BeLP)	66
Quality of Life in Depression Scale (QLDS)	35
Smith-Kline Beecham Quality of Life Scale (SBQoL)	28
Quality of Life Enjoyment and Satisfaction Questionnaire (Q-LES-Q)	93

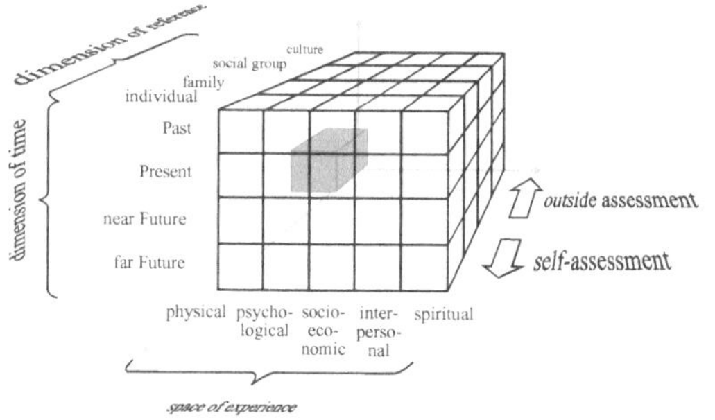

Figure 1. Dimension of quality-of-life according to Küchler.

Frankfurt/Main in 2003. An advertisement from the 'Aktion Mensch' in Figure 2 is to serve as an example for the question of defining of quality-of-life. The media reported about the quality-of-life

Figure 2. Advertisement of the 'Aktion Mensch' (translation: "What means or contains the term quality-of-life?").

atlas for Germany (Korczak, 2003) or the Human Development Index of the UNO 1990.

Just as in those parameters, it is assumed that a metrical representation of quality-of-life is possible in general. In the following, it will be illustrated that quality-of-life parameters are always risk parameters. Subsequently, various hierarchical risk parameters are commented upon.

3 RISK PARAMETERS

3.1 *Frequency of death*

The risk parameter of frequency of death, probability of death respectively corresponds with the classical definition of risk, as it can be found for instance in norms (DIN VDE 31 000). Here, risk is defined as a product of the frequency of occurrence of an incident with damage and the extent of the damage. Various other varieties of the term risk will be mentioned for reasons of completeness.

Frequency of death and probability of death can be regarded as special cases of risk. In this case, the damage is the loss of human lives. Based on necrologies, which were introduced to Australia in the 18th century and to England and Wales in 1837, it is possible to determine the frequency of death for human beings (Hofstetter & Hammitt, 2001).

Figure 3 lists some of the frequencies of death for people in various actions. In Proske 2004, a collection of over 125 values for frequency of death in various situations or actions can be found. When comparing such frequencies of death, one has to consider the basic totality and the date of gathering. In Figure 3 for example, the first two causes are related to partly basic totality of the population. The infant mortality is related only to the group of infants, and the frequency of death of the soldiers is only related to the basic totality of soldiers. Besides these very high values, the relative frequency of death can be represented in very small values, such as 10^{-8} per year for death caused by mass deaths in the history of the earth. As target value for workers, in many countries 10^{-4} is used; as target value for members of the uninvolved public 10^{-6} is used.

The parameter of the frequency of death is usually considered for one year. It does not give evidence of how frequent or for how long the person was exposed to a certain action in one year. In order to improve the quality of the risk comparison, a calibration of time is necessary.

The representation of a mortality risk in relation to a defined reference time of 10^8 or 10^3 hours is called Fatal Accident Rate. Examples for FAR's can be found among others in Proske 2004. Target values for the area of the oil industry are at 15, for member of the uninvolved public at about 0.1.

The frequency of death does not necessarily have to be related to a certain span of time. The frequency of death can also be related to distances,

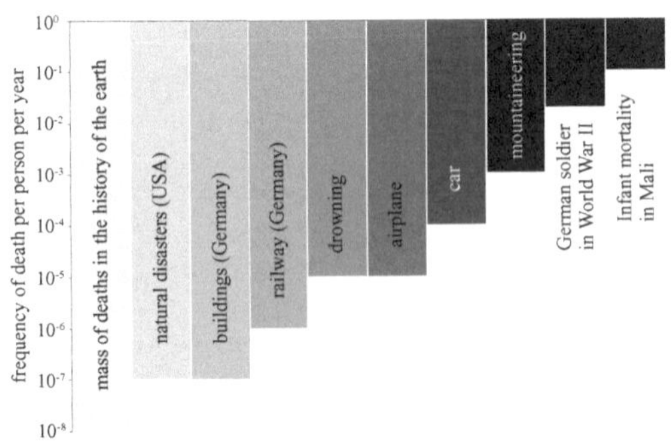

Figure 3. Examples of frequency of death for people for various actions.

number of actions or amounts of substances. The unit-risk-value for air pollution quantities is an example for such a parameter. A unit-risk-value for a certain substance specifies the additionally assumed risk of cancer for humans when inhaling polluted air for 70 years with an amount of air pollution of 1 µg per m^3. The real risk can be calculated by multiplication with the actual time of exposure of the human to the pollution.

3.2 *Family of the F-N-diagrams*

The frequency of death and the fatal accident rates do not consider the extent of a certain incident of damage. The figures will be the same for an accident with one casualty which occurs one thousand times, and an accident with one thousand casualties which occurs only once. Experience has shown that people differentiate between those two cases very much in regards to the subjective judgment of security. If the risk parameter is to be used successfully, it has to reflect the security perception of the population to a sufficient extend.

For the improvement of the subjective risk evaluation, which is also called risk aversion, so-called *F-N*-Diagrams were developed. The first of these diagrams were developed by Farmer in 1967 (Farmer, 1967). The risk research received huge impulses by the building of nuclear power stations. *F-N*-Diagrams became very famous in the so-called Rasmussen-report in the beginning of the 70's of the last century.

F-N-Diagrams are double-logarithmic diagrams, which show on the x-axis the number of casualties and on the y-axis the frequency of accidents with equal or larger numbers of casualties. By this definition, we receive graphs which go down from left to right. We have to consider that there are also so-called *f-N*-graphs, which on the y-axis show the frequency of accidents with N casualties. In this case, a rising graph can be the result.

Since the introduction of the first diagrams, a huge variety of diagrams has been developed, so that it can be called the family of the *F-N*-Diagrams. The basic principle of the diagrams is the same, but the single units on the x-axis can differ. Therefore, there are diagrams, which on the x-axis show the costs of damage, a damage parameter which consists of various other parameters, the number of persons concerned (PAR), the time for removing the damage, the energy which has been used for removing the damage or the radioactive radiation, as has originally been intended. A summary of various representations of *F-N*-Diagrams can be found in Jonkman et al. 2003.

For the development of such diagrams, data of the accidents from the past with specification of the damage and the number of persons concerned are necessary or calculations which provide such data. The data, number of casualties and the frequency of accidents have to be sized according to rising numbers of casualties. After that, the data will be cumulated, which means the frequency of accidents with a number of casualties which equals N or is higher than N will be determined and sized. These data pairs will be represented graphically in the F-N-Diagram. Examples for the development of classical F-N-Diagrams can be found among others in Ball & Floyd 2001, Proske 2004.

The proof of sufficient security will be done graphically in a diagram. Comparison lines, which were developed for a variety of incidents, will be used as a proof. In Ball & Floyd 2001 and Proske 2004, about 20 target graphs are collected. The target graphs divide the area of the diagram in two sub-areas: in an acceptable and an unacceptable area (Figure 4). Some of the target graphs have an additional area, which is under certain circumstances acceptable. This area is known as the ALARP-area (as low as reasonably possible). If the determined graph is inserted into the diagram, one will see whether the graph is in the acceptable area. The average number of casualties or the average costs of damage expected in case of an accident are called the Potential Loss of Life.

3.3 *Lost life years*

The family of the F-N-Diagrams is excellently suitable for the representation of technical and natural risks, since in the cases of those risks high numbers of casualties are possible. In cases of health risks, there is always only one person concerned. Additionally, the age of the person concerned is considered when risks are judged subjectively. The death of a person of 90 years of age, caused by an illness, will be judged subjectively different, than the death of a young person. The risk parameter of the lost life years or lost life days can consider this effect. The parameter is defined as the difference between average life expectation without the analyzed risk and the average life expectation with the analyzed risk. The parameter is widely used in representations of cancer diseases in Germany, but it is also used for other illnesses. Cohen 1991 collected lost life days for various diseases and various social circumstances (Figure 5). Further data can be found in James, 1996, Proske 2004.

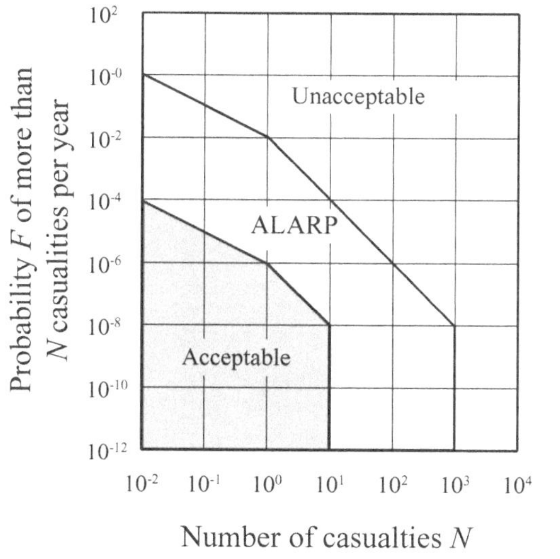

Figure 4. Example of a proving graph (Groningen-graph 1978).

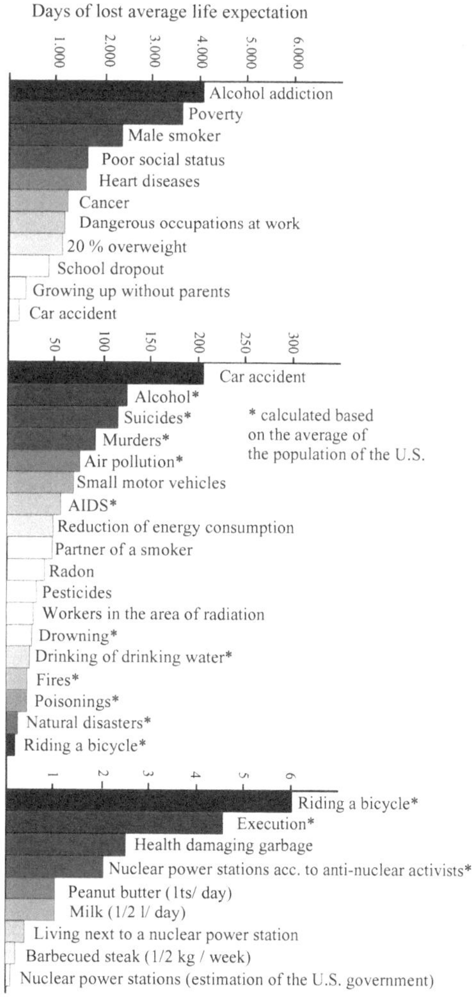

Figure 5. Lost Life Days in various situations according to Cohen.

In addition to the loss of life years, it is possible to collect health reductions during life and to calculate the loss of life time. Here, for instance so-called Quality Adjusted Life Years (*QALY*), Disability Adjusted Life Years (*DALY*) or Health Years Equivalent (*HYE*) is mentioned. Figure 6 is used to clarify those terms. The Figure shows the health profile of a human. The size of the dark are is the *QALY*, whereas the area of the brighter grey is conform to the *DALY*. In addition the difference between the age of the human and the average life expectancy in the Figure complies to the Lost Life Years. One example for the application of this concept, which represents risks in environmental pollution in the Netherlands, can be found in Hofstetter & Hammitt, 2001. In developed countries, the share of the *DALY* amounts to about 10% of the life time, in India about 30%, and in some of the African countries it is almost 50% (Foege, 1994).

3.4 *Quality-of-life parameters*

The Figures for the *DALY*'s clearly show that people who live in regions under poor geopolitical circumstances, are exposed to high risks. This fact also accounts for groups of people who live in industrial countries, in which the conditions of living are equal to those in developing countries. Poverty and poor social status, in industrialized countries also produce the highest losses of life expectation, as Figure 5 shows. The connection between poverty and average life expectation on the level of countries is represented more clearly in Figure 7. Based on the assumption that poverty is an essential introductory figure for life quality, one can draw the conclusion that life quality and life risk are closely connected with each other. Quality-of-life parameters are universal risk parameters, since, according to their structure they can include minor disturbances of life quality as damages.

Life quality parameters as risk parameters can be used to judge the efficiency of risk lowering measures. Risk lowering measures are all safety and security measures, which are installed by human society, such as social security systems, police, jurisdiction, hospitals, technical safety systems or safety systems which are used in order to prevent natural disasters etc.

It is common knowledge that legal standards for safety systems are imbalanced (Joakim et al. 1996; Tengs et al. 1995; Viscusi, 1995). While in some cases, the realization of governmental safety standards is not very successful when at the same time large amounts of financial resources are connected to it; in other cases laws and norms can be realized with very little pecuniary means and show great success in securing humans. Quality-of-life parameters enable us to judge those relations, from the view point of the society as a whole as well as for special cases.

Evaluations of adjuvant therapies in oncology are an example for the evaluation of efficiency. Adjuvant theories include measures for preventing the growth of tumors after complete removal of the primary melanoma. Therefore, for patients with high risk melanomas (size of the tumor ≥ 1.5 mm) after the surgical removal of the primary malignant melanoma the question has to be raised, whether or not the treatment is useful. It can not be predetermined, which of the patients will develop new tumors. Also, the success of the therapy can be determined only with some probability.

As analyses of the patient's quality-of-life (SF-36) showed, the quality-of-life was lowered considerably over a period of several months up to two years for patients who were treated with the adjuvant therapy as a result of the side effects (Figure 8). If one considers that for the therapy success can only be determined with some probability and if we also consider that there is a large amount of side effects, the usage of quality-of-life parameters can provide an answer to the question whether or not a therapy should be done.

In Proske 2004, one example from structural engineering is presented. There, it is analyzed

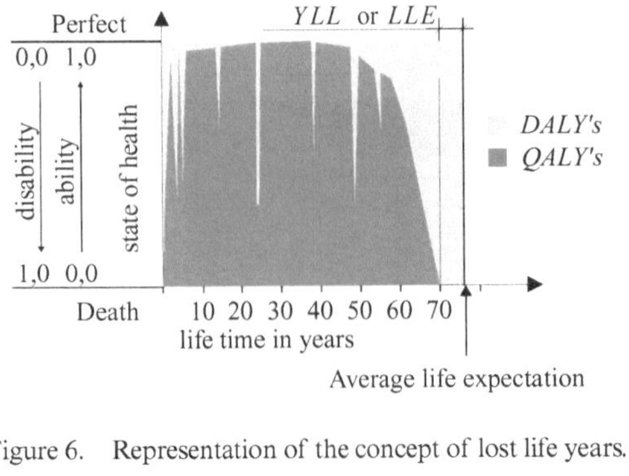

Figure 6. Representation of the concept of lost life years.

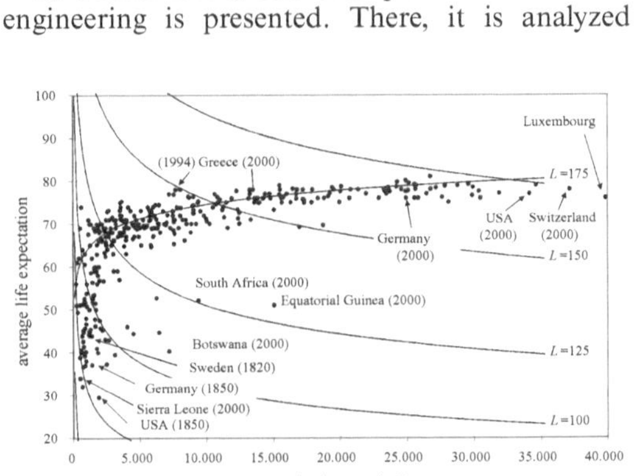

Figure 7. Per-capita-income, average life expectation and life quality index for 170 countries.

with the help of a quality-of-life parameter, which strengthening for bridges is most efficient.

In the sense of actual quality-of-life, the adjuvant theory or the strengthening of bridges should not be analyzed singularly, but in connection to each other. The whole range of medical, technical, and social preventive measures should be comparable. This would in part mean that measures would have to be used parallel to each other; in part it would also mean that competitive measures would have to be used. Quality-of-life parameters must be developed, which function independent of the field of research and which enable us to convert specific quality-of-life parameters. The authors consider the so-called dynamic profile parameters which are reunited with an index parameter a possible solution. Such an indicator is able to consider other quality-of-life parameters as input, for example the quality-of-life parameter Nathwani, Pandey and Lind can consider, for example the health quality-of-life parameter SF-36 in this way, that the life expectancy as input parameter for the quality-of-life parameter Nathwani, Pandey and Lind will be adapted according to some information from the SF-36.

In general, one has to differentiate between profile- and index- quality-of-life parameters. Profile parameters include various introductory figures or groups of introductory figures which can not be summed up, whereas index parameters include all introductory figures into one indicator. Examples for those profile parameters from the field of medicine are the SF-36, the Sickness Impact Profile (SIP) and the Nottingham Health Profile (NHP). Examples for index parameters, also from the field of medicine, are the Karnofsky Index, the EuroQol and the Quality-of-well-being scale. (Skript TU Dresden 2002).

Dynamic profile parameters are based on groups of introductory figures, which can be exchanged according to the issue in question. This modular construction enables the user to develop the system by and by, so that new issues can be treated with those parameters. In order to make those various profile parameters comparable, one has to reunite the dynamic profile parameters into one general index parameter. Higher-ranged index quality-of-life parameters can then be developed out of several of those subject-specific index parameters (Figure 9). Depending on the issue in question, a more or less subject-specific quality-of-life parameter can be selected from a pedigree of quality-of-life parameters.

This development is in part visible in the field of social science, where efforts are made to develop summarized and standardized quality-of-life parameters. Diener's Value-Based Index of National Quality-of-Life (Diener, 1995), Estes'Index of Social Progress (Estes, 1988), the American Demographics Index of Well-Being (Estes, 1988) or the Fordham Index of Social Health should be mentioned here as examples. The Human Development Index of the UNO or the International Well-Being-Index belongs to this group as well. In other areas this development is also brought forward. The quality-of-life index by Nathwani, Pandey and Lind, which takes into account the conversion of life spans while also considering reductions of life quality, is able to integrate quality-of-life parameters from medicine; thus a standardized quality-of-life parameter can be received.

The aim of research should be to bundle up the developments regarding quality-of-life in the individual research fields and to develop comprehensive concepts. Up to now, this has not been done. The development of metrical and interdisciplinary quality-of-life parameters as universal measure for risk is according to the authors of major importance for the society, in order to enable it to evaluate objectively the capability of preventive measures within itself.

Apart from the development of metrical quality-of-life parameters, a graphic representation of the individual components is helpful, in order to check the quality-of-life parameters. As already said,

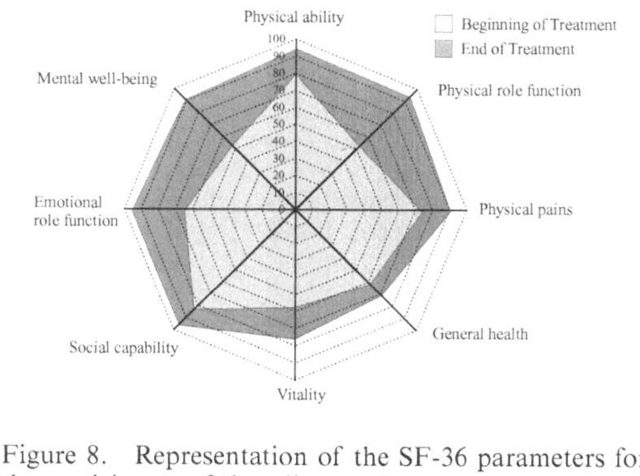

Figure 8. Representation of the SF-36 parameters for the participants of the adjuvant therapy.

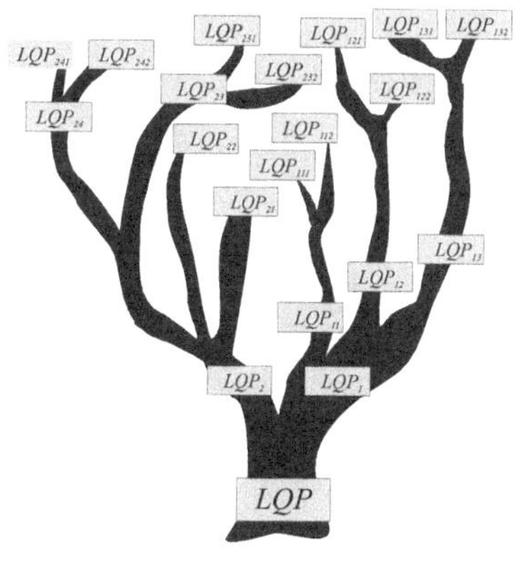

Figure 9. Graduation of quality-of-life parameters (QLP).

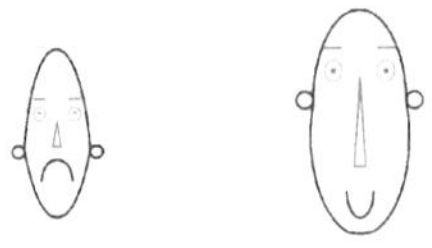

Figure 10. Representation of the SF-36 parameter for the participants of the adjuvant therapy according to the Chernoff-faces.

the relatively high number of introductory figures makes visualization in a diagram quite difficult. Parallel to the developing of computer-based techniques, within the last years a number of ways of visibly representing high dimensional data were developed. The so-called Chernoff-faces (Arnold, 2003; Chernoff, 1973) can be regarded as one form of representation for high dimensional data. In figure 10, the data of the SF-36 from figure 8 is represented again as Chernoff faces.

The usage of human facial characteristics in the visualization of high-dimensional amounts of information is not only grounded on the high differentiable quality of the visualizing of faces and facial expressions, but it also relies on the representation of facial expressions based on a large number of degrees of anatomical freedom and high fine motor capabilities. The visualization of introductory and result figures of quality of life parameters by ways of the human face is after all, what human beings do anyway: to transfer information about their state of health through their facial expressions. The facial expression of human beings enables us extraordinarily to give an answer to the question of the well-being of humans dealt with in the beginning and consequently also to the question of quality-of-life.

REFERENCES

Ahrens, A. & Leininger, N. 2003. *Psychometrische Lebensqualitätsmessung anhand von Fragebögen*. Christian-Albrechts-Universität zu Kiel, Vorlesungsunterlagen 7.7.2003.

Arnold, C. 2003. *Visualisierung multidimensionaler statistischer Daten im Web*. Diplomarbeit. Fachhochschule Hagenberg.

Ball, D.J. & Floyd, P.J. 2001. Societal Risks. Final Report. School of Health, Biological/Environ-mental Sciences. London: Middlesex University.

Bullinger, M. 1996. *Lebensqualität - ein Ziel- und Bewertungskriterium medizinischen Handelns?* Hrsg. H.-J. Möller, R. Engel und P. Hoff, Befunderhebung in der Psychiatrie: Lebensqualität, Negativsymptomatik und andere aktuelle Entwicklungen. Springer-Verlag.

Chernoff, H. 1973. The use of faces to represent points in k-dimensional space graphically. *Journal of the American Statistical Association* 68: 361–368.

Cohen, B.L. 1991. Catalog of Risks extendet and updated. *Health Physics*, September 1991, 61: 317–335.

Diener, E. 1995. *A* Value Based Index for Measuring National Quality of Life. *Social Indicators Research* 36: 107–127.

Estes, R.J. 1988. *Trends in World Development*. New York: Praeger.

Farmer, F.R. 1967. Siting Criteria – a new Approach. *Atom* 128: 152–170.

Foege, W.H. 1994. Preventive Medicine and Public Health. *Journal of American Medical Association* 271(21): 1704–1705.

Frei, A. 2003. *Auswirkungen von depressiven Störungen auf objektive Lebensqualitätsbereiche*. Dissertation. Psychiatrische Universitätsklinik Zürich, August 2003.

Hofstetter, P. & Hammitt, J.K. 2001. *Human Health Metrics for Environmental Decision Support Tools: Lessons from Health Economics and Decision Analysis*. National Risk Management Research Laboratory, Office of Research and Development, US EPA, Cincinnati, Ohio, September 2001.

Hudler, M. & Richter, R. 2002. Cross-National Comparison of the Quality of Life in Europe: Inventory of Surveys and Methods. *Social Indicators Research* 58: 217–228.

James, M.L. 1995–1996. Acceptable Transport Safety. *Research Paper 30, Department of the Parliamentary Library*, http://www.aph.gov.au/library/pubs/rp/1995–96/96rp30.html

Joakim, A.L., Ramsberg, J. & Sjöberg, L. 1996. *The Cost-Effectiveness· of Lifesaving Interventions in Sweden*. Annual Meeting of the Society for Risk Analysis- Europe.

Jonkman, S.N., Van Gelder, P.H.A.J.M. & Vrijling, J.K. 2003. An overview of quantitative risk measures for loss of life and economic damage. *Journal of Hazardous Materials* A 99: 1–30.

Korczak, D. 2003. *Lebensqualitätsatlas*. Opladen.

Küchler, T. & Schreiber, H.W. 1989. Lebensqualität in der Allgemeinchirurgie - Konzepte und praktische Möglichkeiten der Messung. *HÄB* 43: 246–250.

Proske, D. 2004. *Katalog der Risiken – Risiken und ihre Darstellung*. Dresden: Eigenverlag.

Spilker, B., Molinek, F.R., Johnston, K.A., Simpson, R.L. & Tilson, H.H. 1990. Quality of Life, Bibliography and Indexes. *Medical Care*, 28(Suppl 12): D51–77.

Tengs, T.O., Adams, M.E., Pliskin, J.S., Safran, D.G., Siegel, J.E., Weinstein, M.C. & Graham, J.D. 1995. Five-Hundred Life-Saving Interventions and Their Cost-Effectiveness. *Risk Analysis* 15(3): 369–390.

United Nations Development Program: Human Development Report 1990: *Concept and Measurement of human development*. http://hdr.undp.org/reports

Viscusi, K. 1995. *Risk, Regulation and Responsibility: Principle for Australian Risk Policy. Risk. Regulation and Responsibility Promoting reason in workplace and product safety regulation*. Proceedings of a conference held by the Institute of Public Affairs and the Centre for Applied Economics, Sydney, 13 July 1995 http://www.ipa.org.au/Conferences/viscusi.html

WHOQOL-Group: The development of the World Health Organization quality of life assessment instrument: The WHOQOL. In Orley J.; Kuyken W. (Hrsg.), *Quality of Life Assessment: International Perspectives: 41–57*. Heidelberg: Springer.

Risks Challenging Publics, Scientists and Governments – Menoni (ed)
 ISBN 978-0-415-58072-4

Improving information of major technological risks by research in social sciences

A. Van Zanten
Institut Européen des Risques, Honfleur, France

ABSTRACT: Public acceptance of major technological risks is increasingly low. Recent major accidents have even led the public to question the place of industrial activities in modern society. In this context of crisis in public trust, information adapted to the needs and demands of citizens is of great importance. This article aims to show how social sciences can help improve public information and proposes five research topics within the disciplines of economy, sociology and geography. Furthermore, this article aims to explain the role which the "Institut Européen des Risques (IER)" plays for social sciences. The article concludes that the links the IER has established between social sciences and regional decision makers in the Seine Estuary provide favourable local dynamics for research projects.

1 PROBLEM SETTING

That public acceptance of major technological risks has become increasingly low is painfully clear following accidents such as the AZF-disaster in Toulouse (France) and the fireworks disaster in Enschede (the). In the aftermath of these accidents, the news media was filled with articles in which the public questioned the place of hazardous industrial activities in modern society. After all, the public rarely feels it chose to live and work near these activities and rather perceives that these were imposed by industrial companies and the authorities. Accepting the risks that these activities entail has become all the more difficult for citizens since they have become aware of the fact that industrial activities employ few people compared to tertiary activities (service provision). Moreover the public has become progressively more demanding as technological progress reveals new ways in which to reduce risk. Therefore, public debate reflects the opposition between citizens who demand absolute security and the scientific point of view which stipulates that zero risk does not exist.

Information, whether of preventive nature or related to a dramatic event, has become increasingly important in this context in which the public questions governmental decisions and wishes for more transparency and participation in decision making. However, in most cases, it is not until a crisis occurs that public authorities and industrial companies take steps to develop a genuine public information policy. Information about major technological risks is generally accessible through the following means:

1.1 *Public consultation and public meetings*

In France, the creation of a "Local Commission for Information and public Consultation (CLIC)" at any industrial site with one or more upper-tier Seveso establishments is obligatory since the law of the 30th of July 2003. The CLIC must contain representatives from five predefined groups: the State, the communities or local government, the industries, the employees and the local population.

The CLIC ensures that citizens are informed on major industrial risks, creates the settings that ensure that public opinion and concerns are taken into account and integrates the public view in projects which aim to prevent industrial accidents. The CLIC officially advises in public inquiries which precede authorisations for the exploitation of industrial establishments.

1.2 *Public information produced by the authorities and industry*

This is an obligation in Europe since the first Seveso directive, which stipulates the contents of the information that the population concerned by major industrial accidents is to receive.

In France, establishments classified Seveso that have external emergency plans, must edit and distribute public information documents to the citizens concerned by these plans. These documents first of all aim to prepare the population in case of emergency, but are also used to promote the industrial risk culture. Industrial companies have been editing brochures with the required information since 1989.

1.3 *Land-use planning documents*

In Europe land-use planning documents must take into account major industrial risks since the Seveso II Directive (1996). In France, citizens have access to local urbanization plans in their community and may be informed about the major technological risks that they are subject to, through the risk zones that are annexed to these plans.

In addition, recent French legislation (law of the 30th of July 2003) obliges the creation of a Technological Risk Prevention Plan (PPRT) around any upper-tier Seveso establishment before the end of July 2008. These plans implement measures with regards to land-use planning and technical procedures which protect the population from major technological risks. The plan prescribes three possible zones around existing establishments: a zone of expropriation in which the proprietors of real-estate are forced to sell and move out—this zone is established when the risk is very high; a zone in which the proprietor (once aware of the risk to which s/he is exposed) may force the Community (Communauté) to buy his/her property—established when the risk is high but not an absolute emergency; a zone in which the Community has the first right to buy the real-estate—established when the risk is intermediate. The State will ensure that the population is informed about these plans, which it established in collaboration with the Communities and industry.

1.4 *News media*

The importance of news media for public information, as well as its influence on public opinion, has been widely demonstrated and recognized. In France, several major accidents have shown that the discourses found in the news media have also changed. Whereas the responsibility of the authorities following major accidents was not widely questioned some thirty years ago, the responsibility of the authorities has gradually become a more important.

Despite the "easy" access to information about major technological risks by a variety of means, several studies have shown that the public deems this information insufficient or not adapted to their concerns. The Institut Européen des Risques (IER), based in the Seine Estuary region, has identified five research topics which may help improve public information on major technological risks and on which it has initiated research or currently coordinates projects.

2 RESEARCH PROJECTS AND PROPOSALS

2.1 *The economic importance of hazardous industries in the regional economy*

There are many considerations which may lead to either favourable or unfavourable public opinion regarding the place that hazardous industries occupy in society: one the one hand the threats to public health and safety and the environment, and the economic benefits, such as employment, etc. on the other hand. Whereas public information on the risks of industrial activities is more readily available, this is not the case for quantified economic information about these activities. Economic research can therefore contribute to the public debate by providing information about the economic impact of hazardous industries.

The IER together with economists from the University of Rouen (Beaumais et al. 1999), have proposed a research project which aims to measure the weight of hazardous industries in regional economy. The project is likely to be financed by the ministry of Environment in 2006, in which case it will concentrate on the Seine Estuary region. The project may also be financed by risk managers and decision makers of the Arc Manche region[1], as part of a regional collaboration recently initiated, in which case it will concentrate on the entire Arc Manche area.

The research will use the input-output model, based on a matrix representation of exchange between the economic entities in a given region. Every sector consumes goods from other sectors in order to create its product and produces goods which are bought by other sectors. In addition, each supplementary demand for goods in a sector will have consequences for the entire local economic structure. Thus, in order to measure the economic value of hazardous industries, the relation that these industries maintain with other industries and with other activity sectors, must be established. These relationships can be partly captured by input-output models, which have already been applied in several domains such as energy, environment and water management policies.

The results will give an approximation of the direct and indirect importance of hazardous industrial sectors for the economy. In this manner, the

[1]The Arc Manche Region is constituted of the coastal areas of Normandy, Picardie and Nord-Pas de Calais (up to the Belgian border).

model can also calculate the economic consequences of a legislative measure that requires companies to put in place special facilities that help retain accidental pollution or that help contain an explosion. The costs of these measures as well as the supplementary investments in companies that will construct these works can be calculated. Generally, the overall impact of the obligatory facilities is calculated in terms of activity, investments and employment and can reveal winning and losing sectors. Furthermore, the model is able to calculate the impact of a certain public policy and allows comparisons between competing technological options and between competing options for regional economic development (such as tourism versus industry).

It would be interesting to extend this research to different regions in Europe in order to have supplementary data which could help to refine the model and test its underlying hypotheses.

2.2 *Mapping of major technological risks*

Maps based on Geographical Information Systems (GIS) constitute a means for helping citizens understand whether and to what extent they are subject to risk, by mapping out risk zone areas. Risk managers have been using GIS for risk management purposes and are increasingly using them as a means of public information, communication and negotiation. The maps are however still far from being perfect, and more geographical research is necessary to further develop the GIS and to clarify the criteria to be selected for the representation of risk zone areas.

Geographers and lawyers from the Universities of Caen and Rouen[2] are currently involved in a project entitled "Emerging new areas of industrial risks—Methodological support for managers and decision-makers of the Seine estuary through social sciences". The project is coordinated by the IER for the French Ministry of Environment. The downstream section of the Seine estuary, including the industrial basins of Le Havre and Port-Jerôme, provides the experimental area for this multidisciplinary approach. The mission commenced in the autumn of 2004 and is of 30 months duration. Major corporations in Le Havre (TOTAL, ATOFINA, etc.), concerned with the subject and members of the «Industrial Risks» Commission of the Chamber of Commerce and Industry, have become actively involved in the project and have set up working groups with the researchers.

The main objective is to develop a method for the mapping of major technological risks, useful to local authorities and risk managers. The method will take into account future issues caused by changes in the legislation and social demand that may lead to alterations in the delimitations of the areas that are considered "risk zones", i.e.:

- recent French legislation (Technological Risk Prevention Plans), which alters the risk zones surrounding upper-tier Seveso establishments
- delimitations defined by new methods of risk assessment that combine "deterministic" and "probabilistic" approaches, (an international survey is scheduled covering the, Belgium, Italy and Switzerland in cooperation with the Observatory of Risk Management Practices of the IER, in order to have precise references of situations related to the use of the probabilistic model)
- the possible transfer of industrial risk to transport systems, ports, marshalling yards and other intermediate stocking points due to stock reductions in industrial plants

The research will then analyse how to take into account these developments in statutory documents and planning policies, how to redefine the roles and responsibilities in relation to the new risk areas and how to share this information with citizens by means of GIS maps.

This project will provide benchmarking with regards to the choices that are made in several European countries for the representation of major technological risks onto maps. Comparisons could also be made with regards to the manner in which local authorities and risk managers perceive the risks due to the transport of hazardous substances.

2.3 *Mapping of the perception of industrial risks and the impact of public information*

Previous sociological and geographical studies have shown that public perceptions of risks vary from place to place and from population to population. Research conducted in the Seine Estuary region in 2001 for instance (Bonnet, 2003), focussed on the vulnerability of the region with regards to industrial risks. The public considered industrial risks as part of the regional vulnerability and this was a result of different attitudes such as conscious or unconscious ignoring of risk, denial of risk or

[2] Eliane Propeck-Zimmerman and Thierry Saint Gérand, geographers at the GEOSYSCOM laboratory, University of Caen, Lower Normandy and members of the MRSH; Emmanuel Bonnet, geographer at the Laboratory 'Hommes Villes Territoires" of the University of Lille 1 and member of the MRSH; Philippe Guillot, lawyer at the University of Rouen; Béatrice Mazja, lawyer at CENTRE, University of Caen, Lower Normandy and member of the MRSH.

simply ignorance of risk. Interestingly, the survey showed that the populations' perception of risk was not necessarily related to the hazard calculated by the authorities and industry. In several cases, populations rejecting risk the most were in reality the least subject to risk. The survey also revealed an overall poor knowledge of the procedures in case of an emergency.

Public information may be improved by adapting it to the perception of risk in a certain area. Geographical research combined with sociological surveys could help to map risk perceptions and thus enable risk mangers and decision makers to give more appropriate public information.

On the basis of the aforementioned study and other research work, the IER together with geographers and sociologists from several universities of the Arc Manche region[3], have proposed a research project that will allow risk managers to adapt public information on industrial risks to the prevailing risk perception in each area. The project is likely to be financed as part of the regional Arc Manche research programme.

First of all a "standard survey", from which the perception of industrial risk of the population can be deduced, will be constructed and distributed in areas of the Arc Manche region selected by random sampling. The perceptions of risks in the Arc Manche region will be analysed at three different territorial levels. Subsequently, the perceptions from populations, but also from local authorities, risk managers, industry and communities, will be mapped out with the help of a GIS and compared with the "real" risk zones. Interviews will be conducted amongst risk managers and intervening authorities, in order to judge the role they play in the risk perception of the populations.

In this way, the population's perceptions of industrial risks are identified, from which the acceptability of these risks in a given area can be deduced. The project will also provide information about the degree of preparedness by the population in case of an emergency. Those in charge of public information will thus be able to adapt the information they provide to the perception of risk in a given area.

By repeating the survey, it would be possible to follow the progress of the public risk perception in time and therefore allow the evaluation of the success of information campaigns. The mapping of risk perceptions will enable the comparison of different regions with regards to the success of their risk management and information policies.

At the European level it would be interesting to compare the risk perceptions in different regions in relation to public information. This may also help to clarify the cultural and historical factors that may play a role in the establishment of the so-called "risk culture".

2.4 *The role of news media in public information provision*

News media are probably one of the most important means by which citizens are informed about major technological risk and its consequences. They are, however, not necessarily purely informative and may express opinions and hypotheses. An interaction exists between the messages conveyed by news media and public opinion; news media may both express as well as influence public opinion. Furthermore, news media could also influence the opinion of risk managers and decision makers and therefore have an impact on the strategy chosen for public information on major technological risks. Through sociological research these complex interactions could be better understood so that more appropriate public information could be provided to citizens.

For this reason, the IER currently finances a research project together with the Dutch Ministry of Environment, entitled "Comparative analysis of news media and the development of risk communication policy after the AZF-disaster in France and the fireworks disaster in the". The duration of the project is one year, from autumn 2005.

This research by sociologists from the University of Caen and the University of Twente[4], will evaluate the influence news media has had on public opinion in the cases of the two accidents, as well as the extent to which the news media influenced governmental policies with regards to public information following the disasters in both countries.

Coverage from written news media about the disasters will be collected from several newspapers, coded and analysed. The focus will be on the actions of governments and their evaluation in the media coverage and on the appeals to governments with respect to the communication with the public about risks. Public opinion data will be collected from regional and national surveys.

By comparing these disasters, both of which received much attention in the news media, it might become clearer how public information is

[3] Emmanuel Bonnet geographer at the University of Lille 1, Anne-Peggy Hellequin, sociologist at the Université du Littoral Côte d'Opale, and Emmanuel Eliot, geographer at the University of Le Havre.

[4] Anne Lalo, psycho-sociologist at the University of Nice and member of the LASAR centre, University of Caen Lower Normandy, Jan Gutteling and Margôt Kuttschreuter, sociologists at the department of Communication Studies, University of Twente.

influenced by national development and implementation of risk communication with the public and by governmental culture. The comparison will therefore consider:

- media coverage of governmental actions and of appeals for policy changes, the analysis will focus on quantitative (e.g. the extent of coverage) and qualitative aspects (e.g. how are governmental actions evaluated, etc.).
- the possible relationship between media-coverage and public opinion data (gathered from surveys)
- the developments and changes in governmental policies with respect to risk communication since the disasters.

Finally, the project should result in a consortium that will formulate a proposal for a European Research grant on the role of the news media in communication on safety and risk issues.

2.5 *Tools for the evaluation of public information concerning industrial risks*

Sociological research work carried out for the French authorities in 1994 (Lalo, 2002), concluded that information about risks provided in brochures edited by Seveso-classified industrial establishments was not always clear to the public; some documents did not give the right information, whereas others were simply poorly understood. The study, which involved a sociological and linguistic analysis, evaluated the compliance of the information with legal demands, the linguistic clarity of the information as well as the (negative or positive) perception the information was likely to induce amongst the population.

While the conclusion reached by this research is similar to that of several other studies, the study, more importantly, revealed the demand by risk managers, in particular industrial companies, for tools for evaluating the quality of public information documents.

The IER and the University of Nice therefore propose a study to produce a tool based on lexical analyses that allows risk managers to evaluate the information that they prepared for the public. Furthermore, the creation of a database of public information documents is proposed, which will allow risk managers in charge of communication and public information to see what has already been done and to learn from past mistakes or successes. By updating the database with new documents, it would also be possible to evaluate the quantitative and qualitative progress of the public information documents. The database would therefore also be accessible to researchers in social sciences who may carry out such analyses.

The database as well as the evaluation tool could be put at the disposal of risk managers on the Internet. The Internet also opens European perspectives; it could make available public information documents from different countries, in order to allow comparisons and research at European level.

3 CONCLUSION

Research in social sciences can contribute to improving information about major technological risks. The IER allows for an interdisciplinary and transverse approach by linking different social science disciplines. The research initiated in the Seine Estuary Region by the IER and several universities, provides tools for authorities and risk managers by anticipating future developments, which may increase risk exposure or result in public refusal of industrial activities. The links the IER has established between social science research and analysis and regional policy provide favourable conditions for cooperation between scientists and decision-makers.

Research projects initiated in the region therefore benefit from a favourable local situation and are ready to be integrated in European research projects. It would be interesting to compare the means by which the public is informed and public risk perceptions in different countries across Europe, in order to set up appropriate benchmarking at the European level.

REFERENCES

Beaumais, O. & Schubert, K. 1999. La modélisation en équilibre général calculable: un regard sur les interactions économie environnement. *Économie Rurale* 251: 25–32.

Bonnet, E. 2003. *L'estuaire de la Seine: un territoire vulnérable face aux risques industriels*. Mappemonde.

Chevé, M. & Congar, R. 2003. La gestion des risques environnementaux en présence d'incertitude et de controverse scientifiques: une interprétation du principe de precaution. *Revue Économique* 54.

Gurabardhi, Z., Gutteling, J.M. & Kuttschreuter, M. 2004. The development of risk communication. *Science Communication* 25: 323–349.

Lalo, A. 2002. Histoire moderne de la sécurité: le cas d'AZF, Le développement durable—Ambitions.

Zimmermann, 1994. *Risque Technologique Majeur. Conditions de Production et Rôle des Documents Cartographiques dans le Processus d'Identification et de Gestion*. Université Louis Pasteur Strasbourg 1.

Risks Challenging Publics, Scientists and Governments – Menoni (ed)
© 2010 Taylor & Francis Group, London, ISBN 978-0-415-58072-4

Whistleblowing, risk communication as core feature of sustainable management

Björn Rohde-Liebenau
Mediator at RCC Risk Communication Concepts, Hamburg, Germany

ABSTRACT: The author shares some of his experience in systematically structuring risk communication in industry, business and the public health sector. The bottom line is that "whistleblowing" is typically the outrage-based, unstructured form of deviating risk communication, which takes place, where a regular, trustworthy risk *communication* cycle is not in place. Where internal risk communication is gagged, there will be external whistleblowing. The phenomenon of retaliation in the context of whistleblower has been covered extensively in research, particularly in the US. This article shows what goes wrong before retaliation starts, before whistleblowing even starts and what can be done to make better use of the existing resources in a collective effort to identify risks early enough. Essentially risk communication is a resource; coming from a whistleblower simply changes it into the emergency state. It is in the best interest of decision makers to listen…

1 INTRODUCTION

A whistleblower is someone who believes he is observing undesirable or even illegal on-goings at or around the workplace (public or private) and feels it would be impossible or unreasonable to report these to immediate supervisors or others directly responsible—e.g. because he has already unsuccessfully tried to do so. The whistleblower then turns to someone he should normally not approach—either much higher up in the hierarchy or even outside the organisation. His intention is to raise awareness and have the "problem" stop. Where does this sort of behaviour fit into modern management theory? Risk management is certainly the place to consider.

2 WHISTLEBLOWING, RISK, RISK MANAGEMENT AND RISK COMMUNICATION

Advertently or not, a whistleblower interferes with core management functions. One such function might be described as sanctioning behaviour inside the organisation, or put differently: the management's power to decide, what and who is right or wrong. At the heart of the whistleblowing phenomenon is the question, whose responsibility it is to decide what is a "risk", and what to do with it. There can be no doubt that risk management in any sort of organisation is a function, the responsibility for which will always rest with the management, even where processes can be delegated. When management shuts out risk information from inside the organisation, it can be perceived as pretending to "already know everything" or as having assessed all risks to such an extent that there cannot possibly exists new information. This certainly blatantly ignores the nature of risk and eventually constitutes a risk information blockade.

The question what constitutes a "risk" is at the very basis of risk assessment. For the purposes of the whistleblowing research it proved useful to define a risk as *a factor that has a tendency to influence a process in the future in a way that is seen as negative in comparison with expectations.* There is the element of uncertainty in the definition, proprietary to any process in the future. And there is the subjectivity of perception and attribution of (negative) value. While there may nearly be a total consensus on many specific risk inducing factors, it is clear that for some general factors the decision may be largely a matter of perception or viewpoint:

- will this have an influence at all?
- positive or negative?
- of which sort?
- how likely?
- to which degree?
- will the damage, if any, be large or negligible?
- is there a danger of recurrence?
- what actually are the root causes?
- is there something we can do about it?
- where do we start to affect the causal chain?
- what exactly should be the preferred action?
- who decides?

- who is responsible for the implementation?
- how should we monitor the further developments?
- etc.

Risk is innate to any enterprise, because an enterprise can itself be defined as a complex of processes with the expectation to reach certain goals in the future but influenced negatively or positively by factors, known or unknown, influenceable or not. An enterprise in this sense may not only be a commercial undertaking but obviously also scientific research—or even our day to day life. We can say there is no enterprise without risk and no creativity without risk. At the same time we have to admit that all that is safe is only in the past—in the sense that anything in the future cannot be risk-free. Because any (positive or negative) development in the future has roots in the past, it makes sense to monitor it and to see, whether its path can be wilfully influenced by us in the present (the point where past turns into future) so that the chances of negative developments are lowered.

Risk has occasionally been defined as hazard plus outrage.[1] The word hazard, which sounds much like 'danger' but contains a certain accident-like element of lack of control,[2] will not be focussed on now. In a simplifying manner we might even say "risk is risk plus outrage". This might signify, risk becomes activated as a risk, when outrage is added—outrage being what the public confuse with risk in the first place. The added value of such definitions for the whistleblowing issue lies in the element of "outrage", which leads back to the element of expectation but also to trust. Outrage is basically the emotional reaction to "disappointed expectations". Trust is made up of a series of experiences that had met (positive) expectations, which means that outrage is also disappointed trust. This shows that outrage by definition has to be expected where risk is the issue—since risk always involves expectations and at least some of the stakeholders expectations will always be disappointed. Outrage may, of course, incite Whistleblowing. Outrage as a potentially negative factor in a future process will, by definition, be a risk in itself. Whistleblowing, as we will see, may also be a risk in itself, both for the whistleblower as well as for the other agents involved. It is necessary, however, not to fall into the trap of trying to "manage" the outrage instead of the risk, but rather work on the risk in order to be able to address the outrage.[3]

3 RISK

3.1 *Risk management*

Risk management is a management process that has to be initiated when a risk is detected and can only be discarded when the risk ceases to exist. Usually risks change and develop. With risks defined above as "influencing factors", we can see that it is usually more likely that an enterprise will disappear, rather than the influencing factor. That means, a risk management process, once started is not so likely to stop for a good reason. When a risk occasionally does seem to disappear, it is usually replaced by a new one: the environment never goes away—but continuously changes.

Risk management is often depicted as a cycle consisting of information, analysis, strategy (-decision), implementation and monitoring—each in relation to the perceived risk. The cycle may consist of arrows pointing from one element to the next, but usually it is not explicitly stated or not even meant that these arrows symbolise the communication that has to take place before risk management can go from one element to the next. Here, it should also be pointed out, that risk management is a cycle only in the sense of a mill or a grinder: it is not turning ineffectively on the spot, since both the influencing environment and the material being influenced change all the time—and sometimes even the pivot of the grinder...

Enterprises need to manage objects in or from the past only inasmuch as this past can influence the future. If this presumption is correct, it means that management is always necessarily (also) risk management, since it always needs to take into account factors that could adversely affect the future of core processes. Management without a conscious approach to these risks is unthinkable.

Delegating the risk management function to a specialist without retaining control over the process and receiving complete feedback regarding any relevant outcome is unthinkable. In many institutions management, while not delegating risk management, and instead taking it primarily as a formal requirement, has risk management mostly in the head of the manager—who believes he already knows all the risks and necessary factors to make his decisions. This is often particularly the case with enterprise founders, who knew every inch of their enterprise's environment at its founding time. Since gradual environmental changes are often hardly perceivable, managing the growth cycles and their crises receives high priority—albeit without particular regard to their nature as risks.

"Communication" regarding "risks" is then only observable, if ever, between the manager and his inner doubts. This will be the same type

[1] See http://www.psandman.com

[2] French „par hazard" meaning „by chance, accidentally," in English also „haphazard"

[3] The original promoter of risk as hazard plus outrage, Peter Sandman has obviously got that right

of manager who claims being "risk friendly"—in other words willing to take on a lot of risks—which is supposedly why he would not want or need any extra information from others. The "know everything" attitude equals sailing into the storm regardless where the wind comes from: courageous but stupid—certainly neither fast nor effective. An organisation can only sail hard on the risk, if it really knows the risk and if it is always alert to any new information, knowing that "the weather may change any time".

3.2 *Risk communication*

Risk communication has its place before a risk management decision is taken and in the course of the monitoring process. This means it is as continuous as the risk management cycle. Communication theory says we cannot "not communicate". If the manager is completely silent about risks or Risk Management, he may be communicating that he doesn't want any extra information or that he believes the risks are taken care of perfectly well. What he means and what the bystanders understand is also part of the communication process. The arguably resulting ambiguity only shows that this communication is less than efficient, not that there is no communication.

However, there is then very little communication regarding the risks itself—only about attitudes, results and the desire not to be bothered. Any risk communication worth mentioning starts when information pertaining to specific risks and answering questions as on the list in the first paragraph) is exchanged. Ideally the communication should take the character of a dialogue—with questions and answers in either direction, evaluating the information and eventually coming to a preliminary risk assessment, which in turn is the basis for a similar process of seeking adequate means to "manage" the risk. The decision, which means to choose and implement, is the final management decision. Doing nothing may indeed be a responsible decision in the course of a well informed risk management process—but only when it is well informed.

"Risk Communication" about the risk management process itself can also be observed. Often, when experts speak about "Risk Communication", they mean the process of passing information from the management or its spokespersons to the public about some of the outcomes of past risk management processes. Strictly speaking, this should be called "External Risk Communication". It is only functional if indeed there has been a prior risk management process to talk about. That means, there can only be meaningful external Risk Communication, if there has been internal Risk Communication. Otherwise it will soon destroy rather than build trust in the public.

Internal risk communication may also be about past risk management processes, then taking on a function similar to *external* risk communication. This is the case e.g. where Management communicates with the intention to convince employees that everything is being done to protect their health in the workplace. Again, that can only be credible and thus effective, if it was preceded by a meaningful risk management process, including the risk communication necessary internally.

"To communicate" means "to share with" or "to make common". This can be a very one-sided affair: the management shares and "makes common" but discourages any reply. Whereas the term 'risk communication' is often only used in the sense of a monologic, top-down information about certain risk management decisions, truly informed risk management requires open, dialogic communication over risk from its outset. Not just because we as humans are capable of relations and dialogue and because it "feels" better, but rather because of the "problem" of perceptions. Risk management relies on continuous assessment, continuous evaluation, leading to decisions, their implementation and monitoring and eventually back to the beginning. That is essentially cybernetic steering—a necessarily dialogic process, even inside one person.[4] This means risk communication is necessary and ought to be dialogic, even or especially where no new information is provided. The efficiency advantage of dialogue over monologue can be seen in any dispute but also in related technical processes like those underlying "fuzzy logics".

Risk management with a sense of accountability cannot neglect any (new) risk information. It is irresponsible to discourage employees to share risk information they believe to hold. Like any other management function, risk management depends on the highest attainable degree of effectiveness (output oriented) and highest efficiency (input or resource oriented). Clearly redundant, trivial or unprocessable information should be avoided, because it threatens both efficiency and effectiveness. However, the presumption of already knowing everything or already having made the best decision is in reality self-deceit and no good pretence for excluding further risk communication from staff. The best way to deal with the prerogatives of efficiency and effectiveness is to let staff know, which sort of information is sought by the management, which is already available, and how

[4] more on these processes in individuals and in society in Stuart Hampshire „Justice is Conflict"

new information should best be presented. This can be put into rules, which will differ from organisation to organisation and which should be produced with as much staff participation as feasible. Such rules should serve as a filter, not as a blockade.

This can only be achieved, if all risk information from employees receives appropriate feedback—possibly guiding towards a more efficient way of communicating next time, but never generally discouraging the provision of information. Of course, the tendency of management to shut itself off from an overflow of sources of risk information is comprehensible. Complexity makes management decisions difficult and can render them unattainable. It is a management responsibility to organise processes in such a way that only important and relevant information influences decisions. To decide which is which is a management task (which may partially be delegated)—but to exclude all or any internal sources totally would be recklessly faulty management. It is also a severe mistake to exclude new risk communication topics. To motivate communication and to steer it to a reasonable extent, management should itself occasionally introduce new issues, in the sense of "this is what we are looking for, but we are also open to anything that has escaped our attention so far…"

Human communication, by its very nature, is highly error sensitive. Language uses words as representations which necessitate a decoding on the receiving side. This process is "fuzzy". Its intelligence but also its inclination to subjectivity and "mistakes" is owed to the fuzziness of its encoding and decoding. Even between two well trained individuals, who always got along well in the past and who are generally well capable of communicating, an escalation of misunderstandings up to a complete breakdown of effective communication is possible and even likely. Furthermore communication is highly trust related; trust being generated by positive expectations repeatedly met in the past.

That is another reason, why it is recommendable to lead by example. In fact, management always leads by example—but sometimes the example deserved some improvement. So, if staff have the repeated experience that its management is genuinely open to new risk related information and even willing to engage in an exchange on differing risk perceptions, the foundation for the amount of trust necessary to allow meaningful risk communication inside the organisation is laid and external whistleblowing will be less and less likely.

4 WHISTLEBLOWING

In order to understand whistleblowing, a clear picture of the circumstances and conditions under which employees engage in risk communication is necessary—as well as in which of such cases they need protection from retaliation and should therefore be entitled to it. Whistleblowing is often initiated by persons who have an immediate knowledge of underlying facts—i.e. can stand witness for the facts. The same could be said about most others engaged in internal Risk Communication. Without exception all studies have shown, that anyone can become a whistleblower and that there are no outstanding traits that make it more predictable that any one person becomes a whistleblower.[5]

We speak of internal or external whistleblowing and so generally designate the addressee of the information. Such a distinction seems useful, because the risk but also the legitimacy of whistleblowers may be different in either case. While internal risk communication is absolutely indispensable, we have seen above that in "proper" external risk communication, it is usually only possible to communicate the outcomes of informed *internal* risk communication processes. Without such processes and circumventing management, risk communication in the form of external whistleblowing is not entirely desirable and there may often be good reasons to question its legitimacy. Therefore the differences in the circumstances and conditions of internal and external whistleblowing shall be explored further in the following sub-paragraphs.

4.1 *Internal whistleblowing*

As can be seen in Chart I below, internal risk communication means an open exchange of information and dialogue between knowledgeable staff, supervisors, experts and management. In case the immediate supervisor or certain specialists (main route) are not able or willing to engage in this type of internal Risk Communication, there should always be a policy to have a second internal addressee (by-pass I) available. If that is the case and an employee feels he cannot come forth to communicate about certain risks with his immediate supervisor, he may be called an internal Whistleblower. There are many reasons, why even such an activity could lead to major harassment. A basic reasoning behind such harassment is the idea that you ought to be loyal to your "team" or your peers and not so much to the organisation: "Right or wrong—my peer group". This attitude reaps havoc, if it is tolerated inside any organisation that is larger than the "peer group". The primary remedy is to motivate supervisors to be mindful of internal risk communication and to have very clear rules enforced that

---

[5] examples and much further reference in C. Fred **Alford**, Whistleblowers: Broken Lives and Organizational Power, Cornell University Press, 2001

make it common and legitimate to by-pass those who are not willing or able to engage in Risk Communication.[6] Internal whistleblowing becomes necessary and almost unconditionally useful, where open risk communication is blocked, because one or all of the potential addressees or risk communication dialogue partners are unwilling or unable to perform (Chart II).

4.2 *External whistleblowing*

Whistleblowing done externally is usually considered to be dangerous or even damaging. In certain instances, however, external whistleblowing may not only be legitimate but even necessary.

In 2004 the national ombudsman of the German Scientific Research Association (DFG – Deutsche Forschungsgesellschaft) received information about scientific malpractice from 45 persons in universities and public research bodies.[7] These institutions all have rules that permit passing this type of sensitive information to the DFG ombudsman. The ombudsman, however, remains "external" to them. Reporting to the Ombudsperson is therefore "legalised" external Whistleblowing.

Usually such external risk communication is seen as illegal, disloyal and dangerous. It may indeed be dangerous in many respects, but it is not necessarily illegal or disloyal. As we can see in Chart III, it becomes as necessary, legitimate and useful as internal Risk Communication, where *all* internal routes can be perceived as blocked, ineffective or unreasonable. When this is the case may vary in different organisations and may also be a matter of perception. It can and should be defined for some individual instances. But it should be clear that there can and will be cases, in which it will be necessary, legitimate, and useful to use external routes for risk communication. A guiding line may be:

- the more acutely dangerous the risk to be reported is and
- the more blocked internal routes of risk communication may seem,
- the more the public interest outweighs the private interest of the whistleblower or the organisation,

the more necessary, legitimate, and useful will be external Whistleblowing.

The process of setting up specific rules for specific organisations will be more effective to the degree staff is involved. When the rules have proven to be feasible and effective inside the organisation, reporting on these rules and other essential features of the risk management process will help building outside trust. Reporting on the relevant findings and measures chosen in risk management—and consulting over them with the public could be considered a functional equivalent of external Whistleblowing. It is rather what makes external whistleblowing superfluous.

Clearly, cases cannot only be imagined, but do happen, where somebody engages wilfully in extortion, uses real or fake risk information for solely private purposes. Similarly, somebody else sees phantoms; or sees everything right but overlooks the possibility of internal risk communication before going external to blow the whistle there. It is suggested here that nothing *special* should happen with these people. There should be clear rules as to who will sanction what sort of behaviour according to which degree of culpability etc. Effective (humane) resource management will probably call for "recovery" of this human resource rather than "ditching" it.[8]

The rules themselves should not have a chilling effect but rather encourage bringing up relevant risk information, which would otherwise not be brought to the attention of a person willing and able to engage in the management of the respective risk. This means rules must protect any whistleblower from harassment, i.e. illegitimate sanctions and "mobbing" by people who

- have not fully evaluated the case,
- have not given the whistleblower his/her full right to be heard,
- or even refused listening to the risk communication of the whistleblower
- are not responsible for sanctioning, etc.

It is important to keep in mind that even legitimate sanctions against a "pseudo" whistleblower have an effect on other staff. They should not have a chilling effect on legitimate internal whistleblowing.

4.3 *Supportive factors*

Pro-whistleblowing and pro-risk-communication efforts support each other from two different angles Parts of that can be ascribed to organisations like Transparency International,[9] who promote peer review and best practice related to ethics and transparency. Transparency International has realised early on that anti-corruption efforts often depend

[6] This lack of willingness or capacity obviously in itself another risk worth assessing...

[7] Prof. Dr.med. Ulrike Beisiegel, spokesperson of this body, considers the real number of relevant cases much higher: Financial Times Deutschland, August 24, 2005, http://www.ftd.de/rd/19410.html

[8] much more and further reference at www.risk-communication.de/en/index.htm

[9] www.transparency.org

on Whistleblowers. Corruption is known as a secretive crime, where only insiders can help shedding light on the scene. Regular risk communication is only theoretical in these issues is unthinkable for most "observers" because it is simply too dangerous—and often not clear enough that the immediate supervisor is not also involved. Therefore internal or even external whistleblowing will be necessary. This observation applies to all collusive situations. Realising this induces joining efforts to improve the situation from the politico-legal and/or managerial cultural side. Such efforts can be observed in the science and research communities.[10]

Even risk management was once, not too long ago, "forced" onto the businesses.[11] Then it became clear that much of what the new law demands is already practice and that the rest also makes sense. What matters here is the mechanism. Even a formal approach is better than nothing, if it forces the management to think about the issue. Any deeper reflection coupled with peer experience should lead to the conclusion that purely formal measures do not serve a purpose and are therefore absolutely expensive compared to the costs of substantial compliance: simple structural and cultural changes that can lead to enormous savings, a more loyal internal culture, more trust on the stakeholders' side and foremost a better understanding of the risk environment.

Of course, it is also the general tendency toward self-regulating systems that favours better internal risk communication, since it is an essential element of effective self-regulation. Furthermore the promotion of a corporate ethical identity, corporate ethical values, excellence and competition in these fields also suggests to look at the risk communication culture. Quality Management and Human Resource Management may be just as reliant on such a culture as risk management itself. Again, all this obviously does not only apply to commercial operations but also to administration and scientific bodies. Soon it may be just as discrediting to work for or study at a university that does not totally comply with standards of scientific integrity including rules for effective internal Risk Communication, as today it is "impossible" to ride for a bicycle team with a doping scandal.

[10] well established in the Anglo-Saxon world, e.g. with the ORI—Office of Research Integrity—in the U.S., but also institutions like the DFG Ombudsman in Germany. For much further reference and critical analysis see also www.apsab.ch

[11] Gesetz zur Kontrolle und Transparenz im Unternehmensbereich of 27 April 1998, introducing the obligation for a comprehensive risk management system in § 91 II AktG—followed by several other laws strengthening corporate control. Laws to the same effect now exist across Europe.

4.4 *Obstacles to constructive internal risk communication*

There are several distinct obstacles to constructive internal risk communication that can be observed. On the personal level, there are among others:

2.1. the anxiousness of the recipient to loose face, if a "mistake" is detected,
2.2. the age-old tendency to "kill" the conveyor of the negative message,
2.3. the fixed idea that raising risk awareness is disloyal either against the supervisor and/or the entire organisation,
2.4. the reluctance to see the message as opportunity for improvement,
2.5. the human mental difficulty to perceive risk and opportunity at the same time.

These obstacles can be overcome. Otherwise we would have *no* risk communication—and in its absence probably many disasters or much less social and commercial activities could take place. But it is probably for the above reasons that it is still widely accepted to discriminate unfairly against whistleblowers.

5 CONCLUSION

Whistleblowing in itself—especially in its external variety—creates undesirable additional hazards for all involved stakeholders. Some of the more obvious additional risks are directed against reputation, the "human resources" (of the whistleblower and others) in the literal sense, but also against values such as trust, loyalty and eventually the very existence of an organisation. While external whistleblowing may be undesirable from the perspective of state of the art risk management, it seems indispensable, where internal risk communication is ineffective. Besides obvious ethical reasons, this is also an objective reason, why even external whistleblowers need protection against unfair treatment.

A famous classic in sociology is titled "Exit, Voice and Loyalty" Subtitle: "Responses to Decline in Firms, Organizations and States". Hirschman's book is undoubtedly a work of a genius. But it is 35 years old and organisations and their environment have changed. So could Hirschman's idea be applied to whistleblowing? We cannot be so optimistic about *exit* as an alternative any longer. The world and certainly the job market has become a much smaller place. In a sense whistleblowing can be seen as a sign of decline in an organisation. External whistleblowing is not what Hirschman meant by *exit*. Hirschman described whistleblowing phenomena as *voice*. There is not enough voice and it may be debatable, whether or not whistleblowing

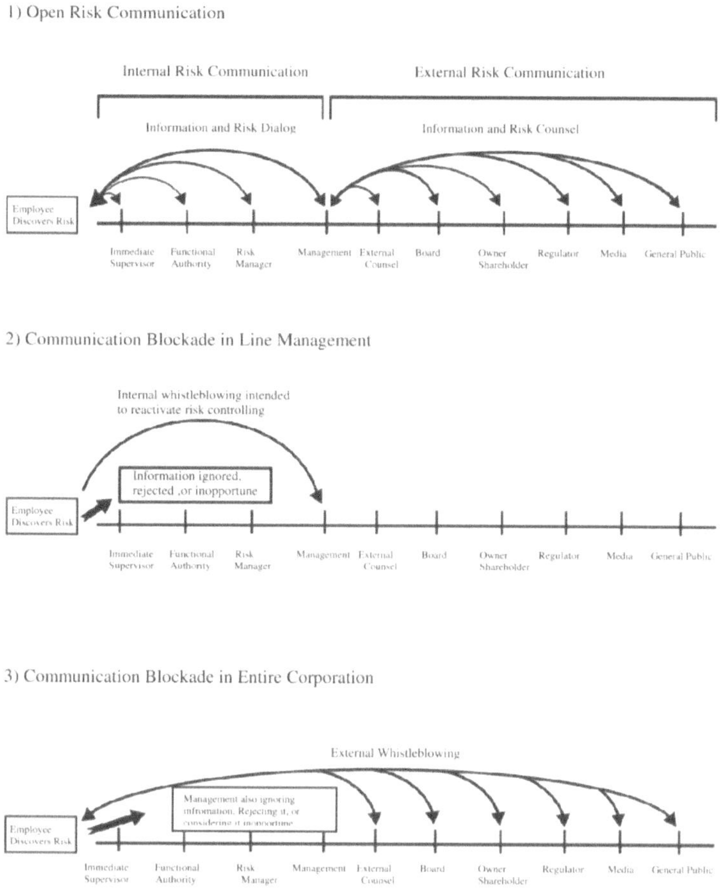

is productive as *voice*. However, whistleblowing as defined in this article is not disloyal—whether or not it is *voice*. If whistleblowing is external risk communication from persons who have reasons to believe that they have no where else to go and a serious risk out of or in relation with their organisation requires action, this attitude is loyal and in fact such whistleblowing may save an organisation from much greater harm.

Loyal, coming from the French "Loi" does not equal "right or wrong—my country" but rather "let's keep the law, even if there are temptations otherwise". *Voice* can be even more loyal when the concern can be voiced internally, but only if there is a trustworthy addressee with the power to effect changes, where necessary. Trust comes from repeatedly met positive expectations. One of the primary expectations in the context of risk communication will be that conscientious risk communication deserves a conscientious assessment and a constructive answer. If this is not experienced repeatedly, no matter what rules will not be able to cultivate trust. However it should also be experienced that harassing whistleblowers or otherwise treating them unfairly, is a serious breach of workplace duties and will be sanctioned accordingly.

Without intra-organisational risk communication of the dialogic type, there will be no Accountability, no Sustainability and no effective Risk Management. Internal Risk Communication, is as precarious as any type of human communication. In a crisis or conflict situation help will probably be indispensable and should be made acceptable- and often that will inevitably be an external help.

REFERENCES

Alford, C.F. 2001. *Whistleblowers: Broken Lives and Organizational Power*. Cornell University Press.

Prof. Dr. med. Beisiegel U., 2005. "Spokesperson of this body, considers the real number of relevant cases much higher." *Financial Times*, Deutschland, August 24, 2005.

Risks Challenging Publics, Scientists and Governments – Menoni (ed)
 ISBN 978-0-415-58072-4

The contribution of economic and public-private cooperation instruments to the prevention and mitigation of natural hazards

G. Pesaro
DIAP – Politecnico di Milano, Milan, Italy

ABSTRACT: Following an economic perspective the damage provoked by a major natural hazard means a loss of development resources for a territorial system as a whole. Public policies should therefore focus on damage prevention in order to minimise as much as possible the loss of territorial resources in case a major natural event occurs, particularly in presence of non-renewable resources. This means to draw attention not only to civil protection and damage management but also to state action goals and tools to enhance prevention and ex ante damage mitigation. The essay offers a brief survey of economic, market based and public-private cooperation instruments, some already tested in the management of natural hazards and some borrowed from successful experiences in the environmental protection field, for each instrument key features, functioning model and conditions and potentials of implementation in hazard and damage prevention field will be discussed.

1 INTRODUCTION

This essay aims to discuss the possible benefits originating by the implementation of a selection of economic instruments to enhance the public systems of management and prevention of natural hazards.

To obtain such goals the proposal is to enrich the existing public intervention systems with economic and participation based tools, able to actively involve an enlarged system of territorial stakeholders and to improve the performance of public action by enhancing capabilities to better face natural hazards and reduce potential damage.

The latter can be considered a loss of resources for the territory as a whole, both because some territorial resources cannot be substituted with others (often financial), like historical cultural heritage or particular natural ecosystems, or because the reimbursements (by the state or the insurance system) and reconstruction financial needs (by public or private territorial stakeholders) absorb economic means, in some cases slowing down territorial development at the local level. Moreover past experiences show that the damage can affect the territorial systems in the medium and long period, with an amplification of the damage impacts over the years. For these reasons the evaluation of different intervention options in hazard and damage prevention should become a more stable public decision support tool, based on the comparison between costs of intervention and benefits resulting from the reduction of the total potential damage for each different option.

If damage caused by a natural disaster can be regarded as a loss of territorial resources, part of which are not reproducible, hazard mitigation goals should primarily refer to ex ante interventions, able to reduce consequences if the disaster occurs. If it is therefore clear that in a short period perspective attention is focused on civil protection and damage management, in a long period perspective prevention goals should be seen as the best action approach. This to reduce the total territorial loss in case of disaster, of course not undervaluing the importance of better organize disaster ex post intervention measures to mitigate damage impacts. But how to gain such a result? And how to intervene on the wide plurality of different territories, natural hazard sources, subjects and activities which characterises contemporary societies?

It seems necessary to increase the general safety degree by widening the policy tool box in order to both influence territorial components and reach a variety of actors. Reference is made to interventions aiming to reduce the dangerousness of hazard sources and/or exposition and vulnerability of territorial objects and stakeholders. The firsts consist in structural interventions on hazard sources by the state or other public bodies, where possible (i.e. hydrological and geological hazards). The seconds, normally non structural interventions, should directly involve territorial stakeholders and activities, pushing communities to a more proactive behaviour facing hazards. In this case interventions can involve a variety of elements: localization of residential and productive settlements, planning and management of built areas development, diffusion of best practises in building construction and maintenance, economic activities features facing different kinds of hazard, social attitudes and hazard and damage perception.

Long term policies have to be considered, as such ambitious intervention goals require not only rather diversified means, sometimes alternative sometimes integrated, but also because time is needed to produce positive effects from a systemic point of view. And it has been already widely recognised that structural measures only, that is those acting mainly on the hazard, are insufficient to provide a satisfactory protection in the long run. To be effective they must be used in combination with non-structural measures, those affecting both exposure and vulnerability (Menoni, 2004).

This is the framework in which public action goals, from the one hand, and economic and public-private cooperation based instruments, on the other, will be analysed. Tools will be compared by functioning features and conditions and reachable goals, this underlining their potentials in terms of prevention goals achievement and the possible role of private sector besides the public one. Some of the selected instruments have already been experimented in the natural hazards field, like insurance and scope taxes. Others come from the natural environment protection field, which shows interesting analogies with natural hazards prevention policies, like subsidized credit, incentives, voluntary instruments, public-private cooperation. In this second case tools are in a sense "borrowed" from successful experiences in the environmental protection field and hypothesis are made for their transfer and adaptation to the natural hazard field.

2 ADAPTING ECONOMIC AND PUBLIC-PRIVATE COOPERATION TOOLS TO NATURAL HAZARD MANAGEMENT

Governance and management of natural hazards mean production of safety, protection of the community by prevention interventions, organization and management of the consequences of natural disasters (civil protection), re-establishment and mitigation of the impacts of damage on the territory. And this of course considering all exposed and damaged components of a community at the same time (non-rivalry), without excluding anyone, no matter what gender or social conditions (non-excludability). The production of safety from natural hazards can therefore be legitimately regarded as a public good.

Such a hypothesis implies the failure of the market in safety production. The economic system alone cannot guarantee the protection from natural hazards to the entire community at the needed conditions: some form of public intervention is required.

If safety from natural hazards is a public good an hypothesis can therefore be made: it should be possible to apply, under appropriate conditions, the tools already available for the governance and management of other public goods, and particularly in the field of natural environment protection.

In the latter many categories of tools are nowadays available to be added to command and control (laws and regulations). These are the result of an evolution of the tool box produced by two main dynamics. First, the increasing complexity characterising environmental depletion phenomena. Second, the need for achieving ambitious results in natural environment and territorial protection. This involving activities and behaviours of an increasing number and variety of territorial stakeholders (from institutional to economic actors and entire communities), whose production and consumption models cause environmental negative impacts. The increased complexity makes it immensely difficult to design and implement laws and widely valid regulations, requiring additional—not alternative - tools and more flexible and suitable solutions to be applied to peculiar situations and territorial actors in a certain period (Pesaro, 2001). Such a territorial framework is relevant also the natural hazard field, since a wide range of actors should be involved in prevention strategy is a rather similar way to the case of environmental protection.

Each instrument can be implemented considering its potentials in obtaining given performance and it can be integrated with others to multiply the systemic effects of intervention and in the meantime reduce the implementation constraints.

In the case herewith examined it is obviously necessary to re-orientate environmental policy instruments to obtain effective results in the natural hazard and damage prevention field. An "exercise" requiring capabilities in recognizing correspondences and differences between the two implementation areas. This both with reference to the nature of hazard and damage themselves and the system of roles, activities and responsibilities of the whole of the territorial stakeholders. Moreover the public approach to safety problem—dimensions and action fields of the state interventions—and the perception of natural hazards by communities are very important and can make the difference.

3 TOOLS ALREADY USED OR TESTED IN NATURAL HAZARD MANAGEMENT

3.1 *Insurance*

In general terms, an insurance system allows a number of subjects to reduce impacts on their welfare due to negative probabilistic events. Insurance can be regarded as a device for indemnifying or guaranteeing an individual against loss produced by negative events, this by pooling economic resources in "common" funds. In case a negative event occurs, reimbursements are made from a fund to which many individuals exposed to the hazard have contributed certain specified amounts (premiums). Subjects reduce their risk by periodically paying amounts calculated on the basis of hazard potential incidence and on the value of exposed objects and activities.

At the territorial level this means to mitigate damage incidence on a community and to reduce negative impacts on collective welfare by providing financial resources to substitute those lost in the event. Insurance is also a device to shorten time needed for recovery from damage.

We want here to underline the potential role of private insurance market working in parallel with the state for reducing the pressure of reimbursements on public expenditure. This in the perspective of the diffusion of private insurance against natural hazards, as the insurance system can effectively operate only if subscribers are numerous and hazard degrees and exposure are different in different territorial contexts, so as to achieve risk-pooling goals.

A way to obtain this is to make insurance for natural hazards compulsory. But in such a case insurance takes the form of a purpose tax (see next paragraph) and functions as a risk-sharing device. This is for instance the French policy option.

Interesting for its contribution to reduce public expenditure, the central problem of the compulsory insurance is that it functions as a purpose tax but does not attain nor the contributive capacity principle (whose having more pay more) that the benefit principle (only the victim get benefits). This because insurance premiums are correlated to objects and activities values but not to any exposition or vulnerability degree (the hazard premium just amounts for the 12% of the fire policy) and because risk-sharing purposes require the entire community's involvement.

Insurance is one of the most known and stable instruments to cope with hazards, but in the logic proposed in this essay it seems not effective in producing the public good "safety". At present it does not contribute to ex ante damage prevention. Premium estimate is still based on the maximum possible damage, that is the loss of the resource (buildings and economic activities under specific conditions) and does not change with reference to different hazard exposure degrees not to possible interventions on vulnerability by clients.

To overcome such a limit insurance companies should recognize different hazard exposure degrees and interventions to reduce exposition and vulnerability by private actors. This rewarding "lower hazard clients" by reducing insurance costs (Kunreuther, 2004), like in case of fire, theft or car insurance policies. If premiums are proportional to hazard exposure and vulnerability degree, subscribers will find interest in increase private safety degree (in the achievable extent) obtaining advantages by reducing both hazard incidence and insurance costs. At such conditions the instrument could contribute in achieving also hazard prevention goals.

Implementation conditions and functioning limits can be observed in the following:

- difficulties in defining annual insurance premiums, specially with reference to uncertainty characterizing potential damage estimation for economic activities;
- need for a very large number of insurance subscribers in different territories to share risk among subjects exposed to different hazard degrees;
- high risk of adverse selection, which makes the natural hazard insurance cover a less interesting market for insurance companies. This because the more exposed subjects are more likely to subscribe, increasing the economic risk degree for companies;
- in case of non-renewable territorial resources, like for example cultural heritage, insurance is not satisfactory at all, as the resource, once lost, cannot be substituted. The insurance can only provide for a kind of compensation but only to a very limited extent compared to the original values involved.

3.2 *Purpose taxes*

From the economic theory perspective taxation is an instrument used by the state to finance public expenditure. There are many different types of taxes and their composition define each national taxation system, mainly related to state main goals and state-market relationship. In the natural hazard field tax revenues finance state activities implying: regulation, civil protection, prevention and ex post damage mitigation.

It is important to distinguish two main categories: general taxation and purpose taxation.

In the first case population pays taxes, normally in a proportional and progressive form. A part of the amount is subsequently employed in activities aiming to produce safety from hazard and damage management (from civil protection to damage reimbursements). In the second case the withdrawal is directly linked to uses. Purpose taxes to cope with natural hazards can be implemented involving only the users of public safety services or the whole community, in this second hypothesis achieving risk and loss sharing goals. Very generally speaking, mixed options between general and purpose taxation depend from each country's welfare and choices of tolerable risk transfer and risk sharing options.

Purpose taxation also aims to report to the community that some individual behaviours, like localization choices or building vulnerability, can imply collective costs. Hazardous localizations or high vulnerable territorial elements mean an increase in prevention intervention, civil protection and damage mitigation means, that is collective costs to be financed by public expenditure. Purpose taxes make clearer the link between the existence of hazard exposure and its costs.

The implementation of purpose taxes in the domain of natural hazard can therefore obtain two results. On the one hand the creation of financial funds to pay for the recovery of territories after a disaster and to prevent them from damage in the future. On the other a disincentive for potentially prejudicial behaviours to public safety quality, since the tax means a reduction of usable income for households and an additional cost for economic activities. An important incentive effect which could be achieved by purpose taxes is therefore to persuade territorial stakeholders to change their behaviours in facing natural hazards: this because persisting in the hazardous path means additional costs due to the tax. As in the case of insurance premiums proportional to hazard degree, territorial stakeholders will find advantages from two points of view: reduction of tax payments and increase in private safety.

Implementation conditions and functioning limits can be observed in the following:

- difficulties in the control of the real use of the tax revenues by public institutions;
- low social acceptability of purpose taxes in natural hazards field. This for two main reasons. First because natural hazards do not—directly—depend on human activities and because being exposed to risk is not every time the result of a choice (i.e. changing in territorial or natural environment features). Second because in many countries tax pressure is already high and new taxes could not easily be implemented;
- encouraging population and economic activities to abandon the most hazardous territories because of high hazard costs can produce potential distortion effects: for instance the increase of existing depopulation trends, the decrease of land management activities and the loss of part of the built cultural heritage (in Italy, for instance, in the Alps and other mountain areas);
- difficulties in defining of tax amounts, as there are still problems in assessing the total risk to which the population and economic activities are exposed.
- difficulties of the tax system to adapt to hazard dynamics. Hazard features as well as typologies and territorial incidence degree can change quite rapidly, while the taxation revision requires time, political and social negotiation.

4 ENVIRONMENTAL POLICY INSTRUMENTS APPLICABLE IN NATURAL HAZARD MANAGEMENT

4.1 *Incentives, subsidies, disincentives and subsidized credit*

In the environmental field, incentives are direct allocations of funds or fiscal advantages offered to the market or households to permanently change their environmental performances. Subsidies are direct allocations of funds to compensate both costs of interventions necessary to reduce environmental depletion and pollution or the abandonment of not environmentally sound behaviours. They can therefore be regarded as forms of economic subsidy by the state which, through its governmental institutions, aims to reduce environmental pressures by private actors, both households and economic actors.

Disincentives, on the contrary, are forms of tax increase or constraints to the access to public services and funds, aiming to reduce advantages deriving from not environmentally sound production and consumption models by territorial actors. Disincentives therefore contribute to internalize social costs and, in a sense, assign a value to the use of environmental resources.

In both cases actors can voluntarily decide if and have to intervene: in the case of incentives they are advantaged because of public financial support; in the case of disincentives they can achieve a stable reduction of costs within a medium/long period (in the form of tax reduction or access to public services and funds).

Such mechanisms could effectively be implemented also in the hazard management field, for instance to promote voluntary reduction of exposure or promote best practices in hazard and vulnerability reduction.

From a collective point of view this means:

- promotion of a hazard prevention culture and enhancement of territorial system's safety by inducing individual/private proactive behaviours;
- potential reduction of the hazard incidence on territorial damage in case of natural disaster;
- enhancement of mitigation interventions in territorial areas with economic problems which in many cases are also the most vulnerable to hazards because of the lower degree of both ordinary building maintenance by owners and of prevention interventions by public institutions;
- enhancement of knowledge regarding the subjects' preferences, community's hazard perception, exposed activities and system's vulnerability. In this case—hazardous conditions and preferences are revealed as in the form of demand for the access to public funds and services in order to reduce exposure or vulnerability and for a detaxation following adequate interventions;
- enhancement of knowledge useful to make estimations of hazards and expected damages.

Flexible and not much invasive, these instruments could be particularly effective where hazard is not very high but frequent. In these cases, minor, simple but permanent interventions can strengthen safety of buildings, home devices and activities in the long period.

At present, incentives and subsidies look more effective than disincentives in obtaining vulnerability reduction performances by households.

Regarding the reduction of exposure, incentives seem partially successful in promoting delocalization from hazardous areas.

Fiscal disincentives or high urbanization charges could, on the other hand, be effective to control future localization choices and to strengthen territorial planning instruments by integrating constraints. In the long period, high localization costs (taxes and urban charges) can, from the one hand, reduce the number of subjects choosing a given site to localize their activities (residential or economic ones), that is exposition to hazards and need for state protection (social costs internalization); on the other hand they can produce financial means to be employed in public intervention for hazard prevention (like purpose taxes).

Differently from incentives and subsidies, implying a state direct contribution, by subsidized credit the state intervenes to reduce loan interests and/or extend repayment periods compared to the normal market conditions. It therefore reduces the costs and/or the burden of private investments from the one hand and the public funding needed to promote action from the other.

Implementation conditions and functioning limits can be observed in the following:

- it is anyhow difficult to make communities and individuals aware of their responsibilities in hazard management and understand the relationship between private/public action and expenditure;
- fiscal disincentives and other public charges (taxation related) can produce social acceptability problems;
- they seem feeble in coping with illegal behaviours related to existing laws on land use and planning;
- their effective implementation depends also on the presence of experts to both guide private initiatives and to control them from the project to the implementation phases.

4.2 *Public-private cooperation instruments*

In the environmental field, the diffusion of voluntary instruments can be considered the result of innovation in environmental policies, requiring a new type of interaction between public and private actors (Pesaro, 2001). As the environmental problems complexity increases, an evolution of interaction model occurs. The resulting policy implementation structure is an effect of the participation and cooperation degree of a number of actors (public bodies at the national and local level, economic actors, citizens and environmental associations etc.). Recent policy evolution shows a tendency of public subjects to actively involve the society as a whole and emphasis is placed on the definition of sets of long term objectives with a wide political and social base.

Voluntary instruments can be therefore defined as instruments offered by public subjects to the territorial system in order to enhance the present policy action and build direct relationships with households and economic actors. Territorial actors decide whether or not to participate in the proposed tool schemes.

Two categories can be distinguished: voluntary adherence schemes and other cooperation and participation schemes.

In voluntary adherence schemes, public actors define action strategies and general operational schemes in order to achieve a well defined set of policy goals. Private subjects decide whether to enter in the proposed schemes, by a decision making process based on their characteristics, interests and action targets. The tool becomes operational through specific adaptation projects, developed by each private actor to meet stated policy goals. Examples in the environmental field are the European Union's EMAS (Environmental Management and Audit Scheme) and the international ISO 14001 (International Standard Organization environmental quality assessment certification) regarding production processes, and the EU Ecolabel, a set of assessment procedures to obtain an environmental quality label on products.

The main strength is the existence of reference procedure schemes already developed by public stakeholders with reference to a range of different possible production processes, products and territorial activities. Moreover during the working phases adhering stakeholders have access to knowledge and technical aid by experts. The final certification can also offer positive external image impacts.

More widely, cooperation, is based on formal and informal interaction forms and promotes consensus building activities also outside "traditional" official sites (parliament halls, courtrooms and administrative offices). One of the more interesting strengths of this instrument is the capability to create new permanent negotiation and interaction sites and to mobilize private resources—economic means, capabilities, knowledge. Cooperation, moreover, contributes to enhance the systemic action by a better monitoring of the territorial system through a regular public-private information exchange.

These instruments could be mainly used to promote risk prevention culture and to enhance and create new knowledge on exposure and vulnerability.

Stable participation and public-private cooperation can therefore support policy decision making processes in the hazard management and prevention field as well as in integrating hazard prevention culture and practice in other sectoral policies.

To obtain the best possible results from these tools, they should be implemented just after a natural disaster has occurred, when public concern on the problem is still high.

Once established, a permanent participation and public-private interaction structure could contribute to maintain high attention on hazard matters and to diffuse hazard prevention and damage mitigation best practices. A stable periodic information diffusion itself about total damage direct and social costs in the short, medium and long period could encourage communities towards more proactive behaviours to cope with hazards.

Finally, hazard prevention matters could be advantageously integrated in existing environmental interaction structures, like Agenda21, where public-private cooperation and participation have already been successfully experimented in structural territorial stakeholders involvement. This would call attention to the links among environmental quality, land fruition models and damage incidence and perhaps push to internalize hazard localization externalities from both environmental and safety perspectives.

Voluntary adhesion schemes seem more effective than other cooperation models in implementing far-reaching interventions, involving a large number of stakeholders. It is therefore easier to involve private subjects in already existing action schemes, where potential solutions and action options are already envisaged to meet with the different needs. The more schemes are able to involve all active actors in prevention and civil protection fields, the more the tool will be effective.

An implementation hypothesis could be the development of projects aimed at making the population aware of all different hazards existing in their territory and to diffuse prevention culture. Public actors could offer professional expertise at no cost to assess the hazard degree in each individual case and suggest possible intervention options to reduce expected damage.

In the case of critical hazardous situations, specially involving economic actors, voluntary agreements could contribute to lower hazard degree by promoting and sustaining private action and investments. For instance in cases where the presence itself of a production activity could significantly amplify total territorial damage in case of natural disaster, or where the temporary or permanently interruption or the closing of a certain economic activity can heavily damage the entire community in the medium/long term perspective (income loss and potential public services interruptions).

Implementation conditions and functioning limits can be observed in the following:

- the adhesion to schemes can reveal the need for major intervention and high expenses to enhance safety: this is an advantage to the entire community, on the one hand, but could however act as a deterrent to the participation of individual households and medium/little firms on the other hand;
- need for stable regular audits to check certifications;
- need for good cooperation and negotiation capabilities by public and private stakeholders need to guarantee the transparency of the decision making process.
- risk of "regulator capture" in presence of asymmetric information: private actors could have more information and knowledge about their real situation and action power then the public counterpart, causing potential asymmetric advantages;
- It can be difficult to persuade communities to pro-active cooperation and involvement, as the expected by products can be either mitigation

expenses on reduced value of goals exposed to hazards.

5 TOWARDS A MIX OF INSTRUMENTS TO ENHANCE HAZARD PREVENTION POLICIES

To overcome implementation conditions and functioning limits, it may be interesting to imagine mix of tools able to strengthen the public action by involving, more or less voluntarily, private actors.

Prior to this, in tables 1 and 2, a summary of possible uses goals and effectiveness conditions are shown.

Mix of different tools for example insurance and incentives, private contribute to reach two main hazard management goals: the enhancement of damage prevention and the transfer of at least a portion of the risk on the private market, reducing damage pressures on public finance.

Hazard insurance subscribers would have access to subsidized credit (or even incentives, as follows) to intervene on private vulnerability only if adhering to programs offering expertise, guidelines and monitoring of final results.

Very interesting results should be reached by mixes of incentives, subsidies or subsidized credit and voluntary instruments. Coupling voluntary intervention schemes and agreements with financial support tools could contribute to overcome tools constraints and to promote proactive behaviours towards hazard prevention. On the one hand the participation to schemes or the willingness to negotiate agreements could be enhanced by advantages in terms of economic means to finance (at least partially) private interventions. On the other hand, public financial effort would be more effective since public subsidies to private stakeholders would be conditioned by the participation to programs controlled by technical and scientific expertise and by the monitoring of the implementation phases.

In general terms it is possible to observe that all voluntary forms of public-private interaction and cooperation can be strengthen by the existence of an economic incentive system.

Moreover, it is very important to underline that voluntary instruments and cooperation appear particularly powerful when coping with specific local situations (Pesaro, 2001). The integration with financial aid would contribute to concentrate funds on specific solutions for specific territories, involving local stakeholders and selecting, making interventions more effective.

The integration between the various instruments mentioned above, and the more traditional ones (command and control, insurance and taxes), can help to diffuse shared-responsibility approaches beside hazard prevention culture, that is a growing awareness of the role of private actors, together with public, in reducing total territorial hazard and expected damage. Consensus since discussion and knowledge dissemination

Table 1. Possible uses of the territorial damage mitigation tools.

	Insurances	Scope taxes	Incentives, subsidies and disincentives	Voluntary schemes and public-private co-operattion
Use goals				
• reduction of vulnerability	(X)[1]		X	X
• reduction of exposed territorial stakeholders and objects	(X)[1]	X	X	
• mitigation of damage impact	X	X	X	X
• internalization of settlements and activities localization externalities		X	X	X
• revelation of preferences, more dangerous conditions and hazard aversion degree of private subjects			X	X
• involvement of private subjects in best practices implementation and information diffusion			X	X
• promotion of prevention culture at territorial system level			X	X
• financing of public activities in the field of hazard management, damage prevention, civil protection organization and management		X		

In the case that insurances, in the future, should arrive to include advantages (premium reductions) for vulnerability or exposure reductions

Table 2. Effectiveness conditions in the functioning of territorial damage mitigation tools.

	Insurances	Scope taxes	Incentives, subsidies and disincentives	Voluntary schemes and public-private co-operattion
• Effectiveness conditions				
• based on different vulnerability conditions	X		X	
• need the direct/active involvement and awareness of territorial governance subjects	X		X	X
• social acceptability		X		
• use for prevention goals (interventions for hazard reduction)		X	X	X
• must be integrated into a damage prevention strategy			X	X
• Regarding specific projects with measurable results			X	X
Effectiveness on settlements localization and shape and on territorial stakeholders' behaviours				
• on existing buildings and territorial activities	(X)[1]	X	X	X
• on the choices for future settlements and activities	(X)[1]	X		X

In the case that case insurances, in the future, should arrive to include advantages (premium reductions) for vulnerability or exposure reductions.

can highlight the role of individual private performances in both increasing hazard and potential damage, if a disaster occurs, and in reducing them by adequate interventions. And this entails: the idea that safety promotion and hazard prevention are duties of the whole community and not only of the state; the consequent idea that it is important to find public but also private funds to reduce the total potential territorial loss in case of disaster the potential increase in the acceptability of new territorial and land use regulation.

From all discussed above it is however clear that the capability to design strategies and the choice for different mixes of tools mainly depend on the knowledge of, at least, the following: typologies of hazards, their severity and frequency; territorial physical characteristics; exposure and vulnerability of a wide range of different territorial stakeholders and activities; short, medium and long period expected damage; short, medium and long period expected results from interventions; technical and scientific expertise in reducing exposition and vulnerability. Moreover the choice for the "right" tool mix also depends on the careful setting of the expected results in the long run and on an intervention profitability analysis, that is on the capability of investments to obtain the expected result (hazard prevention and/or damage mitigation) in the most possible effective and efficient way. New knowledge building and the improvement of existing data sources should be therefore considered as a priority for public institutions with risk mitigation responsibilities.

REFERENCES

Bourg, D. & Schlegel, J.L. 2001. *Parer aux risques de demain. Le principe de précaution*. Paris: Seuil.

Cellerino, R. 2004. *L'Italia delle alluvioni*, Milan: FrancoAngeli.

Cochrane, H. 2003. Economic loss: myth and measurement, in European Commission. *In search off a common methodology on damage estimation*, Workshop Proceedings, European Commission - DG Joint Research Centre, EUR 20997 EN European Communities.

Cole, S. 2003. Protection, risk and disaster in economic networks, in European Commission. *In search off a common methodology on damage estimation*, Workshop Proceedings, European Commission – DG Joint Research Centre, EUR 20997 EN European Communities.

Commission of the European Communities, 2004. *Green Paper on public-private partnerships and Community law on public contracts and concessions*, COM(2004)327, Brussels.

Commission of the European Communities, 2004. *Flood risk management: Flood prevention, protection and mitigation*, Communication from the Commission to the Council, the European Parliament, the European Economic and Social Committee and the Committee of the Regions, COM(2004) 472, Brussels.

FEMA - Federal Emergency Management Agency, 1996. *Report on costs and benefits of natural hazard mitigation*, Washington, in http://www.fema.gov

Hubert, G. & Ledoux, B. 1999. *Le coût du risque. L'évaluation des impacts socio-économiques des inondations*, Paris: Presses de l'école nationale Ponts et chaussées.

International Strategy for Disaster Reduction—ISDR (2004). *Living with Risk. A global review of disaster reduction initiatives. 2004 Version*, United Nations ISDR, Geneva.

IReR, 2004. *Incidenza socio-economica in relazione alla delocalizzazione degli insediamenti e delle attività situate in zone a rischio idrogeologico*, Rapporti di Ricerca IReR, Codice IReR 2003A016, Milan.

Kleindorfer, P.R. & Kunreuther, H. 2001. Managing Catastrophe Risk. *Regulation* 23(4): 26–31.

Kunreuther, H. 2001. Incentives for mitigation investment and more effective risk management: the need for public–private partnerships. *Journal of Hazardous Materials*. 86: 171–185.

Kunreuther. H. 2004. *Public-Private Partnerships for Reducing Environmental Risks: The Role of Insurance, Inspections and Regulations*, Performance Track Participants Association (PTPA) Conference, Baltimore, MD (USA), April 21st.

Kunreuther, H. & Roth, R.J. (eds) (1998). *Paying the price: the status and role of insurance against natural disasters in the United States*. USA: Joseph Henry Press.

Linnerooth-Bayer J. & Amendola A. (2003). Introduction to Special Issue on Flood Risks in Europe. *Risk Analysis* 23(3): 537–543.

Menoni, S. 2004. Different Preventive Strategies Require Diversified Risk Assessment Models, in *Environmental Security and Environmental Management: The Role of Risk Assessment*, Proceedings of the NATO Advanced Research Workshop on the role of risk assessment in environmental security and emergency preparedness in the Mediterranean Region, Eilat, April 15–18.

Menoni, S. 2005. *Costruire la prevenzione - Strategie di riduzione e mitigazione dei rischi territoriali*, Bologna: Pitagora Editrice.

Pesaro, G. 2001. *Environmental Voluntary Agreements: A New Model of Co-operation Between Public and Economic Actors*, Feem Working Papers "Note di Lavoro", VOL/9.2001.

Sandler, T. 2001. *On Financing Global and International Public Goods*, Policy Research Working Paper 2638, The World Bank.

Schneider, S. 1995. *Flirting with disaster. Public Management in crisis situations*. New York/London: M.E. Sharpe.

Slovic, P. (ed) 2000. *The perception of risk*. USA: Earthscan Publications.

Van der Veen, A., Vetere Arellano, A.L. & Nordvik, J-P. (eds) 2003. *In search off a common methodology on damage estimation*, Workshop Proceedings, European Commission - DG JRC, EUR 20997 EN European Communities.

Vetere Arellano, A.L., Nordvik, J-P. & Ranguelov, B. 2003. In search of a common methodology on damage estimation: from an European perspective, in Van der Veen A., Vetere Arellano A.L. and Nordvik J-P. (eds), *In search of a common methodology on damage estimation*, Workshop Proceedings, European Commission - DG JRC, EUR 20997 EN European Communities.

Risks Challenging Publics, Scientists and Governments – Menoni (ed)
© 2010 Taylor & Francis Group, London, ISBN 978-0-415-58072-4

Estimation of economic losses according to the earthquake scenarios for Istanbul

S. Kundak
Istanbul Technical University, Turkey

ABSTRACT: After the earthquake which occurred in Kocaeli and Duzce in 1999, studies adhering seismic vulnerability and risk have gained great acceleration as never before. Despite earthquakes were expected, the consequences were far beyond than expected. Hence, all attention has been drawn to Istanbul which has similar urban development process and similar type of building quality but with a larger population. The historical earthquake records support this kind of concern, as Istanbul suffered large damages during its long history. The aim of this paper, therefore, is to find out economic consequences of a probable earthquake in Istanbul. Earthquake scenarios of Istanbul have been used in order to estimate total damage cost from damages to housing units and interruption of business activities. This research examines on one hand how economic losses are distributed on spatial pattern of Istanbul an on the other hand how elements (housing and business units) at risk have changed since the year 2000.

1 INTRODUCTION

Natural disasters, especially earthquakes, can be devastating to human activities, to social organizations at every level and to economic life. After the first shock, the damage is counted by deaths and injuries. In a while, the destroying effects of disaster appear on the economic structure of the region. Economic losses by severe earthquakes can cause long-term reductions in the growth of a nation's economy and trigger inflation. Therefore, evaluation of economic losses can be considered according to their share in country's economy using either gross domestic product (GDP) or gross national product (GNP) (Chen et al. 1997; Chan et al. 1998; Chen et al. 2002). Coburn and Spence (1992) displayed that while the absolute value of economic losses seem lower or higher in certain events, their share in country's GNP is more significant to understand actual losses in economy. For instance, after the Kocaeli and Duzce earthquakes (1999), which occurred in the most industrialized region of Turkey, total economic loss was estimated around 10 billion USD which represented approximately 4% of GDP in 1999.[1] Moreover, the biggest business enterprise of Turkey, TUPRAS Oil Refinery could recover all damages received in 1999 earthquake in a year and reached again 100% of its capacity in September 21, 2000 (Danış & Görgün, 2005).

Loss estimation techniques are basically developed to assess probable damages to buildings and infrastructure according to engineering perspective. Another point of view covers possible number of deaths and injuries according to the size of an extreme event. The last, but the most complicated one, is economic loss estimation which deals with intricate monetary systems. Studies on economic impacts of earthquakes have been usually divided in two categories: a) loss caused by destruction of built environment (direct loss), and b) loss caused by interruption of economic activities (indirect loss) (Cochrane, 2004; Bendimerad, 2001; Tierney, 1997). The loss estimation of built environment can be made by calculating the cost of reinstatement of all that was destroyed or made unusable by the earthquake. Likewise, the loss of production to the region's economy can eventually be estimated; however, this estimation can not be so precise because of the complexity of fiscal asset of the settlements. Rose (2004) clarifies the complexity of

[1] Total economic losses by two big earthquakes occurred in 1999 differ in each reference, including official records. Different perspectives and estimation methods obviously affect the monetary value of total losses. Hereby, total economic loss is refered by the official database of the General Directorate of Disaster Affaires and its share in GDP is calculated using the data by the Turkish Statistical Institute. Durukal and Erdik (2008) estimates total losses around 16 billion USD with a share of 7% in GDP. In the preliminary report by the World Bank, total losses were estimated between 7–11 billion USD (World Bank, 1999). In the interim report on economic effects of 1999s earthquakes by Bibbee et al (2000), some official statements and estimations from different institutes were given.

economic system defining "stocks" and "flows". This determination helps to better assess probable losses on one hand and on the other hand it prevents any duplication in economic loss modeling.

Economic losses can be evaluated from different dimensions such as: (1) costs from instant losses (i.e.building collapse); (2) costs from losses in business activities (both stocks and flows); and (3) costs from interruption of lifelines (electric shortage, lack of transportation). Costs from instant losses can be defined as the replacement/repair costs or market values of damaged units which are relatively easier and quicker to estimate in hazard scenarios or in the disasters. Interruption of business activities, on the other hand, has two main causes, related with damages on equipments and/or storages which may halt production. The latter represents interdependency of the market where the output of an enterprise is the input of another. This aspect of economic relations requires more complicated analysis in loss estimation which can be named as systemic approach. Furthermore, there are several secondary and tertiary components which make difficult to assess probable economic losses in extreme events. For instance, an important observation, related with business interruption aftermath of the Kocaeli earthquake, is that several small and medium size enterprises suffered from the lack of customers even if there was a slight or no damage in their work place (Yildirim et al. 2003). Another remark from the same event is that the latter significant factor causing business interruption was related with damages to transportation and infrastructural facilities which lead us to the third dimension in economic loss estimation. In 1994, 1995 and 1999, strong earthquakes hit California (USA), Kobe (Japan) and Kocaeli (Turkey) respectively. The common point of these earthquakes is that they affected metropolitan areas where business activities are dominant in urban life. The consequences of these earthquakes showed that damages to lifelines and infrastructural services such as electricity should be taken into consideration in estimating losses by business interruptions.

Loss estimation is not focused only on "what can be lost" but also it is a great tool in providing input to insurance, recovery and mitigation as well. In previous research (i.e. Kunreuter, 2000; Bendimerad, 2001; Sharma, 2001; Champell et al. 2002), it has been shown that due to loss estimation models, urban planners and decision makers would be able to comprehend potential losses and therefore mitigation efforts and consequences would be tangible. At this point, the importance of the accuracy in database is revealed as the main issue. Bendimerad (2001) suggests a tier classification method which prevents any confusion of scales in loss estimation. For instance, once considering a metropolitan area, there are distinguished features at every level such as regional dependency and the role of this metropolitan area in its country; urban pattern and compatibility with soil conditions; and building scale features respectively. Therefore, to better understand probable impacts due to any type of threat, loss estimation analyses should be deployed from macroscopic scale (nation, region) through microscopic scale (street, building).

The aim of this paper is to represent a macroscopic perspective to economic losses in Istanbul, caused by a probable earthquake expected in the Marmara Sea, along the North Anatolian Fault. In the second section of the paper, the seismic background of Istanbul will be introduced with information on historical earthquakes that occurred in this region. Section 3 includes database construction, loss estimation model used in this study and findings. In the last section, results of the study will be evaluated.

2 A RETROSPECTIVE APPROACH TO EARTHQUAKE PERILS AFFECTING ISTANBUL

Turkey is situated on the Himalayan Alpine belt where numerous big earthquakes occurred in the region during the history. Two main fault systems cross the Anatolian land. The Eastern Anatolian fault (app. 700 km) extends between the eastern end point of the North Anatolian Fault (NAF) and the city of Kahraman Maras through the southern direction. 1300 km-long North Anatolian Fault system extends from east side through the west side of Anatolia and parallel to the Black Sea. There are several short but highly active seismic zones in the Central Anatolia and the Aegean Regions as well.

The NAF has some important characteristics to be noted for a better evaluation: (1) the NAF is similar to the San Andreas Fault in the USA, however, it can produce major earthquakes with higher frequency of occurrence; and (2) earthquakes on the NAF follow a specific sequence from eastward through westward, as described by Stein et al. (1997a) as "failing dominoes". Referring to previous research and scientific facts, researchers studying the NAF for a long time were not as surprised as lay people, decision makers and governors in the aftermath of the 1999 earthquakes. Thus, because numerous articles published and several papers presented on the NAF (i.e. Ambraseys, 1970; Barka, 1992; Stein et al. 1997a; Stein et al. 1997b; Ucer et al. 1997; Barka & Kadinsky-Cade, 1988; Barka, 1996; Ansal, 1991; Ansal, 1997; Ansal et al. 1997) at both national and international arena pointed out that the 1999 earthquake was expected as the next failures on the NAF. Celal Sengor[2], giving voice to academia, expressed his primary reaction against the devastating impacts of these earthquakes as: "... When the Izmit Earthquake happened, there were two sorts of feeling arising in me: one was frustration, the other one was anger. Anger with myself, anger with others ... Because sort of we knew it was coming and yet nothing had been done." (BBC, 2003). The actions that had to be done before were strongly connected with the built environment quality and site selection which lead us to urban vulnerability components. The consequences of having poor buildings and poor infrastructural systems (especially water distribution, O'Rourke et al. 2000) have driven attention to the biggest city of Turkey, Istanbul. The main reason is that Istanbul had a similar development process as Kocaeli, with illegal housing, poor building quality encouraged by lack of control and very dense urban patterns. After the Kocaeli earthquake in 1999, the probability of occurrence of a severe earthquake, which would affect large territory including Istanbul, has been calculated looking at earthquake catalogues and tension accumulated on the NAF due to most recent seismic events. In the following 30 years from 1999, it has been estimated about 62% of probability of occurrence of a big earthquake affecting southern Istanbul, whereas about 32% of probability for the following 10 years (Barka, 2000; Parsons et al. 2000).

In Istanbul, earthquake records spanning two millennia indicate that, on average, at least one medium intensity (Io = VII-VIII) earthquake has affected the city every 50 years (Ambrasseys & Finkel, 1991). In its history, Istanbul experienced several earthquakes. According to the historical records, the earliest documented earthquake occurred in 32 B.C (Ambrasseys & Finkel, 1991). The ancient manuscripts and diaries address about an earthquake occurred in 1489, January 16 near Istanbul, with the Io = VIII. However, as there are not many information sources, it is just known that this earthquake caused severe damages on buildings (Cezar, 1963). The most devastating earthquakes occurred in 1509, 1719, 1766 and 1894 along the North Anatolian Fault.

Most of the damage in the 1509 September 10 earthquake was on the north-eastern coast of the Sea of Marmara (Ambrasseys & Jackson, 2000). The intensity in the southern part of Istanbul, the Historical Peninsula, was around VII and greater. The fortifications and the main gates of Istanbul were demolished, domes and minarets of several mosques collapsed, over 1.000 houses were destroyed; some 4.000–5.000 people were killed in the city; the number of injured people was 10.000 (Ambrasseys & Finkel, 1995). According to historical records, the earthquake was felt over a large area from Greece to Egypt. Beside the earthquake, tsunami caused severe damages in the sea shore of the Marmara Sea (Genc & Mazak, 2000). The aftershocks lasted for several months. After the earthquake, the reconstruction process of the city was started. As the majority of losses were suffered by masonry buildings, wooden constructions were encouraged by the local authority.

In 1719, a major earthquake occurred in the eastern part of the Marmara Sea, the estimated epicenter was near to Izmit Bay. The effects of this earthquake expanded through the Marmara region, but the main damages were in the Izmit and Istanbul areas. In the earthquake, 6.000 people were killed, buildings were severely damaged, 40 mosques and 27 towers were totally collapsed (Ambrasseys & Finkel, 1995). Taken into consideration its epicenter and its effects on the settlements, it can be thought that the earthquake of 1719 was suchlike the 1999 earthquake occurred in the same zone (Kundak, 2006).

After 257 years from the major earthquake of 1509, in 1766 May 22, a fault segment between Izmit and Gallipoli (through the eastern part to the western part of the Marmara Sea) broke. The earthquake caused tsunami which affected all the seashore of Marmara. The aftershocks were numerous and lasted for a long time (Ambrasseys & Finkel, 1995). According to the records, the earthquake lasted for 2 minutes and just 4 minutes after another shock hit the area once more. In the second day after the earthquake, it was reported that the number of death was around 4.000–5.000. Beside the damage in the city center, the villages in the northern part of the city suffered severe losses as well. City walls of Istanbul collapsed and Topkapi Palace was severely damaged so that the Ottoman Sultan had to live in a tent in the garden of palace for a while (Ambrasseys & Finkel, 1995).

The earthquake of 1894 July 10 was felt in a large area and caused high damages in Istanbul. As well as all the city once more suffered from the big earthquake, the most tragic event was the collapse of the Grand Bazaar. The earthquake occurred in the day time when many people working or shopping there. Because of the shakings the big doors of the Grand Bazaar were closed, first the domes and then the walls of the Bazaar collapsed. Many people were killed in this event (Ambrasseys, 2001). After the earthquake, according to the assessment by the local authority, damages were described into 5 regions: first

[2]Professor at the Istanbul Technical University, Department of Geological Engineering. He is the first Turkish academician to get elected as a foreign associate of National Academy of Sciences

region covered Izmit-Adapazari region where all buildings had collapsed; second region covered northern part of the Sea of Marmara where only low construction quality buildings collapsed; third region was the southern part of the Sea of Marmara where there were no damages to buildings but the house equipments moved or damaged during the earthquake; forth region referred whole the Anatolia and Balkan region where it was felt and fifth region described Europe, Asia and a part of Africa where the earthquake was felt just by seismographs (Genc & Mazak, 2000).

Since the 19th century when the last large earthquake occurred, Istanbul grew faster than ever seen in its history. At the beginning of the 1950's, the development of Turkish economy reinforced the dominant economic role of Istanbul. In this period, the rapid population growth due to migration from rural part of the country caused rising density and expending urban area. However, the planning processes remained insufficient against this "rapid development" and Istanbul gained a complex and uncontrolled urban pattern. Today, with its 12 million inhabitants, Istanbul is the most populated city in Turkey. Moreover, Istanbul undertakes several leading roles in cultural, financial, commercial, tourism and service functions. This feature of the city certainly reflects on nation's economy. Istanbul's contribution to tax revenues reaches 42%, its contribution to the budget is 34% and its share in GDP exceeds 20%.

Expansion of urban land in Istanbul showed linear development to the South, from the eastern side to western side, parallel to the NAF. Both population and building density increased in the fringes of the city. Newly developed sub-centers and industrial areas enabled to change mono-centric structure of Istanbul to poly-centric structure. This development process facilitates inner-city flows and protects forest land in the northern part of the city, though earthquake vulnerability has increased in Istanbul. When the 1999 Kocaeli earthquake hit the Marmara Region, in Istanbul, Avcilar (in south-west) and Tuzla (in south-east) were the most affected districts with collapsed buildings. In Istanbul 1–2% of the buildings were damaged, 454 people were killed and 3600 people were injured (Erdik et al. 2000).

Just after the 1999 earthquakes, government (both central and local level) took initiative on conducting multidisciplinary research to develop earthquake risk analysis and risk reduction schemes for Istanbul. In 2002, the Istanbul Metropolitan Municipality (IMM) and the Japan International Cooperation Agency (JICA) undertook a comprehensive project in order to both estimate the earthquake magnitude which will be produced by NAF and to assess the impacts of a severe earthquake on the Istanbul Metropolitan Area[3]. The case study covered 27 districts of Istanbul and additionally 3 outer districts of Istanbul. The main objectives of the study were: (1) to develop a comprehensive database of Istanbul to help to deal with hazards; (2) to investigate buildings inventory according to earthquake hazard; (3) to assess the risk and the elements at risk in the city; (4) to develop outputs for mitigation projects.

Four earthquake scenarios were developed in this study: Model A (7,5), Model B (7,4), Model C (7,7) and Model D (6,9). After developing these scenarios, the study focused on Model A which is considered the most probable scenario and on Model C which is the worst case scenario in order to assess probable losses to buildings and human life. The damage ratios for buildings were classified in three groups: heavily, moderate and slightly damage. According to the Model A, heavily damaged buildings will be around 51,000 and moderately damaged buildings will be 114,000. The death toll is estimated around 73,000 people and the number of injured around 120,000. According to the Model C, the number of heavily damaged buildings is 59,000 and the number of moderate damage buildings is 128,000. According to this scenario, the number of death might be as high as 87,000 and the number of hospitalized injured people 135,000 (IMM and JICA, 2002).

Historical earthquakes and development process of Istanbul give some clues for a retrospective approach and to produce scenarios on what would happen when a major earthquake hits again this metropolitan area. Damage due to past earthquakes was mostly distributed in the historical peninsula where the city is densely developed. Most of the loss in human lives and injuries was caused by collapse of masonry buildings and some local fires related with ignition of wooden buildings. Today, Istanbul is 20 times bigger as for the land occupied and 13 times more populated comparing to the 19th century. Furthermore, as a consequences of the industrialization process that took place in the last century, new kind of manufacturing with large complex, more hazardous materials and more environmentally incompatible arose, introducing technological hazards and na-tech in the late 1900s. Therefore, in the light of scientific findings in seismology, geology and engineering and in the presence of current urban development, we may suppose that Istanbul, and of course Turkey, is facing a very serious earthquake threat much larger than ever been before.

---

[3]This project will be referred as "IMM & JICA Project" in the rest of the paper

3 A MACROSCOPIC PERSPECTIVE TO PROBABLE ECONOMIC LOSSES FOR ISTANBUL

Estimation techniques imply uncertainties due to several concealed variables. Uncertainty can be seen as the lack of knowledge and information or as a consequence of lack of comprehensive research. Hence, new technologies and lessons learnt from recent disasters are crucial in improving performance of loss estimations. For instance, in 1964 Nigata and 1995 Kobe earthquakes, liquefaction have drawn attention; likewise the 2001 El Salvador earthquake triggered landslides which caused extensive destructions. These cases are mile stones to better understand liquefaction mechanism and earthquake triggered hazards. On the other hand, of course there are still inconsistencies between the expected (or estimated) and observed aftermath of disasters. The field surveys after the Kocaeli earthquake showed that there had been some over-estimations in slightly damaged buildings and under-estimations in severely damaged and collapsed buildings (Bird et al. 2004; Spence et al. 2003). Nevertheless, despite loss estimation models carry out some deviations, they are essential in defining risks.

Taking into account uncertainties, over- and under- estimations, some economic losses likely to be caused by a "major earthquake" have been discussed by focusing on destruction of houses and interruption of business activities. Two earthquake models developed in the IMM & JICA Project and their three-leveled damage ratios (highly-moderate-partly damaged) have been used in order to express losses in the built environment. Therefore, in this study, economic losses have been grouped in three different damage levels as a consequence of two different earthquake magnitudes, using the census of the years 2000 and 2008.

3.1 *Database Construction*

In this study, inner city neighborhoods of Istanbul have been extracting information on houses and business stock vulnerability. The building damage ratios of the two earthquake scenarios with magnitude of 7.5 (most probable-case scenario-M7.5) and 7.7 (worst-case scenario-M7.7) for each neighborhood have been included in the database. As an indicator of economic exposure of neighborhoods, GDP is used.

The previous version of this study covered 615 neighborhoods of Istanbul which represent the inner city (Kundak, 2004). While the earthquake scenarios were produced in 2002, the database was obtained by the census of the year 2000. Recently, the Turkish Government has implemented new adjustment to spatially organize the boundaries of existing neighborhoods and districts, and established new districts in the outskirts of the city. Therefore, while the total number of districts has increased from 32 to 39, the inner city neighborhoods has decreased to 548 in 2008. On the other hand, in the last 8 years, GDP of Turkey and Istanbul has changed dramatically which would affect the economic exposure of case study neighborhoods.

Damage ratios of earthquakes with M7.5 and M7.7 had been calculated for each neighborhood unit in the IMM & JICA Project. This parameter is the function of geology, geomorphology, seismicity, soil and earthquake energy attenuation characteristics. Damage classification is grouped into three categories: a) partly damaged, b) moderate damaged; and c) heavily damaged. Partly damaged buildings are still useable buildings but they require reinforcement because their stability against earthquake might be reduced. Moderate damaged buildings are standing buildings but they are not safe for living inside before restoration. Heavily damaged buildings are totally or nearly collapsed buildings that require reinstatement. According to the past earthquake experiences in Turkey, the approximate cost had been calculated. Housing that has collapsed or is too heavily damaged to be inhabitable will need to be demolished and rebuilt at an estimated cost of US$20.000/unit. Furthermore, contents cost of housing unit which refer to all the equipment of a house is also added. According to insurance compensations of an average house, it is about US$20.000/unit. Housing with moderate damage is estimated to cost US$ 8.000/unit for repairs and light damage reparable at US$3.000/unit (World Bank Report, 1999).

The data set representing losses caused by business interruption includes number of business units and the share in GDP (2001 & 2008) of each neighborhood. This data enable to calculate indirect losses caused by earthquake. After major earthquakes, some work places can not continue their production for a while because of loss in their employees or buildings. Moreover, in many case, major earthquakes cause deep and long-term monetary losses in the fiscal asset of both region and country in the mean of tax contribution, value added and share in GDP and GNP. As mentioned in the first part of this paper, these losses can cause long-term reductions in the growth of a nation's economy and trigger inflation.

3.2 *Model Construction*

Loss estimation models are flexible that one may transform basic equations depending on the nature of the problem and the purpose of the study. In the

Istanbul case, to obtain a comprehensive picture, database and model are designed in macroscopic scale. Estimated economic loss in each neighborhood ($Loss_N$) is represented by the sum of total damage cost of housing units ($Loss_{H_N}$) and total damage cost from interruption of business activity ($Loss_{B_N}$).

$$Loss_N = Loss_{H_N} + Loss_{B_N}$$

Damage cost of housing units is expressed as:

$$Loss_{H_N} = DCH_{h_N} + DCH_{m_N} + DCH_{p_N} + CCH_{h_N}$$

where;
DCH_{h_N} = Damage cost of heavily damaged houses
DCH_{m_N} = Damage cost of moderate damaged houses
DCH_{p_N} = Damage cost of partly damaged houses
CCH_{h_N} = Contents cost of heavily damaged houses

Damage cost from interruption of business activity is expressed as:

$$LOSS_{B_N} = D_{B_N} \times GDP_N$$

where;
D_{B_N} = Percentage of heavily damaged work places
GDP_N = GDP of neighborhood

3.3 *Findings*

The findings of this study show on the one hand how probable economic losses are distributed in the inner city and on the other hand, they reinforce the statement that vulnerability is definitely time dependent due to the changes on urban economy, household structure and urban patterns, especially in the case of metropolitan areas.

As mentioned previously, in this study two earthquake scenarios have been used to estimate probable economic losses in housing units and business interruption and to make a comparison how economic vulnerability of Istanbul has been changed in the last 8 years. The number of housing units has increased about 7% whereas the increase in business units has reached 32%. Hereby, as the main concern of this paper is probable losses due to earthquake, it would be better to evaluate the spatial distribution of these increases in the whole city (Figures 1 & 2). In both Figures, two critical data were given: the first one shows the spectral acceleration according to the Model C (Mw = 7,7) presented as gal and the second one indicates neighborhoods where dramatic changes in housing and business units occurred. Neighborhoods with an high increase in housing units are concentrated in the western part of the Historical Peninsula (district named Fatih), historical part of the Asian Side (mostly Uskudar district) and at the fringes of the city. Fatih has always been the highest population and building density district of Istanbul. In the last 8 years, as well, despite of the fact that there were limited open and green spaces, we observe that the district has been getting denser than before. Highly dense urban pattern decreases the quality of life, but more important, considering soil conditions of this area, the risk is getting higher. This kind of development would likely make difficult to handle emergencies, not only due to earthquakes but also to other natural and technological hazards. The same situation is observed in Uskudar district at the Asian Side of the city. However, due to better soil conditions we can assume that though there has been a dramatic increase in housing units, risks are not getting higher as they are in Fatih district. The third concentration point where housing units have doubled or more, is the city fringes. In the late 1990s, the notion of gated community following mass housing entered into the real estate scene. They were developed mostly near to the main transportation nodes where accessibility to down town is provided. After the 1999 earthquake, these gated communities and other mass houses gained another "mission" than providing houses for new comers but also providing "safer" houses against expected earthquakes. Therefore these new residential areas have been mostly located on stiff soil or rock and relatively far from the NAF (Figure 1).

Since 2000, the major shift has occurred in the number of business units in Istanbul. Despite deep economic impacts of previous earthquakes and economic crises, business establishment in Istanbul never slowed down. The spots indicated as future sub-centers of Istanbul in the Master Plan of 1994 were developed accordingly. Once again, considering the spatial distribution of these new enterprises we can see that they are concentrated in the southern part of the city, closed to the fault line and on soft soil. The remarkable note is that more than 100 neighborhoods which are on the zones where the ground motion is expected higher than the other parts of the city, the number of business units doubled or even tripled (Figure 2). On the other hand, in the last 8 years GDP of Turkey has increased and consequently GDP of Istanbul has improved from $31 billions to $133 billions with a share in country's GDP from 20% to 17,8%. This development can be evaluated from different angles. In a bright side, economic development in the other parts of the country improves comparing with the past and therefore the dominance of Istanbul as the biggest economic asset is getting lower. Yet, considering the absolute value of annual

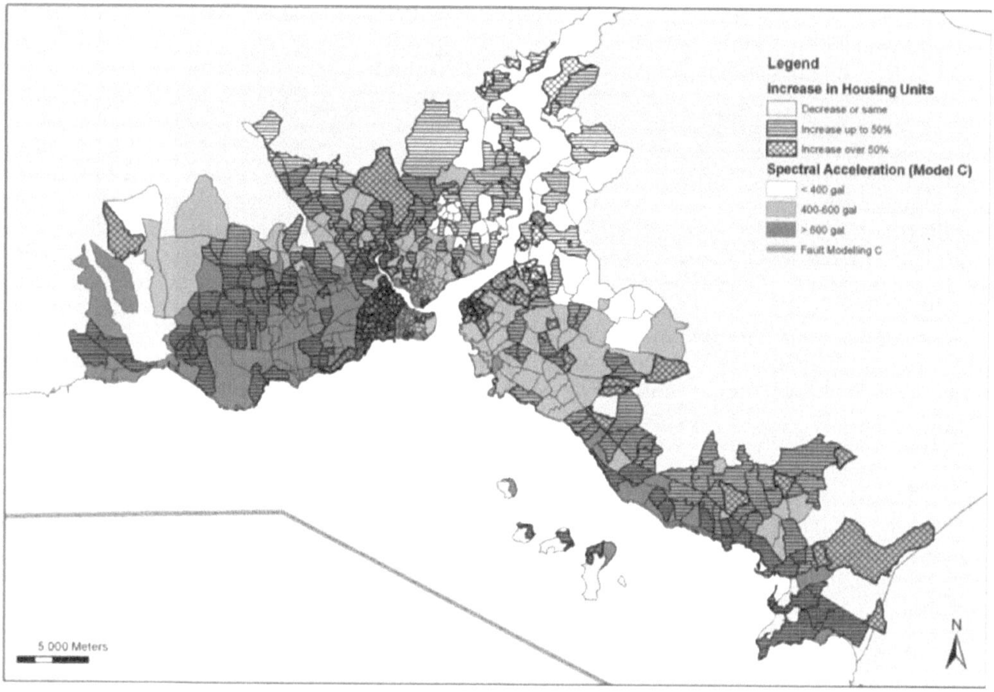

Figure 1. Spatial distribution of the increase in housing units.

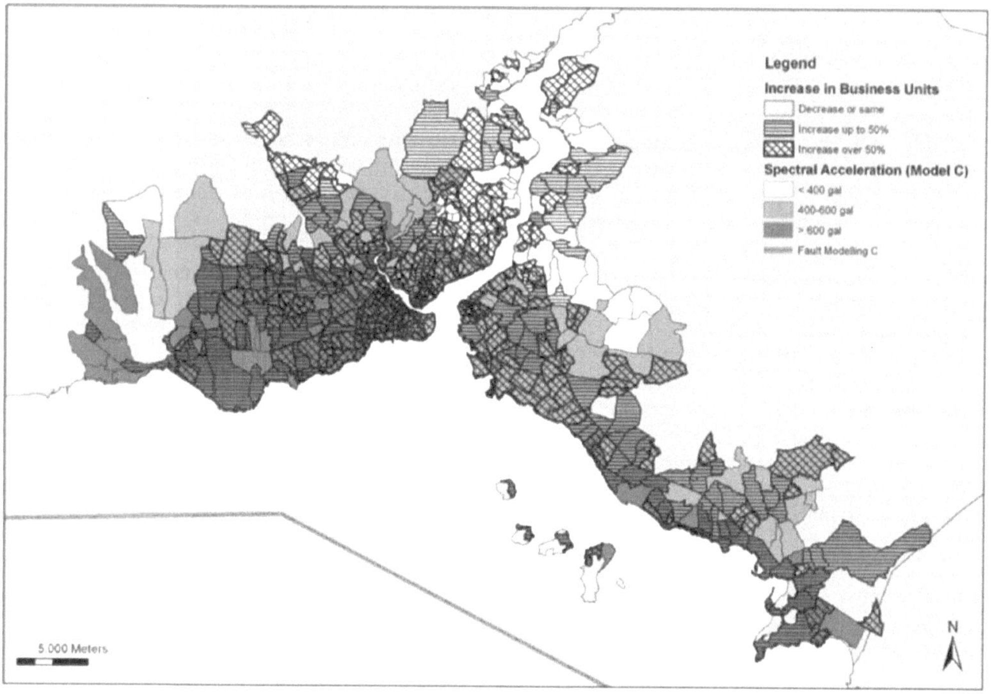

Figure 2. Spatial distribution of the increase in business units.

GDP of Istanbul, the economic exposure nearly quadrupled. As mentioned in the previous parts of this study, estimation of economic losses considering each business units as single entities would lead to under-estimation of probable losses. Since it is well known that all economic activities are strongly dependent one from the other by the means of providing services and goods.

Regarding the monetary losses calculated hereby, the major shift is indicated in the case of interruption of business activities. Therefore the absolute value of losses increased for the Model A in 2000 with $26 billions to 2008 $56 billions and for the Model C in 2000 with $30 billions to 2008 $64 billions (Tables 1 & 2). Another important finding can be followed in the latter Tables that the population living in risky areas of the city has considerably increased since 2000. For instance, in the total section of the Table 2 for the Model C, we can notify that 42,2% of the total population in 2000 lives in the first two regions (where losses are expected less that $40 millions), whereas in 2008 the ratio decreases to 20,4%. Not always with the same ratio, but with similar trends, housing and business units are concentrating in hazardous areas (according to soil structure and distance from the fault) and therefore they contribute to increase risks.

4 CONCLUSION

This paper addresses the links between both the progression of urban and economic development of Istanbul from the point of view of earthquake related losses. The expected economic losses represented in this study include only housing and work places indicators and are equivalent to nearly 8% of country's GDP of the year 2008. Other losses in infrastructure and urban facilities can increase these costs. In the concluding part of the previous version of this study using the database of the year 2000, it was suggested the need of a comprehensive planning process by means of spatial reorganization and administrative adjustment which would increase resilience of the city against earthquake. Updating the study does not lead to different conclusions. The rapid increase in the Southern part of the city by the means of both population growth and building construction is revealed as the major component in increasing earthquake risk. On the other hand, urban regeneration projects, adaptation of new building codes and mass housing projects conducted by government might help in decreasing risks at least theoretically. However in practice, there are still crucial problems to solve in order to achieve resilient city and resilient community:

a. Nature versus Safety: Istanbul has been developed in a linear way near to the NAF. In the Northern part of the city there are water reservoirs and forest areas and in the eastern and western regions there are agricultural land. After the 1999 earthquakes, large housing projects have been concentrated in these areas and they are becoming a new threat to natural resources of the city. Moreover, such a development in which all infrastructural facilities are provided, encourages illegal housing development in the adjacent areas, as it was experienced in the past.
b. Renewal versus Gentrification: Urban renewal is a common method in urban planning to provide livable settlements in old and decayed areas. New implementations of this method, today, lie through urban regeneration and gentrification which target low income decayed areas to turn into rehabilitated areas affordable for medium to high income people. In the near past, urban renewals in Istanbul have targeted both improving quality of life in decayed areas and decreasing earthquake risk. However, besides the physical re-adjustments that have taken place, social consequences of the implementation have overrode the positive outcomes. Low income families who used to live in the city center have been driven towards outer parts of the city and obviously they refused to thus in a very harsh way. In order to enhance resilience of the city, it is necessary to take precautions by renewal and rehabilitation, while social aspects should be taken into consideration.
c. External versus Autonomous Building Consultancy: Since 2001, new buildings should get external and independent consultancy during the every stage of construction (from blue prints through occupancy permit). Private entities, according to this new law, would make sure of buildings passing all controls by consultancy firms (in 19 provinces of Turkey). This law takes apart governmental structures, urban facilities and residential buildings of the Housing Development Administration of Turkey. These structures are exempted from this regulation as the governmental bodies have their own self-control. However, once considering progression of vulnerability in urban areas due to wrong decisions and mismanagement, this autonomous approach is likely to cause problems in the future. On the other hand, big construction companies establish their own consultancy firms to make the consultancy process faster. Therefore, the present implementation of this law mostly covers buildings constructed by individuals, which represents small numbers

Table 1. Distribution of probable economic losses on housing and business activities according to earthquake Model A.

	MODEL A (2000)							MODEL A (2008)						
		Total loss		Number of neigborhood		Number of housing units			Total loss		Number of neigborhood		Number of housing units	
		#	%	#	%	#	%		#	%	#	%	#	%
HOUSING	<20	2.685	15,4	334	54,0	930.049	29,3	<20	2.056	10,7	257	46,9	856.201	25,3
	20–40	4.292	24,6	147	23,8	900.446	28,3	20–40	3.450	18,0	119	21,7	857.320	25,3
	40–60	3.054	17,5	63	10,2	539.712	17,0	40–60	3.699	19,4	76	13,9	612.520	18,0
	60–80	2.117	12,1	31	5,0	275.499	8,7	60–80	2.361	12,3	34	6,2	318.688	9,4
	80–100	1.502	8,6	17	2,8	178.142	5,6	80–100	2.374	12,4	27	4,9	277.708	8,2
	>100	3.816	21,9	26	4,2	353.404	11,1	>100	5.195	27,2	35	6,4	467.678	13,8
	TOTAL	17.466	100	618	100	3.177.252	100	TOTAL	19.135	100	548	100	3.390.115	100
		Total loss		Number of neigborhood		Number of business units			Total loss		Number of neigborhood		Number of business units	
		#	%	#	%	#	%		#	%	#	%	#	%
BUSINESS	<20	2.720	32,0	510	82,6	380.070	60,3	<20	2.036	5,5	214	39,1	152.424	18,3
	20–40	1.679	19,7	59	9,5	99.534	15,8	20–40	3.507	9,5	121	22,1	164.704	19,7
	40–60	1.149	13,5	24	3,9	67.745	10,8	40–60	2.399	6,5	50	9,1	83.013	10,0
	60–80	844	9,9	12	1,9	31.097	4,9	60–80	2.414	6,5	35	6,4	70.004	8,4
	80–100	178	2,1	2	0,3	6.074	1,0	80–100	2.892	7,8	32	5,9	61.263	7,4
	100–200	1.185	14,0	8	1,3	28.845	4,6	100–200	7.729	20,8	53	9,5	154.931	18,6
	>200	749	8,8	3	0,5	16.475	2,6	>200	16.096	43,4	43	7,9	146.308	17,6
	TOTAL	8.504	100	618	100	629.840	100	TOTAL	37.073	100	548	100	832.647	100
		Total loss		Number of neigborhood		Population			Total loss		Number of neigborhood		Population	
		#	%	#	%	#	%		#	%	#	%	#	%
TOTAL	<20	2.481	9,5	253	40,9	2.017.605	23,1	<20	965	1,7	95	17,3	918.669	9,2
	20–40	4.531	17,4	153	24,8	2.179.258	24,9	20–40	2.827	5,0	98	17,9	1.437.542	14,3
	40–60	4.254	16,3	87	14,1	1.483.203	17,0	40–60	4.495	8,0	92	16,8	1.759.180	17,5
	60–80	3.078	11,8	44	7,1	862.102	9,9	60–80	3.984	7,2	57	10,4	1.117.338	11,1
	80–100	2.426	9,3	27	4,4	669.116	7,7	80–100	3.654	6,5	41	7,5	971.608	9,7
	100–200	5.585	21,4	40	6,5	1.052.018	12,0	100–200	14.066	25,0	96	17,5	2.074.356	20,7
	>200	3.687	14,3	14	2.2	473.671	5,5	>200	26.217	46,6	69	12,6	1.750.120	17,5
	TOTAL	26.042	100	618	100	8.736.973	100	TOTAL	56.208	100	100	100	10.028.813	100

Table 2. Distribution of probable economic losses on housing and business activities according to earthquake Model C.

	MODEL C (2000)							MODEL C (2008)						
HOUSING		Total loss		Number of neigborhood		Number of housing units			Total loss		Number of neigborhood		Number of housing units	
		#	%	#	%	#	%		#	%	#	%	#	%
	<20	2.562,9	12,8	310	50,2	817.553	25,7	<20	1.954	8,9	237	43,2	729.130	21,5
	20–40	3.993,1	19,9	138	22,3	788.324	24,8	20–40	3.335	15,2	115	21,0	774.968	22,9
	40–60	3.768,7	18,8	78	12,6	570.609	18,0	40–60	3.661	16,7	75	13,7	595.408	17,6
	60–80	2.501,9	12,5	36	5,8	364.892	11,5	60–80	2.516	11,5	38	6,9	322.597	9,5
	80–100	1.867,2	9,3	21	3,4	184.387	5,8	80–100	3.010	13,7	34	6,2	339.584	10,0
	>100	5.383,7	26,8	35	5,7	451.487	14,2	>100	7.454	34,0	49	9,0	628.428	18,5
	TOTAL	20.077,5	100	618	100	3.177.252	100	TOTAL	21.930	100	548	100	3.390.115	100
BUSINESS		Total loss		Number of neigborhood		Number of business units			Total loss		Number of neigborhood		Number of business units	
		#	%	#	%	#	%		#	%	#	%	#	%
	<20	2.924	29,8	496	80,3	357.289	56,7	<20	1.850	4,4	190	34,7	119.609	41,4
	20–40	1.718	17,5	61	9,9	99.019	15,7	20–40	3.355	7,9	116	21,2	152.930	18,4
	40–60	1.568	16,0	32	5,3	85.373	13,5	40–60	3.020	7,2	63	11,5	101.814	12,2
	60–80	694	7,1	10	1,6	20.662	3,3	60–80	2.430	5,7	35	6,4	66.301	7,9
	80–100	604	6,2	7	1,1	22.444	3,6	80–100	2.510	5,9	28	5,1	46.407	5,6
	100–200	853	8,7	6	0,9	17.418	2,7	100–200	9.675	22,9	67	12,2	182.385	21,9
	>200	1.433	14,7	6	0,9	28.635	4,5	>200	19.431	46,0	49	8,9	163.201	19,6
	TOTAL	9.794	100,0	618	100,0	629.840	100,0	TOTAL	42.271	100	548	100	832.647	100
TOTAL		Total loss		Number of neigborhood		Population			Total loss		Number of neigborhood		Population	
		#	%	#	%	#	%		#	%	#	%	#	%
	<20	2.239	7,5	224	36,2	1.688.120	19,3	<20	782	1,2	79	14,4	745.213	7,4
	20–40	4.496	15,0	152	24,6	2.001.127	22,9	20–40	2.679	4,2	92	16,8	1.305.580	13,0
	40–60	4.626	15,5	93	15,0	1.571.204	18,0	40–60	3.907	6,1	79	14,4	1.321.428	13,2
	60–80	3.120	10,4	45	7,3	880.564	10,1	60–80	4.049	6,3	60	10,9	1.246.918	12,4
	80–100	3.031	10,1	34	5,5	610.043	7,0	80–100	3.956	6,2	45	8,2	863.866	8,6
	100–200	6.183	20,7	47	7,6	1.308.846	14,9	100–200	15.465	24,0	107	19,5	2.395.799	23,9
	>200	6.179	20.7	23	3,7	677.069	7,8	>200	33.362	52,0	86	15,8	2.150.009	21,5
	TOTAL	29.874	100	618	100	8.736.973	100	TOTAL	64.200	100	548	100	10.028.813	100

comparing with those of government and big construction companies.

New facts declared by scientific researches on Istanbul do not obviously alter the development process of Istanbul. Since 1999, risky zones are been a problem for earthquake mitigation. The main question arises in how to "clean" these areas. Today approximately 1.000.000 people are living in the most risky areas which are on river basins, very soft soil, landslide prone areas and filled up land. Once, these zones will be wiped out from low quality buildings and structures, they will probably be used as open, green and recreational spaces. But as observed in the past, due to economic pressure on these lands, there will always be a sort of threat of transforming them into built-up area. At this point, a strong component of risk mitigation sets in as a success condition: risk perception of the community and the decision makers. The economic losses figured out in this paper are consequences of underestimation of risks and overriding seismic activities. In the 10th anniversary of the big earthquakes in Turkey (2009), despite there has been no dramatic changes in the community's risk perception, at least it is well known how a big earthquake can change all sets in seconds. Therefore, dissemination activities on disaster mitigation efforts would be great tools to increase earthquake awareness in the entire city and consequently guide in reducing physical, social and economic exposure of Istanbul.

REFERENCES

Ambraseys, N.N. 1970. Some Characteristics Features of the Anatolian Fault Zone. *Tectonophysics* 9.

Ambraseys, N.N. & Finkel, C.F. 1991. Long-term seismicity of İstanbul and of the Marmara Sea region. *Terra Nova* 3: 527–539.

Ambraseys, N.N. & Finkel, C.F. 1995. *The Seismicity of Turkey and Adjacent Areas*. İstanbul: Eren Yayıncılık.

Ambraseys, N.N. & Jackson, J.A. 2000. Seismicity of the Sea of Marmara (Turkey) since 1500. *Geophys. J. Int.* 141: F1-F6.

Ambraseys, N.N. 2001. The Earthquake of 10 July 1894 in the Gulf of Izmit (Turkey) and its relation to the earthquake of 17 August 1999. *Journal of Seismology* 5: 117–128.

Ansal, A. 1991. Istanbul'da Deprem. *Istanbul ve Deprem Sempozyumu:* 7–27. Insaat Mühendisleri Odasi Istanbul Subesi.

Ansal, A. 1997. Istanbul için Tasarim Deprem Özelliklerinin Belirlenmesi. *Prof. Dr. Rifat Yarar Sempozyumu Kitabi* 1: 233–244.

Ansal, A., Iyisan, R., Duman, F. & Eyidogan, H. 1997. Istanbul'da Alinan Kuvvetli Yer Hareketi Kayitlarinin Özellikleri. *4. Ulusal Deprem Mühendisligi Konferansi*: 22–29 ODTÜ, Ankara.

Barka, A.A. 1992. The North Anatolian Fault Zone. *Annales Tectonicae* VI:164–195.

Barka, A.A. 1996. Slip distribution along the North Anatolian fault associated with the large earthquakes of the period 1939–1967. *Bulletin of the Seismological Society of America* 86: 1238–1254.

Barka, A.A. 2000. The Next Expected Marmara Earthquake, *International Istanbul Earthquake Meeting*. TUYAP Fuarcilik, Istanbul, Turkey.

Barka, A.A. & Kadinsky-Cade, K. 1988. Strike-slip fault geometry in Turkey and its influence on earthquake activity. *Tectonics* 7: 663–684.

BBC 2003. *Earthquake Storms*. Documentary Movie by BBC.

Bendimerad, F. 2001. Loss Estimation: A Powerful Tool for Risk Assessment and Mitigation. *Soil Dynamics and Earthquake Engineering* 21:467–472.

Bibbee, A., Gonenc, R., Jacobs, S., Konvitz, J. & Price, R. 2000. *Economic Effects of the 1999 Turkish Earthquakes: An Interim Report*. OECD, Economics Department Working Papers No. 247.

Bird, J.F., Bommer, J.J., Bray, J.D., Sancio, R. & Spence, R.J.S. 2004. Comparing Loss Estimation with Observed Damage in a Zone of Ground Failure: A Study of the 1999 Kocaeli Earthquake in Turkey. *Bulletin of Earthquake Engineering*, 2: 329–360.

Campbell, K.W., Thenhaus, P.C., Barnhard, T.P. & Hampson, D.B. 2002. Seismic hazard model for loss estimation and risk management in Taiwan. *Soil Dynamics and Earthquake Engineering* 22: 743–754.

Cezar, M. 1963. *Osmanlı Devrinde İstanbul Yapılarında Tahribat Yapan Yangınlar ve Tabii Afetler*, İstanbul.

Chan, L.S., Chen, Y., Chen, Q., Chen, L., Liu, J., Dong, W. & Shah, H. 1998. Assessment of global seismic loss based on macroeconomic indicators. *Natural Hazards* 17: 269–283.

Chen, Q., Chen, Y., Liu, J. & Chen, L. 1997. Quick and approximate estimation of earthquake loss based on macroscopic index of exposure and population distribution. *Natural Hazards* 15: 217–229.

Chen, Y., Chen, L., Federico, G., Ota, K. & Li, J. 2002. Seismic Hazard and Loss Estimation for Central America. *Natural Hazards* 25: 161–175.

Chen, Y., Chen, X, Fuzhengxian, Y.Z. & Yan, M. 1996. Estimating Losses from Future Earthquakes in China. In V. SCHENK (ed.), *Earthquake Hazard and Risk: 153–168*. Kluwer Academic Publishers.

Coburn, A. & Spence, R. 1992. *Earthquake Protection*. John Wiley & Sons.

Cochrane, H. 2004. Economic Loss: Myth and Measurement. *Disaster Prevention and Management* 13(4): 290–296.

Danis, H. & Gorgun, M. 2005. Marmara Depremi ve TUPRAS Yangini. Deprem Sempozyumu, 23–25 March 2005: 1362–1369. Kocaeli – Kocaeli University.

Durukal, E. & Erdik, M. 2008. Physical and Economic Losses Sustained by the Industry in the 1999 Kocaeli, Turkey Earthquake. *Natural Hazards* 46: 153–178.

Erdik, M., Durukal, E., Biro, Y. & Birgören, G. 2000. İstanbul'da Binalar için Deprem Riski ve Risk Azaltımına Yönelik Somut bir Öneri, *İkinci İstanbul ve Deprem Sempozyumu*, TMMOB İnşaat Mühendisleri

Odası İstanbul Şubesi, 27 Mayıs 2000, İstanbul Teknik Üniversitesi Maçka Kampüsü, 131–149.
Genç, M. & Mazak, M. 2000. *İstanbul depremleri: fotoğraflarla ve belgelerle 1894 depremi.* Istanbul: IGDAŞ.
IMM (Istanbul Metropolitan Municipality) & JICA (Japan International Cooperation Agency) 2002. *A Disaster Prevention/Mitigation Basic Plan for Istanbul.*
Kundak, S. 2006. *Economic Vulnerability Assessment In Housing, Case Study: Istanbul.* Graduate thesis Centre D'Etudes des Risques Geologiques, Universite de Geneve.
Kundak, S. 2004. Economic Loss Estimation for Earthquake Hazard in Istanbul, *44th European Congress of the European Regional Science Association, Regions and Fiscal Federalism*, 25–29 August 2004, Porto, Portugal.
Kunreuther, H. 2000. Public-Private Partnerships for Reducing Seismic Risk Losses. *Euro Conference on Global Change and Catastrophe Risk Management, Earthquake Risks in Europe, IIASA, July 6–9 2000, Laxenburg Austria.*
O'Rourke, T.D., Erdogan, F.H., Savage, W.U., Lund, L.V. & Tang, A. 2000. Water, Gas, Electric Power and Telecommunications Peformance. *Earthquake Spectra* 16(S1): 377–402.
Parsons, T., Toda, S., Stein, R.S., Barka, A.A. & Dieterich, J.H. 2000. Heightened odds of large earthquakes near Istanbul: an interaction-based probability calculation. *Science* 288: 661–665.
Rose, A. 2004. Introduction. In Okuyama, Y., Chang, S.E. (eds.), *Modeling Spatial and Economic Impacts of Disasters*:13–36. Springer.
Sharma, V.K. 2001. Gujarat earthquake, some emerging issues. *Disaster Prevention and Management* 10(5): 349–355.
Spence, R., Bommer, J., Del Re, D., Bird, J., Aydinoglu, N. & Tabuchi, S. 2003. Comparing Loss Estimation with Observed Damage: A Study of the 1999 Kocaeli Earthquake in Turkey. *Bulletin of Earthquake Engineering* 1: 83–113.
Stein, R.S., Barka, A.A. & Dieterich, J.H. 1997a. Progressive Failure on the North Anatolian Fault Since 1939 by Earthquake Stress Triggering. *Geophysical Journal International* 128: 594–604.
Stein, R.S., Dieterich, J.H. & Barka, A.A. 1997b. Role of stress triggering in earthquake migration on the North Anatolian fault. *Physics and Chemistry of the Earth* 44: 78–91.
Tierney, K.J., 1997. Business Impacts of the Northridge Earthquake, *Journal of Contingencies and Crisis Management* 5(2): 87–97.
Ucer, B., Eyidogan, H., Barka, A., Gurbuz, C. & S. Barıs, 1997. Seismic investigations of the Marmara Region, In C. Shindler and M. Pfister (eds.) *Active Tectonics of Northwestern Anatolia-The Marmara Poly-Project.* Hochschulverlag AG an der ETH Zurich, 89–99.
World Bank Report 1999. *Turkey: Marmara Earthquake Assessment.*
Yildirim, E., Fidan, F., Sututemiz, N. & Yilmaz, T. 2003. Adapazari'ndaki Ozel Sektore Ait Isyerlerinde 17 Agustos Oncesi ve Sonrasinda Deprem Hazirliklari ve bu Hazirliklari Etkileyen Faktorler. *5. Ulusal Deprem Muhendisligi Kongresi, 27 May 2003.* Istanbul Technical University.

Risks Challenging Publics, Scientists and Governments – Menoni (ed)
© 2010 Taylor & Francis Group, London, ISBN 978-0-415-58072-4

Evaluation of the possible urban planning and insurance policies to create a risk reduction culture in Turkey

A. Taylan
Konya Selcuk University, Konya, Turkey

ABSTRACT: The paper evaluates the history and current system of loss financing in Turkey and shows the necessary partnership between urban planning and earthquake insurance systems. This can provide the basis to create a risk reduction culture among households in terms of land-use and insurance policies. The cooperation of urban planning and TCIP (Turkish Catastrophic Insurance Pool) should be in the context of an urban risk management approach (URM) that is the most suitable framework to manage earthquake risks as it provides the capability to directly intervene in the physical environment. TCIP can provide financial incentives and credits to homeowners and municipalities. So, homeowners are expected to invest reduces the risk of their houses. However, individuals do not always behave rationally. For this reason, the study recommends the investigation of factors influencing households' current decision process purchasing insurance and taking preventive measures.

1 INTRODUCTION

The comprehensive disaster management consists of two main parts as pre- and post disaster activities, which are sequential activities according to the occurrence of disasters. In this circle, the preparedness and risk reduction activities take place before a disaster strike. The activities after disasters should be followed are the emergency management, rehabilitation, and reconstruction. Till the 1999 Kocaeli and Düzce earthquakes in Turkey, the conventional disaster policy focused on the *adhoc* activities after each disaster, which gives the role to the State as the main compensator of the earthquake losses based on the Disaster Law (7269). This situation of social organization before 1999 earthquakes can be described as the fatalistic society, when the attributes of resources are as random and post disaster contributions (Balamir, 2001).

After the experience of great losses in both human lives and economy in 1999, Turkey faced over 18,000 losses of lives, 50,000 injuries, 30,000 destroyed dwellings and 50,000 businesses premises. This meant over 13,000,000,000 US$ loss in Nations' capital (Balamir, 2002). The low insurance penetration ratio in residential areas led State to undergone large amount of financial losses with the Nation's capital and foreign loans (Selcuk, 2001). With these losses, it is seen that earthquake disaster in urban areas cannot be covered by the State in the future.

The most crucial changes in the legal and institutional environment of the disaster management system were for the improvement of supervision during construction[1] and in the financial system of disaster losses. The introduction of the *Obligatory Earthquake Insurance* (OEI) constituted a self-mechanism to accumulate premiums before disaster and create a pool for using after disaster strike. Therefore, State abandoned also from paying for the future earthquake losses. These changes were also accepted indicator of moving from fatalist society to with the shift to risk management, which gives importance to preparedness and the pre-disaster processes.

In this direction, more regular, formal and continues flows in the finance of disaster losses should be charged. However, the available resources to form disaster resistant communities should also provide some incentives for mitigation, retrofitting and inspection, which are not included in the OEI implementation. The inclusion of such kind of measures can lead to risk reduction in the society (Balamir, 2001).

2 EARTHQUAKE LOSS FINANCING IN TURKEY

Before 1999, there were two financial mechanisms for compensating the disaster losses as State and private insurance companies. When the business and industry losses were mostly paid by the insurance companies and their reinsurance coverage,

[1] Others are "Control of Construction Processes" and "Improved Proficiency in Construction".

the small business and housing areas brought the country into financially difficult situation. The most problems were seen in the housing areas. In addition to the fact that the housing stock in Turkey is produced physically vulnerable during the rapid urbanization since 1960s, the penetration ratio of insurance was around 10%. On the other hand, the State had insufficient sources for reconstruction.

Although Turkey tried different methods to control earthquake insurance, the loss financing after disasters was basically dependent on the State's aid, mostly because of the populistic political background. Between 1904 and 1939, earthquake insurance was supplied as an allied peril in fire insurance. In 1939, it was excluded from fire insurance because of the huge losses in Erzincan Earthquake. It was again included to the security coverage in 1960. With free market tariff in the insurance industry in 1990, the earthquake insurance policy prices decreased in spite of high risk.

The sequential earthquakes in 1993 led the Treasury to determine a certain tariff for earthquake insurance and search the ways of making it compulsory (Selcuk, 2001). After the great economic impact of 1999 earthquakes, a new system came into agenda and as a part of the Marmara Earthquake Emergency Reconstruction (MEER) project of the World Bank in 1999. There had to be established a new system, which will increase the purchase of earthquake insurance by homeowners through making it affordable and bringing some enforcements (Wilczynski, 1999).

The Obligatory Earthquake Insurance regulation was began to be regulated with the Decree Law (No: 587; 27.12.1999) as a country scale regulation. With the regulation, there was established a public-private institution as the "Natural Disasters Insurance Authority (NDIA)", the Turkish name is DASK, to execute the finance of the future earthquake losses through Turkish Catastrophe Insurance Pool (TCIP). The main compensator of the legally built residential buildings became DASK, which offers Obligatory Earthquake Insurance (OEI). It also works with the insurance companies in the market to sell the policies and pay them according to their efforts. The insurance purchasers of OEI were defined as the owners of the legally built residential buildings (recorded in the Title Deed in the boundaries of municipalities) and parts of them as independent units according to the Flat Ownership Law (No. 634)[2].

DASK is not offered to the buildings constructed after 27/12/1999, either, if they have not any license. In other words, the illegally built houses, which were built on their own land, but did not get any license or title deed before this date, whether they are included to the Flat Ownership Law, can buy policy from DASK. Except the housing units in the buildings, the professional offices or small businesses, which have the independent and privately owned sections at the residential buildings, are also obliged to buy OEI. Besides, private buildings, which are built by State or by credit given or extended by the State after previous disasters, are also obliged to buy OEI.

If the homeowners do not buy obligatory earthquake insurance, although they are dependent on DASK, could not receive any assistance after a disaster. They are neither allowed to buy any other earthquake insurance from market, nor to get the assistance of the State. There is not any more responsibility of the State for reconstruction, after OEI has changed the Disaster Law. Besides, the other type of illegal house buildings, which were built on the land of Treasury or other persons without any permission, has not any financing resource in the case of a disaster. There is a high uncertainty of the risk with them and unwillingness to reward and legitimize them with insurance.

In addition to DASK, there are two more resources as private insurance companies and State that take place in the earthquake loss financing system in Turkey, after 1999. The buildings that are used only for commercial or industrial purposes like business centers, administrative buildings, and education buildings, whether they are included in the "Flat Ownership" Law No. 634, can buy earthquake insurance as an allied peril to the commercial, industrial insurance policies from the private insurance companies in the market. However, the small and medium sized enterprises (SMEs) are continued to be neglected. They are expected to purchase insurance in the market, while most of them cannot afford or ignore to buy insurance.

Finally, the public buildings like hospitals and schools, the buildings in rural areas, the buildings which were built after disasters in the past by State are dependent on State with the Disaster Law (7269).

2.1 *The organizational structure of OEI system and DASK*

There are three main bodies constituting the structure of DASK as *the General Directorate of Insurance*, *the Board of Directors*, and *the Operational Manager*. DASK works under the control of General Directorate of Insurance, which is a part of the under secretariat of the Treasury and responsible for the design and announcement of the principles, general conditions, tariffs, and procedures of program.

[2] A legislative piece that governs the way condominiums are managed.

Board of directors monitors the performance of DASK and provides technical support.

The members can only be appointed by the Ministry after the determination of the Treasury. These can be the representatives of institutions such as Prime Ministry, Under secretariat of the Treasury (General directorate of Insurance), Ministry of the Public Works and Settlements, Capital Markets Board, The Association of Insurance and Reinsurance Companies of Turkey, Operational Manager Company of DASK and an academic person with background and experience in one of the related fields (as civil engineering, geophysical engineering, geological engineering or equivalent).

Operational manager of DASK can be contracted out by the under secretariat of Treasury. The agreement is made with Milli Re for the 5 years, which is the largest reinsurer in Turkey. The second agreement is made with Garanti Re in 2005. Insurance companies that are permitted to work in Turkey under the Law of Insurance Arrangement (No: 7123), are selling the insurance policies to real estate owners on behalf of DASK and they are responsible to transmit all risks and premiums to the Operational Manager. There are currently 30 insurance companies authorized to issue the policies.

Operational Manager pays a commission amount to insurance companies according to the sold policy amount as the 12,5% of the insurance premium in Istanbul and 17,5% in other cities. In addition to the OEI policies, insurance companies are offering 2 more earthquake insurance in the market that can be bought voluntarily. The earthquake insurance as an allied peril for home and fire insurance for the compensation amount above 85,000 YTL which is offered by DASK. Another policy is offered for the housing assets Only the homeowners, who are allowed to buy OEI can buy these insurance policy for their homes.

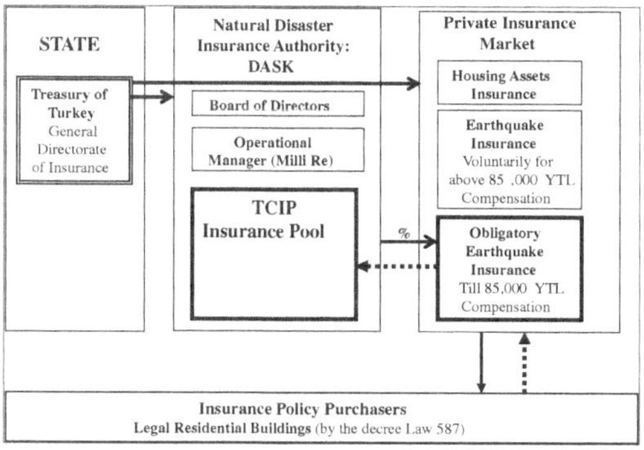

Figure 1. The Organizational Structure of the Earthquake Insurance Market for Residential Areas in Turkey.

2.2 *The current situation of DASK*

DASK aimed to provide affordable earthquake insurance for private property owners in the urban residential areas as a stand-alone product including the secondary perils of earthquakes such as fire, explosion and land-slides. It is a country scale and obligatory regulation. It intends to lessen the financial burden of the State in the future disasters in terms of accumulating premiums in the pool for future disasters. It shares the risk within the country and transfers it to international capital markets via reinsurance, so that the State has pledged not to compensate the earthquake losses in urban areas any more. Nevertheless, one of the establishment aims of DASK is to use insurance mechanism to promote better construction practices and to produce healthy buildings, although it is not reached so far.

The dwelling units dependent on DASK should be the buildings which were constructed according to the Flat Ownership Law (No. 634). The amount of the dwelling units in the whole housing stock is assumed as around 12,980,000. However, the whole housing stock in Turkey was not constructed in legal ways, because of the rapid urbanization and shortage in the housing supply. Only the licensed houses are comprised by the Flat Ownership Law, which constitute the 62.5% of the whole stock (UH, 2003). On the other hand, it does not mean that this legal stock is less vulnerable than the illegal one. Firstly, the populist policies caused to legitimizations in the illegally built areas. Secondly, the supervision of the buildings during construction of legal ones was inadequate. This situation of the housing stock does not make only difficult to identify and assess the risks, but also to implement the Flat Ownership Law (asking the building title deed[3]) to the homeowners, who want to purchase OEI. If the purchased insurance policies were only from the licensed stock, the penetration ratio would be nearly 29%. However, DASK insures the unlicensed stock and accepts the title deeds of land as enough documents during insurance purchase[4]. For this reason, the real penetration ratio is around 18 and 19% (Figures 2 & 3). The nearly maintained ratio is, however, shows that there is required a radical change in the system.

Not obeying the rules of Flat Ownership during the policy selling, lead to appear three

[3] According to Flat Ownership Law, the owners of the land have title deeds. They can get construction licence before construction and using permit after the construction is completed. Only the buildings that have using permit are licensed have building title deeds.

[4] TCIP, www.dask.gov.tr (29.10.2006)

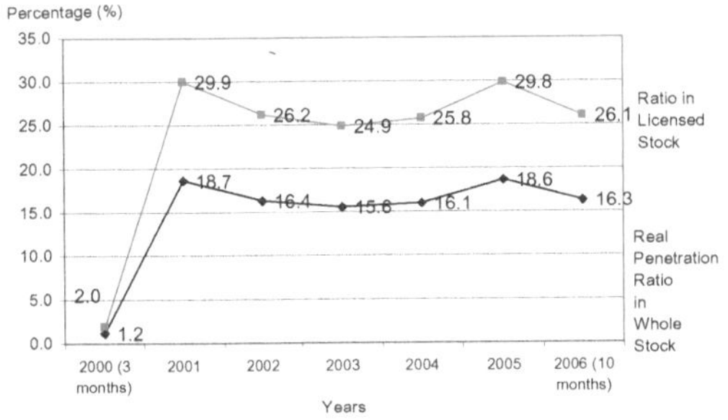

Figure 2. The Ratio of Insured Housing Units into the Housing Stock and Licensed Housing Stock.
Source: Reproduced from DASK (2006) and SIS (2000).

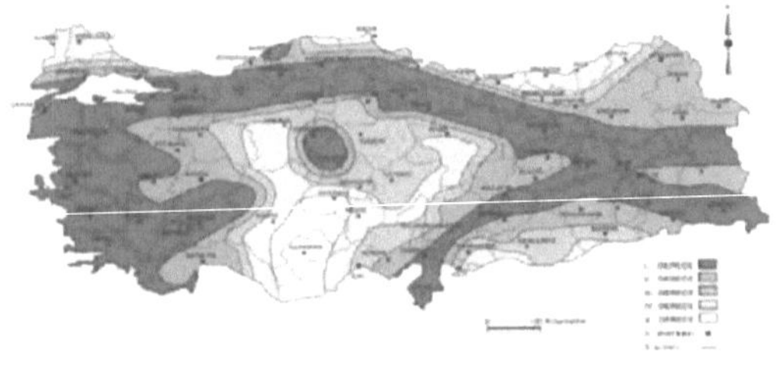

Map 1. The Earthquake Hazard Map of Turkey.
Source: General Directorate of Disaster Affairs, Turkey.

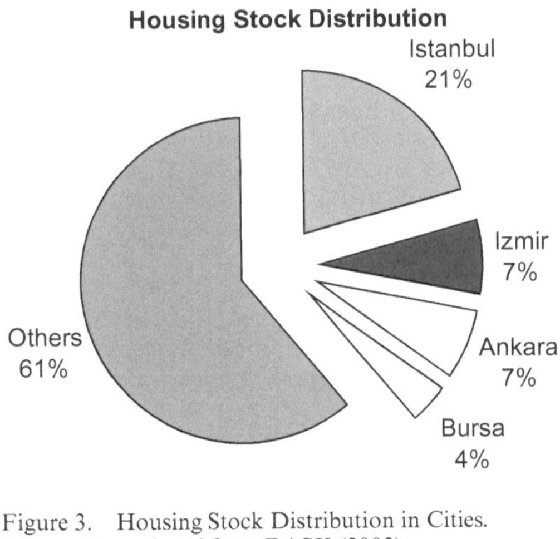

Figure 3. Housing Stock Distribution in Cities.
Source: Reproduced from DASK (2003).

aspects of the current situation as the property ownership rights and social fairness. If the unlicensed dwellings are continued to be insured, they would gain the right to be paid after a possible earthquake.

In other words, they would be legitimized after disaster strikes and get the right to re-build. On the other hand, if they are not comprised by the

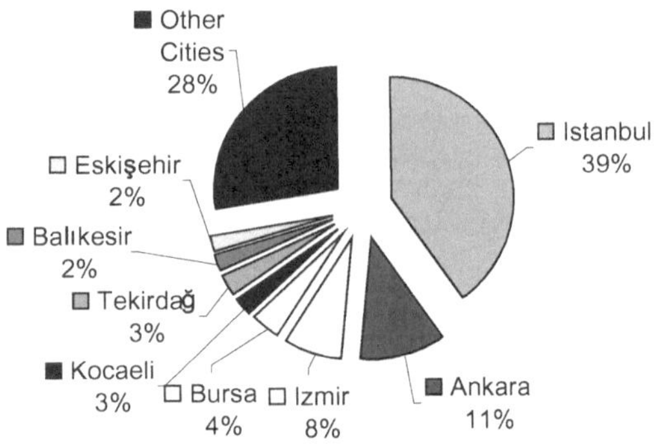

Figure 4. OEI Policy Distributions in Provinces.
Source: Reproduced from DASK (2003).

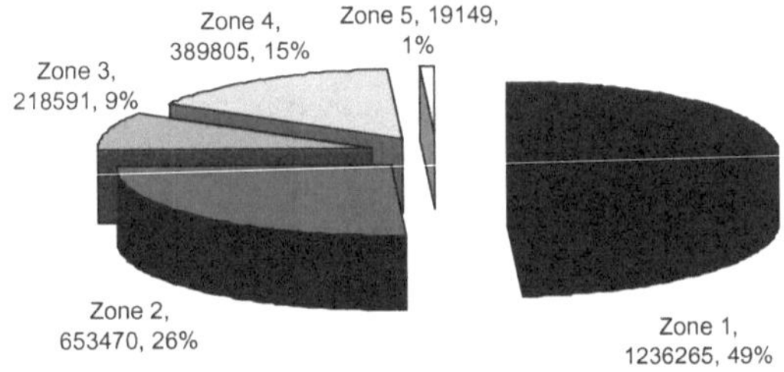

Figure 5. The Distribution of Insurance Policies in Hazard Zones, Turkey.
Source: Reproduced from DASK (2006).

system, they would not have any other right as being financed by the State. Finally, if the penetration ratio is not increased within the licensed stock, only way to increase it remained as including these most vulnerable stock to the insurance system.

Total compensation amount was 73,3 billion YTL (US$ 50 billion), when the payment power is around US$ 1 million with the support of reinsurance (Gencosmanoglu, 2005). The losses paid until 2005 cost 8,3 million TL. Istanbul has 21% share in the housing stock, when Ankara and Izmir have each 7% and the other provinces 65% in total (Figure 3). The policy amounts were distributed nearly in the same sequence in the year 2003. Istanbul had the most written insurance policies with the ratio 39% in the total sold OEI insurance policies in the year 2003. Izmir had 8% share in the policy distribution, when Ankara had 11%, although Izmir was settled at the 1st degree hazard zone (Figure 4). Considering the 70% of the population living in the 1st degree hazard zones, the distribution of the insurance ratio is 49% in the first degree hazard zone is not amazing[5] (Figure 5).

[5] TCIP, www.dask.gov.tr (05.12.2003)

2.3 *The threats and requirements for DASK*

In regard of the obligatory implementation of the insurance, the penetration ratio around 18% since the beginning of the OEI may lead the insolvency of the DASK. The possible problems that are faced in the earthquake hazards and insurance mechanisms, especially in the case of lower reinsurance amount, are the ambiguity of risk low probability high consequence event, high correlations of claims from a single event, asymmetric information problems (moral hazard and adverse selection).

a. *Ambiguity of risk* leads difficulty in identifying what losses may occur. Specific site conditions and construction standards can cause wide variations. The losses from a single event can be tremendous, because earthquakes are *low probability/high consequence events* (Kunreuther, 1998).

   Identifying and assessing the risks are difficult due to rapid urbanization and construction process in urban areas of Turkey. Both the licensed and unlicensed housing stock is vulnerable to earthquake.

b. Catastrophic losses may occur in the case of being exposed of all or many units of the insured group to the same loss causing event and at the same time. When an earthquake damaged all its insured units simultaneously, insurer can suffer from a catastrophic loss (Athearn, 1969). Because earthquakes are geographically focused events, the concentration of claims in a single area makes it harder to diversify risk and may cause the *high correlation of claims from a single event* (Ganz, 1998). An insurer may face this problem, if all eggs are in the same basket (Kunreuther, 1998).

   Under the ambiguity of risks condition in the housing stock, the distribution of the insurance policies in the hazard zones shows that the policies in the first degree hazard zone constitute 49%, when the second Zone has 29% of the insured policies. There is also expected a big magnitude earthquake in the Istanbul city-region, where the most of the capital of Turkey is invested. For this reason, DASK should diversify its portfolio across the country and also in the urban areas using detailed hazard and risk maps based on the detailed risk estimation.

c. There are information asymmetry problems as adverse selection and moral hazard in the insurance mechanisms. *Adverse selection* results from the tendency of people that have greater probability of loss to seek insurance than the average (Athearn, 1969). People at high-risk will buy insurance immediately, if the average price that is determined by an insurance company is low for them. People at low-risk will avoid buying insurance because of the high prices (Kohler, 1982).

   The threat is that as a result of the people in hazardous areas are buying insurance more than the others, there insurance policies may be sold more than the high-risk exposure people and DASK may suffer from high losses in one area.

d. Although *moral hazard* in natural hazards is limited, because the individuals could not control time and magnitude of natural disasters, moral hazard problem may lead insured users of the dwellings a lack of incentive to take preventive action (Li, 1998). The insured homeowners feel safer and have less interest in adopting mitigation measures, due to the security provided by hazard insurance for reconstruction after an earthquake.

   In Turkey, it is common to add additional storages to the buildings without any permission, which brings more burdens to the structural system of the buildings. There is not enough supervision. The land title deed is enough to purchase insurance. Moreover, there is not any provision or culture for risk reduction.

e. To overcome the insolvency risk, transferring the risk to reinsurance market or using other risk transferring tools, requires investing huge amounts in global reinsurance markets, which makes the pool dependent on the global reinsurance markets and which requires selling more insurance policies. The possible solutions for increasing the amount in the pool and making it a sustainable resource, which is not dependent on the global reinsurance market for financing earthquakes, requires to be included into a general risk management framework that regards linking insurance with risk mitigation and assessment (Kunreuther, 2004; Gurenko, 2004).

   The penetration ratio of DASK is still around 18%, which makes difficult to accumulate enough in the pool. It is not under any comprehensive risk management that aims to reduce the risks.

f. Besides, the implementation problems due to the *inconsistency with the flat ownership law* (No: 634), may lead to experience complications involved in the ownership rights after disasters. If disaster strikes the building, and if all homeowners in the same building are not insured, it is impossible to retrofit some part of the buildings, when other parts remain due to lack of insurance (Gülkan, 2002). Moreover, it is not fair to reward the unlicensed buildings with the reason that they bought insurance, while the licensed buildings did not buy, under the conditions of insufficient knowledge and awareness in the society.

3 THE CONTEXT OF THE DISASTER RISK MANAGEMENT IN URBAN AREAS

The general framework for risk management and to link insurance with risk mitigation could be in the context of "urban risk management' (URM), which is a more comprehensive method of disaster management and offers cooperation with other actors that take place in the risk mitigation. For Balamir (2000), the priorities in the URM such as risk avoidance, assessment, risk reduction and transfer and share, regards the *ex-ante* and *ex-post* stage of disaster management in relation with urban planning. Balamir (2002) also emphasized that the characteristics of the formation of and intervention in the vulnerability of the social and physical urban environment against earthquake hazard, requires settlement scale for risk mitigation interventions. Risk mitigation requires socio-spatial interventions ability to control the development of society.

Balamir (2002) adds that urban planning has abilities based on its techniques and methods to intervene and change the physical environment based on the land use decisions, which is also legitimized with the Development Law (No:3194). Besides, urban planning has the management ability to implement such plan decisions with its close relationships to the local managers and with its strategic consideration of all stakeholders to find a common way for the public interest.

Urban Planning uses risk based plans that consider all risk sectors in urban areas in a hierarchical plan order as contingency plans, master plans and local action plans. Specifically, local action plans includes public safety projects in terms of the risk mitigation programs. Risk related land-use decisions may be evacuation, renovation, rehabilitation, regeneration and retrofitting of the buildings depending on the earthquake risk level and considering the city as a whole with its socio-economic characteristics. However, to implement these plans, it needs financial resources and incentives.

Moreover, strategic planning methodology, which regards all involved stakeholders. Actually, urban planning is a social and political process which is dealt with many actors in different sectors in urban areas such as economic, social, and political and hazard management in urban areas (Friedman, 1987). As Britton (1995) agreed, insurance and urban planning should cooperate in the case of disaster risk, because insurance sector and insurers are also dealt with the same sectors. In fact, because OEI is the most important insurance actor involved with earthquake insurance in residential areas in Turkey, it should cooperate with urban planning.

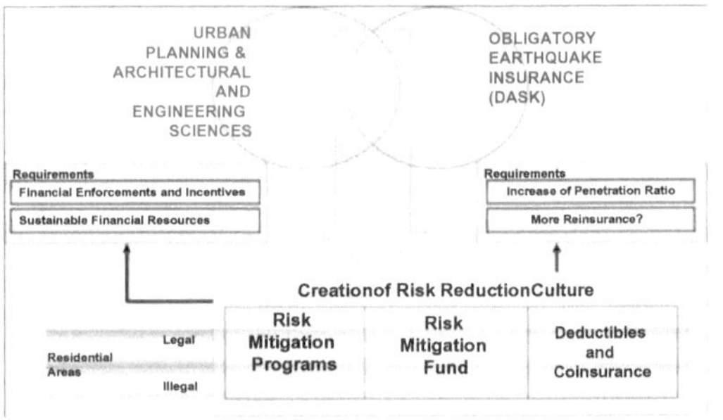

Figure 6. The Conceptual Framework of the Cooperation of OEI and Urban Planning to Create Risk Reduction Culture.

3.1 *The cooperation of urban planning and OEI in the context of URM*

DASK is one of the most important stakeholders dealing with earthquake risk intervention, which should be in cooperation during the intervention for risk mitigation with other stakeholders. Hence, TCIP may develop a coinsurance methodology and deductibles from insurance premiums so that as an incentive instrument, it may encourage homeowners for mitigating their risks. From the side of the DASK, the solutions of its insolvency risks are risk assessment, risk mitigation, risk communication, and design of the deductible and coinsurance mechanisms.

Because DASK does not have the technical capabilities to do everything by itself, it needs to cooperate with other stakeholders in the frame of URM. On the other side, the requirements for the solutions of the problems in the insurance side provide the financial resource and incentives required to implement risk mitigation programs by the local administrations in the frame of URM.

Moreover, the solution of those problems and threats of TCIP is in the cooperation with urban planning, architectural and engineering sciences to increase the policy amount with risk assessment and risk mitigation in the frame of URM. TCIP can be a financial source contributing to another fund that may serve insured homeowners and the central and local governments for risk mitigation.

OEI may contribute to such a partnership with its financial risk assessment techniques, deductibles and with small amounts (i.e. 5% from the annual income of the pool) from the insurance pool to another risk reduction pool (Balamir, 2004). While deductibles can serve as an incentive instrument to encourage risk mitigation, the support for risk reduction can create an important financial resource to realize the urban regeneration projects in action areas of contingency risk plans.

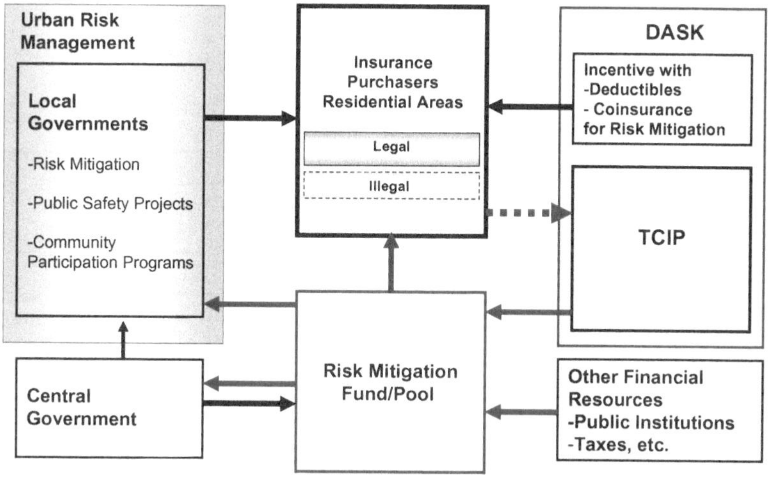

Figure 7. The Conceptual Flowchart for Urban Risk Management with Insurance.

3.2 *Urban disaster risk management with insurance*

In fact, with the low written insurance policy amount, OEI cannot be expected to be solely financial source for risk mitigation, at least till the high increase in the policy amounts. There can be constituted a Risk Mitigation Fund (RMF), which will have other resources such as taxes and transfer in some proportions from the income of public institutions. RMF could give credits to the homeowners to participate in the risk mitigation programs RMF will also give financial support to local administrations in terms of their efforts to provide risk reduction projects (Balamir, 2004; Earthquake Assembly Report, 2004; Kral, 2004).

On the other side, local administrations can take an important role in the increase of the written earthquake insurance policies (Balamir, 2004) through creating risk communication programs and some other incentives (i.e. creating checkpoints in the payments of the real estate taxes, electricity and water bills) within their administrative zones, so that TCIP can sell more insurance policies. Another gain by TCIP through this cooperation, can be to insure the residual risk, after the most of the risks are reduced, which gives the opportunity not to invest huge amounts from the pool in reinsurance. Consequently, this provides not only to increase the insurance policies, but also to increase the capacity of the pool keeping huge amounts in the pool and after a while, to contribute to the RMF more than at the beginning (Balamir, 2004).

Besides, designing of deductible and coinsurance mechanisms according to the risk levels will provide TCIP monitoring the ambiguity of risks and preventing information asymmetry problems, such as moral hazard and adverse selection. This kind of programs requires detailed risk assessment to constitute a variety in the insurance premium amounts in terms of the geo-scientific, civil engineering and urban planning risk researches (Kunreuther, 1998), which are in the URM framework in cooperation with local administrators (Balamir, 2000). With the cooperation for risk assessment with local administrations, the less risky buildings can pay less insurance premium amount, while the high risky buildings will pay more. In turn, such cooperation will provide an incentive to the URM implementing and persuading homeowners to take risk mitigation measures and participate in the risk mitigation programs.

However, the buildings covered by OEI are only the legal residential buildings in the zones of the local administrations. Although it is known that nearly half of the buildings in the residential areas (especially in Istanbul) are illegally built which do not have any disaster assistance according to the current implementation. During urban risk mitigation programs, which could not be in the single building scale, it is difficult to exclude those illegal buildings (i.e. one of the two neighbor buildings can be legal, while the other can be illegal). For this reason, creation of local safety programs that provide legality to these buildings and give the opportunity to buy earthquake insurance, after participating in local risk mitigation programs, will provide to rehabilitate and control the residential building stock.

4 EVALUATION

The cooperation between DASK and other stakeholders risk assessment, risk mitigation, risk

Table 1. The Comparison of the Household Income, Insurance Premium and Housing Expenditures According to the 20% Income Groups.

	20% Income Groups					
	1.% 20	2.% 20	3.% 20	4.% 20	5.% 20	Average
Average Annual Household Income* (Billion TL)	3,67	6,22	8,77	12,57	29,11	12,07
Proportion of the OEI Premium** to the Average Annual Household Income*	1,7	1,0	0,7	0,5	0,2	0,5
Proportion of the Housing and Rent Expenditures*** of the Households to the Average Annual Income*	29,7	30,2	29,7	28,6	26,7	28,3

The table is produced or the steel or reinforced concrete framed structures in the first degree hazard zone.
* State Institute of Statistics of Turkey: The Urban Household Income according to the 20% Income Groups by the year 2003.
**According to the Tariff Implementation of DASK, the construction value of property for 1 m2 Steel or Reinforced Concrete Framed Structures is accepted as 310 Million TL and the annual insurance premium in the first degree earthquake hazard areas amounted as 62 Million TL. (DASK, www.dask.gov.tr (05.12.2003))
*** State Institute of Statistics of Turkey: The Consumption Pattern according to the 20% Income Groups by the year 2003.

financing and sharing takes place in the URM framework, may provide a risk reduction culture with effective risk mitigation policies especially in the residential urban areas. However, the participation of the households living in urban areas in these policy and programs to be envisaged requires investigation of the decision process of the households to reduce risks and purchase earthquake insurance. The reasons of not purchasing by households may be several.

Firstly, when OEI is not actually obligatory due to its uncompleted legal process, it does not offer any effective incentive to encourage the household to buy insurance. It is carried out with the decree law (No: 587) which is not authorized to include monetary punishments according to the Turkish Laws. Turkish Parliament has not passed the draft law since 2002[6]. The most affective sanction is accepted that the State has pledged not to pay the losses any more. Besides, OEI policy is required during the house buying and selling process in Title Deed. In addition, there is also an incentive as the 10% discount rate for at least 8 independent units in one building.

The resistance for not putting OEI in force with real sanctions -or not passing the draft law in the Turkish parliament- led to rise another question: Is the compulsory implementation a right policy, or not? In fact, the non-obligatory earthquake insurance implementation creates a voluntary market, which can be another policy preference. However, there are also discussions about the affect of possible monetary punishment on the penetration ratio of insurance[7].

Instead, there are intended sanctions as the requirement of OEI for the subscription of water, natural gas, and electricity (Gencosmanoglu, 2005). Nevertheless, these are insufficient and there are required real incentives to increase insurance ratio. In addition, because it is not obligatory and State is continuing to pay the disaster losses after the Bingöl and Afyon Sultandağı Earthquakes, people are expecting disaster relief from State, which decreases buying earthquake insurance.

On the other side, there is not enough study showing the influence of such factors as social, cultural, economic and psychological factors. In the case of unaffordable premiums, there can be developed other financial instruments for lowest income homeowners such as micro-insurance mechanism as sub-programs of OEI. If the risk perception of the homeowners is really low, there can be developed risk communication programs in the public.

According to the survey of Fişek (2002), income level and education are influencing insurance and risk mitigation decisions, while risk perception has no effect. However, the 2003 Household statistics (SIS, 2003) shows that the insurance premium constitutes only in the range of 0,2 (for the highest 20% income group) and 1,7 (for the lowest 20% income group), when the proportion of the housing and rent expenditure to the average annual income is in the range between 30,2% and 26,7% of the 5 income level groups.

[6] In fact, due to the government change after the new elections, the draft law couldn't be discussed at the parliament and it is remained as belonging to the old government and it is expired.

[7] (1) Requirement of OEI for Water, Natural gases, Electricity; (2) Authority given to the Apartment Manager (Law: 634); (3) Monetary Punishment after 3 years

5 FURTHER STUDY AND SUGGESTIONS

To develop suitable and implemental risk reduction policies and culture, there is required to search the factors influencing the decision process of Households for risk mitigation. The evaluation of the factors influencing the Households' Decision for purchasing (obligatory) earthquake insurance and taking preventive measures can give direction to the policies. The further research will focus on the possible factors that affect the decision process of households in risk mitigation and buying insurance. Therefore, there can be envisaged programs that integrate risk mitigation and insurance, which will create a risk reduction culture in the society. These factors can be *demographic and socio-economic factors*, *psychological factors* (the hazard anxiety, risk perception, self-efficacy, action-coping and perceived behavioral control) *social and cultural factors* (social context, expected government relief, cultural characteristics, protection responsibility, trust in society and institutions, sense of community), and *the features of risk reduction tools*. To search the factors affecting the decision process of households to purchase insurance and to take risk mitigation measures can provide different type of risk communication and persuasion programs.

ACKNOWLEDGEMENT

This paper was written while the author was visiting the Chair of Insurance Science, University of Karlsruhe in Germany, as a Scientific Guest Researcher and DAAD (Deutscher Akademik Austauschdienst) Scholarship holder. The author is grateful for the financial support provided by the Chair of Insurance in the University of Karlsruhe during the scientific searches within the frame of DAAD Sandwich Programme. She greatly appreciates the helpful comments of her advisor Prof. Murat Balamir (Middle East Technical University, Ankara, Turkey) and co-advisor Prof. Ute Werner (University of Karlsruhe, Karlsruhe, Germany). In addition, she would like to thank to the audiences for their useful suggestions and support. All the opinions and interpretations expressed in the paper are, however, the responsibility of the author.

REFERENCES

Athearn, J.L. 1969. *Risk and Insurance, Meredith Corporation*, United States of America.

Balamir, M. 2004. The current developments and the expectations in the earthquake issues. *Journal of the Chambers of City Planners and Union* 27: 15–23.

Balamir, M. 2002. Urban Risk Management: Methods and Tools for Safety Design, in E. Komut (ed.), *Natural Disasters: Design for Safety*: 26–54. Union of International Architects-Chamber of Architects of Turkey, Kardelen Ofset (in Turkish).

Balamir, M. 2001. *Disaster Policies and Social Organisation*, unpublished paper presented at the Disasters and Social Crisis sessions of the European Sociological Association Conference, Helsinki 27 August-3 September, Finland.

Balamir, M. 2000. Is Turkey in the Process of Restructuring its Earthquake Strategy? *Mimarlık* 295: 44–47 (in Turkish).

Fişek, G.O., Yeniçeri, N., Müderrisoğlu, S. & Özkarar, G. 2002. *Risk Perception and Attitudes Towards Mitigation*, IIASA-DPRI Meeting Integrated Disaster Risk management: Megacity Vulnerability and Resilience Laxenburg, Austria, July 29–31.

Ganz, S. 1998. *Public Policy Options and Earthquake Insurance.* Paper represented in Western Earthquake Insurance Summit presented by Western States Seismic Policy Council, www.wsspc.org.

Gencosmanoglu, E. 2005. *Compulsory Earthquake Insurance*, presented in the Seventh Round Table Meeting on Disaster Management in Turkey, Middle East Technical University, Convention Center (Hall - C).

Gurenko, E.N. 2004. Building Effective Catastrophe Insurance Programmes at the Country Level: A Risk Management Perspective, in Gurenko, E.N. (ed.), *Catastrophe Risk and Reinsurance: A Country Risk Management Perspective*. London: Risk Books.

Gülkan, P. 2002. *Setting the Stage for Urban Risk Mitigation: Seismic Risks and Compulsory Insurance Policy Issues in Turkey*, Second Annual IIASA-DPRI Meeting, Integrated Disaster Risk Management: Megacity Vulnerability And Resilience, July 29–31, Laxenburg, Austria.

Kıral, O. 2004. The Evaluation, Suggestions and Counterviews for the Earthquake Insurance Commission Report in the Earthquake Assembly. *Journal of the Chambers of City Planners and Union* 2004/2.

Kohler, H. 1982. *Intermediate Microeconomics: Theory and Applications*. USA.

Kunreuther, H., Deodatis, G. & Smyth, A. 2004. Integrating Mitigation with Risk Transfer Instruments, in E-N Gurenko (ed.), *Catastrophe Risk and Reinsurance: A Country Risk Management Perspective*. London: Risk Books.

Kunreuther, H. 1998. *Insurance as an Integrating Policy Tool for Disaster Management: The Role of Public-Private Partnerships*, paper represented in Western Earthquake Insurance Summit presented by Western States Seismic Policy Council, www.wsspc.org.

Li, W. 1998. *Hazard Insurance and Mitigation Measures For Natural Hazards*, paper represented in Western Earthquake Insurance Summit presented by Western States Seismic Policy Council, www.wsspc.org.

Selcuk, S. 2000–01. *The Feasibility Report for the Natural Disaster Insurance of Turkey*, METU, Disaster Management Implementation and Research Center, Report No: METU/DMC (Turkish).

Selcuk, S., Yazıcı, S. & Gülkan, P. 2001. Compulsory Earthquake Insurance Scheme for Residences in Turkey and its Financial Implications. *Journal of International Insurance* Vol. 2 Issue 4, Kluwer.

Wilczynski, P.M., Chhibber, A. & Goldberg, J.R. 1999. *Turkey – Marmara Earthquake Emergency Reconstruction Project*, The World Bank Project Appraisal Document Report.

Li, W. 1998. *Hazard Insurance and Mitigation Measures For Natural Hazards, paper represented in Western Earthquake Insurance Summit*, presented by Western States Seismic Policy Council, www.wsspc.org.

DASK, www.dask.gov.tr (05.12.2003)

DASK, www.dask.gov.tr (29.10.2006)

Earthquake Insurance Commission Report in the Earthquake Assembly, 2004, Ministry of Public Works of the Turkish Republic, Ankara.

General Directorate of Disaster Affairs, Ministry of Public Works and Settlements of Turkish Republic, Turkey.

The Undersecretariat of Housing (UH), 2003. The Housing Need Survey for 2000–2010 in Turkey, Ministry of Public Works and Settlements of Turkish Republic.

State Institute of Statistics of Turkey, 2003. The Urban Household Income according to the 20% Income Groups.

State Institute of Statistics of Turkey, 2000. The Census of Buildings.

Risk mitigation criteria in land use planning and critical infrastructures siting and design

Risks Challenging Publics, Scientists and Governments – Menoni (ed)
© 2010 Taylor & Francis Group, London, ISBN 978-0-415-58072-4

The "seismic behaviour" of urban complex systems

A. Galderisi & A. Ceudech
DiPiST – Department of Urban and Regional Planning - University of Naples Federico II, Naples, Italy

ABSTRACT: Damages due to earthquakes are varied and heterogeneous, even though the scientific literature has been mainly focusing on physical ones and the consequent loss of human lives. Very few studies have pointed out the multiple dimension of damage, highlighting the heterogeneous kind of damages occurring in the different time periods after an earthquake. This study, looking at the city as a complex system and grounding on the analogy between the behaviour of a complex system hit by an external stress and the city hit by an earthquake, focuses on the "seismic behaviour" of urban systems. In detail, a conceptual framework for describing and forecasting such behaviour is provided. Moreover, the role of urban planning in seismic risk mitigation has been highlighted, pointing out that both structural interventions on buildings and infrastructures and actions facing the seismic problem in its "urban" dimension are required.

1 THE SEISMIC BEHAVIOUR OF URBAN COMPLEX SYSTEMS

The main question that this research work try to answer, basing on the analysis of some international case studies, is: "What happens when an earthquake strikes a city?"

To answer such question, a comparison between two case-studies has been worked out: Los Angeles-Northridge (17th January 1994) and Kobe (17th January 1995) earthquakes. The two earthquakes had similar magnitude (6.9 for Kobe and 6.7 for Northridge) and occurred about at the same hour (both of them occurred before the dawn). Nevertheless, they hit two metropolitan areas very different with respect to urban patterns, distribution of activity and infrastructures on the territory and so on. Thus, in spite of the similarity between the two natural events, the consequences were very different[1]. The Northridge disaster was evaluated in about 15 billion of dollars, one of the most expensive natural disasters ever occurred in the United States, while the Japanese one was evaluated in about 147 billion of dollars (Maddox, 2002).

Such lack of proportion between the damages occurred in the two urban areas was due to different factors. First of all, it was due to the characteristics of urban patterns. Northridge urban texture is regular, often orthogonal, highly accessible; in Kobe it is messy and developed in one main direction. Furthermore, Kobe is located between the sea and the mountains with few transversal connections and accesses from the outside; furthermore its ancient core is characterized by a dense texture of little and scarcely accessible alleys. These two different urban patterns determined different conditions of accessibility for the rescue teams.

Secondly, fires –caused by the break-up of gas supply network system– occurred in Kobe after the earthquake although about 1.300 well trained fire officers were employed in the emergency. The situation was made even worse by the presence of many wooden old houses, of narrow alleys and roads which determined serious time lags before fires were extinguished and by the interruption of water supply.

Summing up, the comparison of the two case-studies highlight that *when* an earthquake strikes huge urban areas, different damages occur, only partially related to the geologic characterization of the grounds. Damages depend on the event but even on the features of the hit areas, such as the age and the technological and structural characteristics of buildings and infrastructures, the morphology of urban patterns, the building and population density, the decay of built stock, the accessibility and permeability of urban areas. Moreover, the Kobe event clearly highlights the many interactions among different events determining different damages which can rise during and after an earthquake (EQE International, 2002; USGS, 1996). It highlights the complex dimension of a seismic disaster in a large metropolitan context. The seismic impact is not only physical, but also functional,

[1] The Northridge earthquake provoked 57 died and about 1.500 injured, while the Kobe earthquake produced 5.500 died and about 35.000 injured. Buildings with damages from moderates to severe were approximately 12.500 in Northridge, against the 180.000 of Kobe.

organizational, social. Moreover, the damage, in its wider and multiple meaning, is determined by the interaction of phenomena that may occur in different times after the earthquake.

The multiple dimension of damages due to an earthquake has been analyzed also in the context of the 1980 earthquake in Naples. In detail, in this case many damages or, better, holdups, apart from the physical ones, occurred in the first 12 hours after the earthquake: traffic jams, difficulty to access to emergency facilities, lack of safe open spaces, lack of network services, difficulty of the rescue activities (Galderisi & Ceudech, 2005).

Briefly, the investigated events show that in the earthquake instant, physical damages to buildings and consequent loss of human lives occur. Then, during the first 24 hours after the earthquake (emergency phase) other kinds of damages not entirely related to the physical ones (such as difficulty to provide health care services; difficulty to shelter people; traffic jams; difficulty to provide basic network services as water, gas, telephone, etc.), may occur. Then, at medium and long term, economic and social damages (such as loss of activities, unemployment, etc.) generally come out.

Damage is a multidimensional concept which cannot be reduced to the physical effects only. Nevertheless, till now physical damage and human sufferings have been studied in depth, while "functional damages" which may occur in the immediate event aftermath, have been scarcely investigated, even if they seem to be as relevant as the physical ones, inducing long term effects, with economic and social consequences, up to the "death" of the city.

Even if the idea that cities are complex systems, consisting of elements strictly interrelated, is now widely shared, till now, only few studies, looking at the city as a whole, have analyzed the "multiple" dimension of the damage (physical, functional, economic, etc.), mainly due to the complex interactions among urban elements.

Interpreting the city as a complex system, it is possible to achieve a better understanding of the consequences due to the impact of an earthquake. The analogy between the behaviour of a complex system hit by an external stress and the behaviour of a city hit by an earthquake could be very helpful.

System theory has been deepened since the end of the 1940s in many disciplinary fields, such as biology, cybernetics, etc. A first systematization of the studies was provided by the "General System Theory" (von Bertanlaffy, 1971). During the 1960s and the 1970s, system theory was applied to the analysis of territorial and urban phenomena (Alexander, 1967; Mc Loughlin, 1973). Starting from the 1970s, many theories related to the complex system behaviour have been worked out[2], highlighting the role of non-linear relationships among the system elements, their adaptive features and the role played by the "chance" in their evolution (Thom, 1972; Prigogine, 1976).

With respect to the theories related to complex systems and to their behaviour, it is possible to state that a complex system hit by an external stress can follow three different directions: restore a dynamic equilibrium; evolve towards a new organization, or to die. Given initial similar level of stress, those different outcomes are determined, by the "coping capacity" of the system or, in other words, by the "resilience" of the system. But how to define the resilience of a complex system?

The concept of resilience has been introduced at the beginning of the 1970s to point out the capacity of natural systems to absorb perturbations, preserving their functioning and structure. Thus, the resilience depends on the "load" that a natural system can absorb without changing its structure and processes. Hence, a resilient system is a system able to absorb perturbations, restoring its equilibrium (Holling, 1973).

Subsequently, the resilience concept has been deepened with reference to complex systems, highlighting the capacity of such systems, with respect to an external stress, to renew and re-organize themselves (Gunderson & Holling, 2002), reaching a new state of equilibrium. Thus, the resilience has been defined as the amount of disturbance that a system can absorb and still remain within the same state or domain of attraction, or as the capability of the system to renew and reorganize itself when change occurs (Carpenter et al. 2001; Folke et al. 2002). It is useful to highlight that in a resilient system, as stated by Folke et al. (2002), change has the potential to create an opportunity for development and innovation, shifting the system from the current equilibrium to another one.

The topic of resilience has been widely faced during the World Conference on Disaster Reduction held in Kobe in January 2005. In the Conference final Document, resilience has been defined as the capacity of a system, community or society, potentially exposed to hazard to adapt to change, by resisting or modifying itself, in order to maintain or to reach an acceptable level of functioning and structure. Briefly, it is possible to state that a resilient system can absorb the shock, restoring a dynamic equilibrium or evolving towards a different organization.

Strictly related to the concept of resilience is the concept of vulnerability. The term vulnerability points out the predisposition of people, goods or

[2]Complexity Paradigm, Catastrophe Theory, Dissipative Structure Theory.

Figure 1. Landscape of the dead town of Poggioreale (Source: R. Carmassi, I paesi abbandonati, Parametro, 251, Edizioni C.E.L.I., Faenza, 2004).

activities to be damaged by an external stress: from this point of view, it represents a measure of fragility, of the incapacity of people or community, buildings, infrastructures or territory, to absorb the stress. Vulnerability of a system, as stated by Fortune and Peters (1995), is the obverse of its ability to absorb disturbances, which in turn can be considered to be an indication of its resilience. Folke et al. (2002) consider vulnerability as the flip-side of resilience too, highlighting that in a vulnerable system even small changes may be devastating. The strict correspondence between the concepts of resilience and vulnerability is clear with reference to the most recent definitions of vulnerability. These latter, even though referred to the community, widen the concept from the propensity or susceptibility to be damaged, to the incapacity to anticipate, to face, to withstand and to recover from the impact of a calamitous event (Wisner et al. 2004).

Thus, according to the proposed analogy between cities and complex systems, it is possible to assume that a city hit by an earthquake will follow three evolution trajectories, depending on the resilience of the city itself. These different trajectories represent the different "seismic behaviour" of cities. The first possibility is to absorb change, restoring in a short time its equilibrium, though inevitably different from the previous one. The second chance is that the system will shift to a different state, using change as an opportunity for renewal and redevelopment. For example after the Messina and Reggio Calabria Earthquake, in 1908[3], the city of Reggio Calabria was rebuilt with a regular and orthogonal schema which completely changed the original settlement. Finally, the third chance for the urban system is to lose its functioning and structure, and eventually die. Such an evolution affected many small towns seriously damaged during the Val di Noto Earthquake, in 1693[4] (Figure 1).

With respect to mentioned above meanings of the resilience concept, it is possible to state that, also the concept of urban resilience/vulnerability, as the damage one, is a multidimensional concept.

The capacity of a city to respond to a given earthquake, or its seismic behaviour, depends on urban resilience which consists of many components, all of them equally relevant: the resistance of buildings and infrastructures; the capacity of the city to cope with the demand for activities and services of the population in the post-event phase; the organizational capacity of institutional bodies; the economic and cultural development of urban context. All these different aspects have to be taken into account in order to analyze the urban resilience. Urban planning can contribute to the analysis and mitigation of all the components of urban resilience/vulnerability. Nevertheless, the main task of urban planning can be identified in the capacity of urban systems to cope with the demand for activities and services of the population in the post-event phase. This component represents the "functional component" of the resilience/vulnerability of the urban system and it's the most directly related to the spatial and functional organization of the city (Galderisi, 2004).

In order to better understand this component, it is possible to state that as the resistance of physical elements depends on the morphological and structural features of buildings or infrastructures which allow them to respond to the stress determined by the earthquake, so the capacity of the urban system to cope with the demand in the post-event phase

[3] The Messina earthquake of December 28th 1908, which induced a violent tsunami which hit the Sicilian and Calabrian coasts. It was, together the Val di Noto earthquake of 1693, one of the most catastrophic Italian seismic events. Its maximum intensity was in Reggio Calabria, where the percentage of building damages and the loss of human lives were certainly the most elevated. The damage of Messina was greater due to its greater population density (90.000 inhabitants against the 27.000 of Reggio Calabria). Probably, the number of victims is included among 60.000 and 120.000.

[4] The Val di Noto earthquake occurred in January 11th 1693, caused the total destruction of more than 45 towns and a total amount of around 60.000 died. The intensity of the earthquake was equal or superior to the IX degree MCS and the hit territory was around 5600 Kmq.

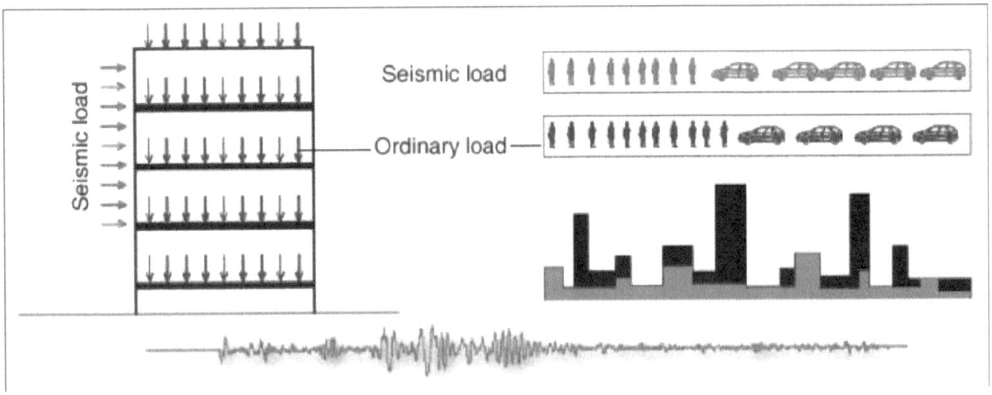

Figure 2. Resistance of buildings/capacity of the city to cope with the activity demand.

depends on the spatial and functional organization of the city which allows the city to respond to the activity demand (Figure 2).

2 THE SEISMIC BEHAVIOUR OF URBAN SYSTEMS: THE FUNCTIONAL COMPONENET OF THE RESILIENCE/VULNERABILITY

Analyses related to the functional damage and the post-event coping activities have been carried since the mid-1970s. For example a model of disaster recovery activity was worked out at the end of Seventies, to describe the coping activities required after an earthquake and the different time periods in which each of them comes out (Haas, 1977; Vale & Campanella, 2005).

According to these studies, the consequences of the 1980 earthquake in the historical city of Naples have been analyzed, describing through curves the evolution of the relation between activity demand and supply have newspaper chronicles, were used as the main intermediate main source, as official reports do not consider functional damages at all.

In the historical city of Naples (Figure 3), the earthquake of 1980 produced widespread and low damages to the built stock even though only a few building collapsed. Despite the relatively low level of physical damage, the seismic event determined a relevant functional crisis due to the lack of functioning in some urban areas, with consequences for large zones of the city which, in many cases, lasted for a long time after the emergency. The historical city of Naples is also characterized by heterogeneous typological features, an elevated population density, the presence of urban activities and relevant tourist flows. Moreover, such as many of historical layers, it is characterized by elevated density compactness and low accessibility due to the irregular secondary road-network with narrow and steep roads. All those features, especially during the periods of maximum touristic flow, determine sometimes also in "peace time" crisis and/or functional congestion due to the inefficacy of the system to supply ordinary activity demand.

Figure 3. The historical city of Naples.

The singling out of the main functional damages occurred in Naples after the 1980 earthquake has been set up following the steps listed below:

- collection and classification of the information related to the event occurred consequences in the first week after the earthquake;
- definition of demand and supply components;
- definition of the functional damage by acquiring scores to the unsatisfied demand components. The scores have been assigned on the basis of qualitative judgments based on newspaper descriptions, frequency of the topic inside the articles and the "weight" of the article (position and length of the article in the newspaper).

For what concerns the choice of the demand and supply components, the research focused on the demand aspects more related to the city spatial and functional features. Thus, with reference to the first 24 hours after the earthquake, we analyzed the following demand components: the access from hit areas to emergency collection points and to health-care services, and the access of rescue teams to hit areas (Figure 4).

With respect to the supply components, the location and distribution of emergency collection points and emergency facilities and the adequacy of road networks have been considered.

With reference to the first week after the earthquake, to the mentioned components of demand and supply, the demand for temporary housing and the supply of temporary accommodations has been added. Obviously, demand components for services and activities are mainly related to population density and distribution, while supply of spaces and activities mainly depends on morphological features of the urban patterns and on activities location and distribution.

Basing on the score attribution for each temporal unit (one day), the curve of total functional damage (which represents the unsatisfied demand) of the historical city of Naples in the first week after the event has been plotted (Figure 5). As it can be seen in the diagram, the trend of the functional shows a damage peak in the first 12 hours (the maximum value is around the 70% of the maximum possible damage). Then, it suddenly goes down to a value of about 50% of the first day value. Subsequently, there is a wave increasing course with some peaks, to the second, fifth and eighth post-event day and some points of minimum, to the third and sixth day. Such trend highlights that the functional damage components determine the crisis of the system with different roles and weights and, above all, they arise in different times. The peak-point of the first 12 hours is due to traffic jams, difficulties of rescue and emergency activities, the lack of emergency collection points and the break-down of network services, such as telephone and electricity.

Time	**Demand**	**Supply**
First 24 hours after the earthquake	Access from hit areas to: Emergency collection points Health care services Access of rescue teams to hit areas	Location and distribution of emergency collection points Location and distribution of emergency facilities Adequacy of road networks
First Week	Temporary housing	Temporary accomodations
Mainly Urban Features	Population consinstency and distribution	Morphological features of urban fabrics and activity distribution

Figure 4. Components of demand and supply.

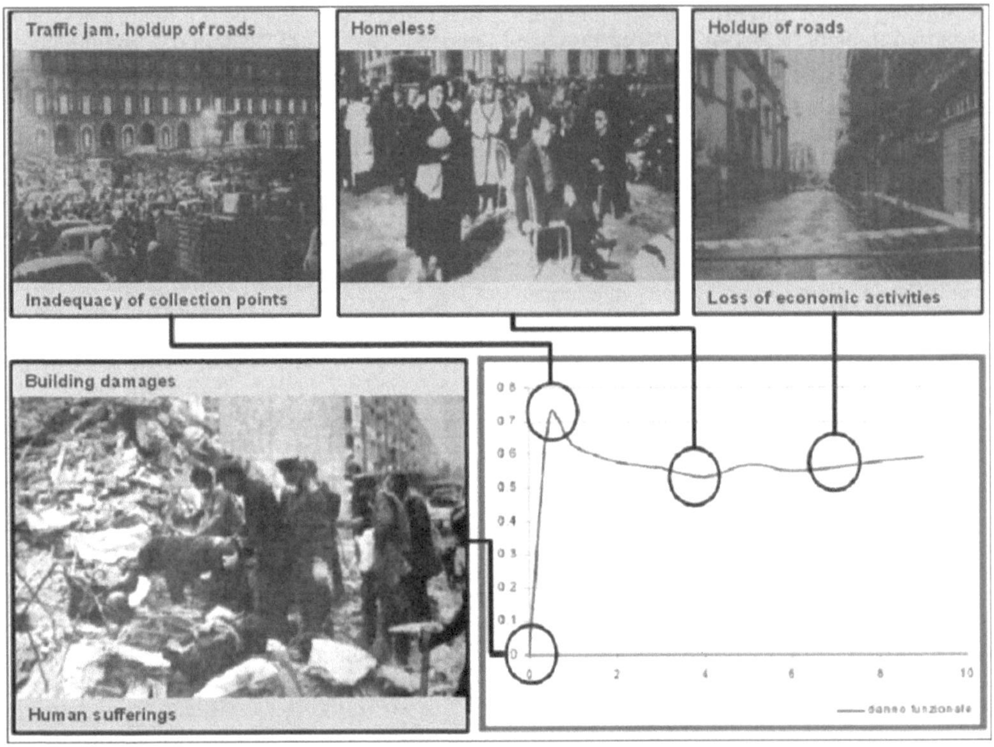

Figure 5. Functional damage of Naples after the 1980 earthquake.

Briefly, the analyses related to the functional component of urban resilience allow us to highlight that:

- the capacity of a city to cope with the demand (for services and activities) in the post-event phase, can be represented through curves which have a "wave" course;
- the curves show thresholds which represent changes of the system state;
- the curves referred to each demand (or to each coping activity) have "different thresholds of entrance";
- the unsatisfied demand (difference between demand and supply) represents the "functional damage" within the system.

3 HOW TO FORECAST THE SEISMIC BEHAVIOUR OF THE CITY

Up to now, only back analyses of seismic behaviour have been worked out. In order to apply the same interpretative model to the forecast of seismic behaviour of a given urban context hit by an earthquake, we have to take into account that complex systems are non-linear; hence, their evolution trajectory is not predictable on the basis of initial conditions, as complex systems are adaptive and capable for self-organization; as outlined by Louise Comfort (1995), "self-organizing systems spontaneously reallocate resources and rearrange their activities to create a better fit between their internal operation and their immediate external environment".

The features of complex systems make even more difficult to provide forecasts related to the urban resilience. In detail, the capacity for self-organization lays heavily on the possibility for cities to cope with the demand of activities and services which arise after the earthquake.

Nevertheless, chosen a seismic event, it is possible to single out the different components of the demand and to draw out some curves, even though simplified, related to the demand evolution. Such a representation allows to identify the most relevant demand risen by functional damage and to compare their evolution with ordinary supply. Thus, urban planning actions aimed at improving ordinary supply or redirecting the demand can be outlined.

For this task, scenario techniques can be applied with efficacy.

Generally, in risk analysis, scenario techniques are applied to choose a particular hazard occurrences, looking only partially at what happens after the event. Setting up scenarios means to accept the ignorance and to build one or more hypothetical orders among different urban phenomena as an answer to the question of: what may happen if, a given event occurred. Scenarios are then built relying on signals, forecasts, hints, tendencies that may be found in the present and/or in the past. (Secchi, 2000). Scenario are also an important tool for the communication of risk to the population and to share information among the experts in charge of its development. From this point of view, the possibility to communicate what concretely could happen in case of seismic event can contribute to improve risk awareness of the community. Nevertheless, until today predictive damage scenarios mainly referred to physical damages and to victims. As main references in applications of the HAZUS (Hazards United States) software developed by the U.S. FEMA (Federal Emergency Management Agency). Can be cited this software, using heterogeneous data, deselecting the hazard characteristics and those of the settlements, provides an estimate of expected damage in terms of human losses, number of homeless, damage to buildings and infrastructures.

The forecast of damages referable to the loss of efficiency of urban systems or, in other words, of the damage due to the overwhelming cities' capacity to cope demand for activities and services generated by the earthquake, has not been studied enough yet. Only recently, some Japanese (Kajitani, Okada, Tatano & Motohisa, 2002) and Italian (Nuti & Vanzi, 1999) authors worked out predictive damage scenarios aimed at outlining functional damages not directly linked to physical ones, taking into account aspects related to the post-event emergency phase, with particular reference to the existing relationships between buildings and infrastructures and situations of functional crisis determined by the lack of performances in emergency.

With respect to these studies, a damage scenario for an area included in the historical city of Naples (Figure 3) has been set up. The chosen event is a VII Mercalli degree earthquake, like the 1980 earthquake, and the occurring time is 20 pm–8 am. With reference to the targets potentially hit by a seismic event and the temporal phases of the emergency, some possible chains of losses have been investigated.

The scenario structure is based on a tree-diagram developed in a row-column framework. In the rows there are different groups of urban targets which can be damaged by the earthquake, for example building stock, people, infrastructural networks, activities and so on. In the columns different times are considered: the earthquake impact and the first emergency phase (24 hours after the earthquake), the short term, the medium long term. For each target, the amount of damage has been determined, considering that each type of damage occurs in a different time (Figure 6).

Chosen the event and the occurrence time, the outlined scenario includes:

- the distribution of damage to the building stock;
- the assessment of people injured or died;
- the assessment of loss of efficiency of the urban system (functional damage).

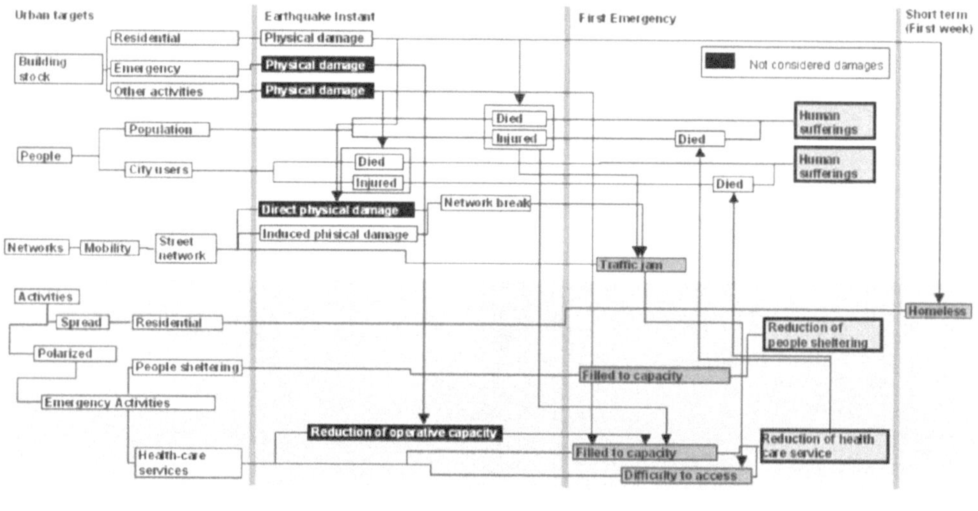

Figure 6. Scenario structure.

The starting point for the definition of the components of functional damage is the setting up of a scenario referred to the earthquake impact when immediate damage to building and to people occur. Such a scenario has been referred to territorial units which represent the grid base of the procedure. The historical town has been divided in 48 territorial units identified according to homogenous features of urban patterns, building age, activities and morphology (Figure 3).

As far as the damage to buildings is concerned, it was decided to equal to the damage to the built stock to that caused by the 1980 earthquake. Data related to the number of damaged buildings and the severity of damages were available in each administrative district of Naples (Rippa & Vinale, 1983). Therefore, a new distribution of building damage has been set up related to the territorial units in which the historical town has been subdivided. A reduction coefficient has been applied to the damage level of each territorial unit in order to take into account the strengthening of buildings carried out after the 1980 earthquake, but such coefficient does not determine relevant variations of the obtained results. Population distribution has been defined using the 1991 Istat census data.

Human losses assessment has been carried out through the determination of the number of died and injured as percentage of the population living or working in each damaged of each territorial units (Figure 7).

Then, we focused on the most relevant components of the functional damage which may occur from the first emergency till the first week after the earthquake, trying to single out the peak-points of the trend of these components. In detail, we analyzed the access of the people to safe collection points, the efficiency of emergency health-care services connected to the capacity and accessibility of these activities (considered in their full efficiency) and, finally, the efficiency of the residential system.

The loss of efficiency of the urban system referred to the access of the population to collection points immediately accessible has been defined through the assessment of the size of open public spaces, of their influence areas and of the open spaces remaining capacity. Hence, the areas characterized by not served population has been calculated (Figure 8).

The assessment of efficiency reduction of urban system related to the emergency health-care activities has been carried out on the basis of the total number of injured that would burden each hospital in the area and the different level of accessibility to such equipments determined by traffic jams along the road network (Figure 9). The distribution of injured along the road network has been carried out through the definition of an index which takes into account the number of injured on each segment of the net and its length.

Finally, the assessment of the efficiency reduction of residential system has been carried out through the estimate of the number of homeless, calculated with reference to uninhabitable buildings. Also for the assessment of homeless, the

Figure 7. Human sufferings and efficiency reduction of residential system.

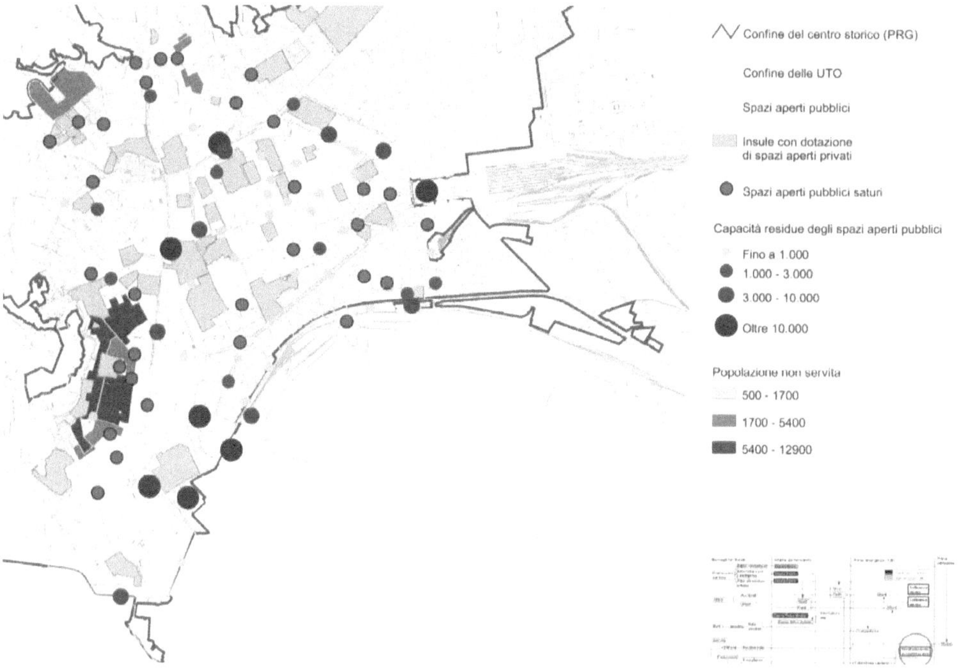

Figure 8. Collection points: not served population.

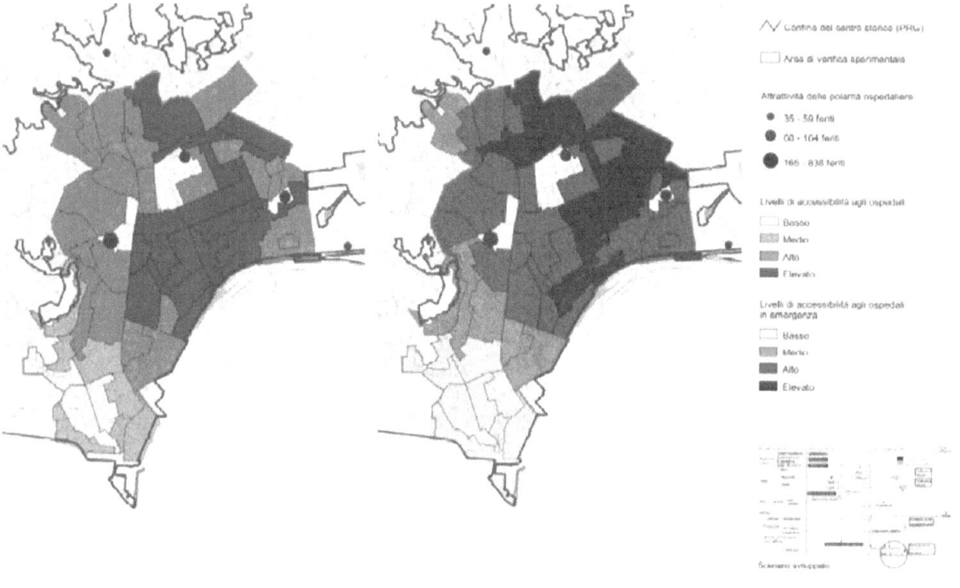

Figure 9. Efficiency reduction of emergency health-care activities

reduction coefficient has been applied in order to take into account the interventions carried out on the built stock, but such coefficient doesn't determine relevant variations of the obtained results.

4 CONCLUSION

As stated above, it is possible to apply the demand/supply interpretative model to the forecast of

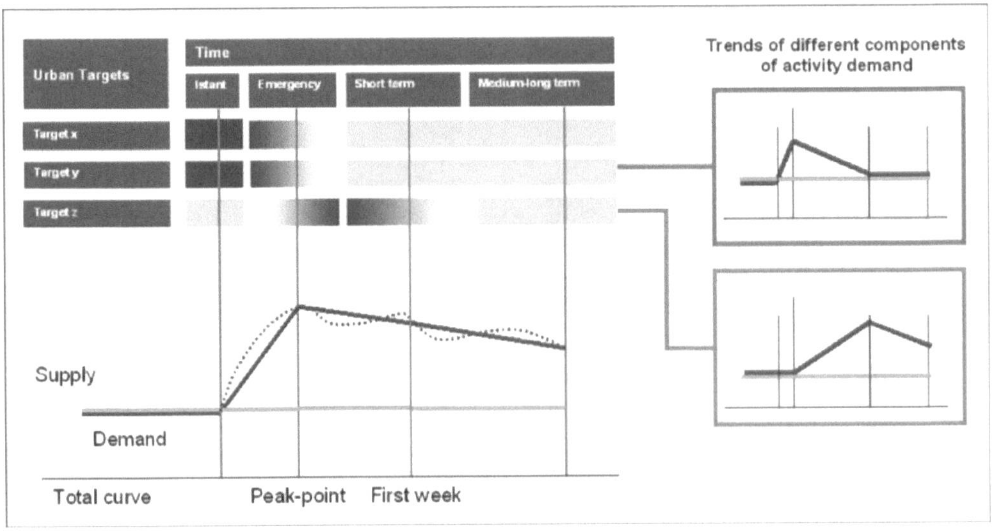

Figure 10. Scenario and trends of activity demands.

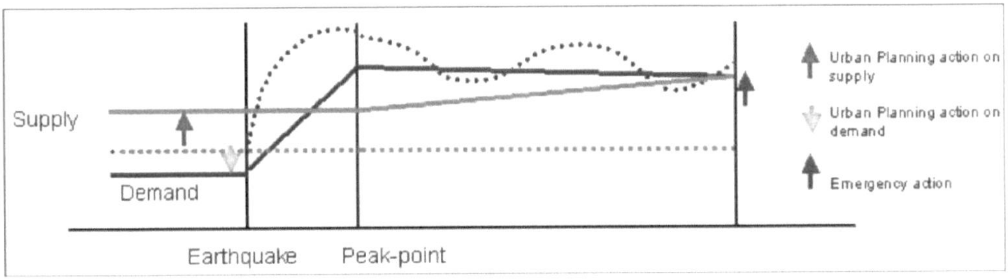

Figure 11. Urban planning and emergency actions on demand and supply of activities.

seismic behaviour of an urban system hit by a given earthquake, or at least some components of this behaviour, taking into account that the evolution trajectory of a complex system is not predictable on the basis of systems initial conditions. Case-study analyses highlight that functional crisis of the hit city depends on the equilibrium between demand and supply of activities and services. Scenario techniques, with respect to a given seismic hazard, can enable us to forecast the maximum intensity of each demand components and to compare them with the ordinary supply levels during the time after the earthquake (Figure 10). Thus, the simplified diagram of the trend of each activity demand at every temporal threshold can be obtained: the difference between the total demand values and the ordinary supply values represents the total amount of functional damage. This amount is, in turn, a simplified trend of the capacity of the city to respond to the earthquake or, in other words, of one of the determinants of the urban system seismic behaviour.

Scenarios represent useful support both for defining urban planning actions aimed at improving supply and/or re-direct demand and defining and scheduling emergency actions, in order to improve the capacity of cities to cope with the demand for activities and services of the population in the post-event phase, which represent, as stated before, one of the components of urban resilience (Figure 11).

The analysis of urban systems seismic behaviour shows that interventions on the built stock one not enough to mitigate seismic risk in cities. The implementation of effective seismic risk mitigation policies requires specific tools, strategies and actions aimed at facing problems in their "urban" dimension. This highlights the key-role of urban planning for the knowledge and the mitigation of the seismic risk in urban contexts, which have been for a long time considered as specific matter of structural and technological disciplines.

REFERENCES

Alexander, C. 1967. *Note sulla sintesi della forma*. Milano: Il Saggiatore.

Carpenter, S.R., Walker, B., Anderies, J.M. & Abel, N. 2001. From metaphor to measurement: Resilience of what to what? *Ecosystems*, 4.

Comfort, L.K. 1999. *Shared Risk: Complex Systems in Seismic Response*. Pergamon.

Di Sopra, L. 1981. Gli aspetti spazio-temporali dei disastri: approccio urbanistico. In B. Cattarinussi, C. Pelanda (eds.), *Disastro e azione umana. Introduzione multidisciplinare allo studio del comportamento sociale in ambienti estremi*. Milano: Franco Angeli.

EQE International, 2002. *The January 17, 1995 Kobe Earthquake* [Online], Available: http://www.eqe.com/publications/kobe/kobe.htm.

Folke, C., Carpenter, S. et al. 2002. *Resilience and Sustainable Development: Building Adaptive Capacity in a World of Transformation*, Scientific Background Paper on Resilience for the process of the World Summit on Sustainable Development on behalf of the Environmental Advisory Council to the Swedish Government [Online], (Available: http://www.sou.gov.se/mvb/pdf/resiliens.pdf.

Fortune, J. & Peters, G. 1995. *Learning from failure*. Great Britain: John Wiley & Sons.

Galderisi, A. 2004. *Città e terremoti. Metodi e tecniche per la mitigazione del rischio sismico*. Roma: Gangemi editore.

Galderisi, A. & Ceudech, A. 2005. Il terremoto del 23 novembre 1980 a Napoli: la ricostruzione del danno funzionale nella prima settimana post-evento. In S. Lagomarsino, P. Ugolini (eds.), *Rischio sismico, territorio e centri storici*. Milano: Franco Angeli.

Gunderson, L.H. & Holling, C.S. (eds.) 2002. *Panarchy: Understanding Transformations in Human and Natural Systems*. Washington, DC: Island Press.

Haas, J., Kates, R. & Bowden, M. 1977. *Reconstruction following disasters*. Cambridge University Press, MIT.

Holling, C.S. 1973. Resilience and stability of ecological systems, *Ann. Rev. of Ecol. and Syst*. 4.

Kajitani, Y., Okada, N., Tatano, H. & Motohisa, H. 2002. Statistical Analysis of Spatial Temporal Distributions of Human Activities after Earthquake Disaster. In Proceedings of the Second Annual IIASA-DPRI Meeting for Integrated Disaster Risk Management [Online], Available: http://www.iiasa.ac.at/Research/RMS/dpri2002/Papers/kajitani.pdf

Maddox, D. 2002. The Great Kobe Earthquake of January 17, 1995, Natural Hazards [Online], Available: http://www2.msstate.edu/~dem35/aboutme/kobe.pdf

Mc Loughlin, J.B. 1973. *La pianificazione urbana e regionale*. Padova: Marsilio.

Nuti, C. & Vanzi, I. 1999. Criteria for evaluation of seismic risk of regional hospital system; a case study: the Abruzzo Region. In Design And Retrofitting of Hospitals in Seismic Areas, Proceedings of the Workshop [Online], Available: http://host.uniroma3.it/dipartimenti/dis/ricerca/Hospitals/Articoli_pdf/nv.01.eng.pdf

Prigogine, I. 1976. Order through Fluctuations: self-Organization and Social System. In E. Jantsch & C.H. Waddington (eds.), *Evolution and Consciousness*. London: Addison-Wesley.

Rippa, F. & Vinale, F. 1983. Effetti del terremoto del 23 novembre 1980 sul patrimonio edilizio di Napoli. In Associazione geotecnica italiana, Proceedings of XV National Congress, Spoleto.

Secchi, B. 2000. *Prima Lezione di Urbanistica*. Bari: Laterza.

Thom, R. 1972. *Stabilité structurelle et morphogénèse*. Paris: Intereditions.

UN/ISDR, *Terminology: Basic terms of disaster risk reduction* [Online], Available: http://www.unisdr.org/eng/library/.

U.S. Geological Survey, 1996. *USGC Response to an Urban Earthquake. Northridge '94*, Open-file Report 96–263, Denver.

Vale, L.J. & Campanella, T.J. (eds.) 2005. *The Resilient City*. Oxford University Press.

Von Bertanlaffy, L. 1971. *La teoria generale dei sistemi*. Milano: Isedi.

Wisner, B., Blaikie, P., Cannon, T. & Davis, I. 2004. *At Risk. Second Edition. Natural hazards, people's vulnerability and disasters*. London: Routledge.

Risks Challenging Publics, Scientists and Governments – Menoni (ed)
© 2010 Taylor & Francis Group, London, ISBN 978-0-415-58072-4

Spatial planning processes, territorial planning law and flood risk in the region of Valencia (Spain)

J. Olcina Cantos
University of Alacant, Spain

ABSTRACT: The Region of Valencia suffers from significant flood risk. The territory is experiencing the effects of a very high population and extreme urban development pressure, as well as being one of the most important areas of expansion for commercial agriculture and tourist-residential activity. Traditional measures for the prevention of flood risk have focused on the construction of fluvial infrastructure works, but all they have all been insufficient in reducing flood risk effectively. In January 2003, the Region of Valencia approved a Territorial Planning Scheme for the prevention of flood risk (PATRICOVA) that includes a series of measures for dealing with flood risk in future land use planning. This paper offers a critical review of the real reasons for the increase in flood risk suffered by this sector of the Spanish Mediterranean coast in the last thirty years, as well as making a detailed study of the repercussions of the application of the Territorial Planning Scheme.

1 THE REGION OF VALENCIA, A MAJOR FLOOD RISK REGION IN SPAIN

The territory of Valencia suffers from an exceptional range of extraordinary natural events which make certain areas very vulnerable and where economic activities and settlements are affected by extreme events. If we understand natural hazards to be the territorial results of human activities that are not compatible with the characteristics of the physical environment, then the Region of Valencia offers many examples of land uses that have not taken into account, either through ignorance or recklessness, its physical reality. This has made Valencia one of the most risky regions in Spain.

Seismic activity, landslides, intense and torrential rain, high winds and hailstorms, abnormally high and low temperatures and the temporary lack of precipitation are the most significant extraordinary natural events that affect the Region of Valencia. The extensive usage of the territory and the very diversity of the extraordinary natural events mean that the Region of Valencia is severely at risk.

Historical records (sedimentological analyses, Reports, Liturgical Acts, Mayoral Reports, Local Chronicles and pro pluviam and pro serenitate prayers) show the extreme vulnerability of Valencian towns to extraordinary natural events. Many authors have analysed the frequency of the appearance of these phenomena and their social, economic and territorial consequences in studies of the historical behaviour of climate in the Regional of Valencia. Using the results of these studies and consulting contemporary information (records of capacity, meteorological records, damage reports, press reports, field work) it is by no means exaggerated to state that Valencia is the most dangerous region in Spain as regards the exposure and vulnerability to extraordinary natural events.

To all this we must add the fact that the Valencian society is especially sensitive to natural hazards, as a result of having suffered numerous catastrophes in the second half of the 20th century, especially, in the 1980s and 1990s. The list of major disasters still very much present in collective memory begins in 1957 with the flood of the River Túria in Valencia and includes, in chronological order, the chronic drought suffered in southern Alacant in 1966–67, leading to the approval of the Tajo-Segura Water Transfer Scheme, the floods in October 1972 and 1973 in La Vega Baja del Segura, the collapse of the Tous dam in October 1982, the "wave of cold" in January 1985, the incredible hailstorm of the Day of Santiago in 1986, the overflowing of the Rivers Xúquer and Segura in November 1987, the high winds of February and the floods in September 1989 in many Valencian municipalities, the severe drought in the early 1990s, the flooding of the city of Alacant and surroundings in September 1997 and rivers bursting their banks and flooding large areas of the provinces of Castellón and Valencia as a result of the torrential rain in October 2000.

All of this led to the implementation of varied mitigation strategies, including curative and preventive measures that seek an ever more difficult balance between human occupation of the territory and the dynamic natural behaviour of the physical environment. An awareness of the natural hazards existing in the Region of Valencia is essential when it comes to proposing changes in land use and other territorial planning measures that respect the environment.

The most common natural hazard and that which causes most damage is undoubtedly the intense torrential rainfall, which releases significant amounts of rain in a very short period of time, causing the sudden rise of both large and small watercourses. The Region of Valencia has been the setting for many events involving torrential rain and flooding in the past, given the frequent occurrence of these events and the extremely high occupation of vulnerable territories. This region has also seen some of the highest daily rainfall figures in Europe and the even the world. The Region of Valencia is therefore a privileged laboratory for the study of flooding.

Some points should be clarified here. Firstly, the difference between events involving "heavy" rain with precipitation values of 50 mm per hour and totals which do not exceed 100 mm in 24 hours, and events involving "torrential" rain, meaning abundant heavy rain with values exceeding 50 mm per hour and totals which exceed 100 mm in 24 hours. They both cause damage to the affected areas, but the latter are more serious. In order to characterise flood risk areas, we must also take into account the most important event that has occurred in a particular area in a period of at least 50 years, analysing the rainfall figures, the circulating flows and depths reached by the flood water. It is thus necessary to carry out a historical study of local floods in order to define detailed territorial risk factors.

When it comes to the "record" of daily rainfall figures, the importance of torrential rain events in the Region of Valencia is underlined by the 871 mm that fell in Xàbia on October 2nd 1957 and the 817 mm in Oliva on November 3rd 1987. These are two of the highest daily rainfall values in Europe and are not far behind the figures recorded for some monsoons and tropical rainstorms.

Three factors lie behind the torrential rains and floods that lash the land of Valencia:

1. Atmospheric: Involving a combination of atmospheric situations that favour the development of torrential rain, all with the common denominator of the appearance of cold fronts and depressions at altitude, created by the expansion of masses of cold air of diverse origin (arctic, maritime polar, continental polar) and circulation of Northeast or Southeast windsat ground level (cyclogenetic developments from Algeria or the Balearic Islands, situations at the southern edge of a blocking cyclone and even low barometric pressure events). These conditions favour the creation of powerful convective nuclei that become "mesoscale convective clusters". Only the genesis of these cloud formations can explain rainfall greater than 200 mm per day and hourly rates of 100 mm and more (see Figure 1).
2. Geographical: Here we must highlight the presence of warm seas off the Valencian coastline. The maritime sectors of Alboran-Algeria and the Balearic Islands are the warmest in the western Mediterranean from August to October inclusive. The waters are thus potential cloud generators, with high vapour pressure figures during these months. Without the presence of warm seas we would be unable to explain daily rainfall as high as that recorded in the Region of Valencia. The proximity of Africa is another important factor as regards torrential rainfall and flood processes for two reasons: the formation of cyclogenetic developments in Algeria is linked with the presence of the Atlas-Riff mountain range in North Africa (vorticity) and with the incorporation of condensation nuclei with the mesoscale cloud formations. To this we must add the presence of the relief of the Betic and Iberian mountain ranges close to the coast, which act as triggers for the maritime flows that penetrate inland. Finally, we have a network of Mediterranean regime fluvial collectors with torrential behaviour. These are allochthonous (Xúquer, Túria, Segura and Millars) and autochthonous rivers (Palancia, Serpis), as well as short rivers/dry riverbeds, watercourses and ravines with significant load capacity (Carraixet, Poyo, Ovejas, Abanilla). They all cross areas with intense farming activity, where crops have invaded both large and small riverbeds and where urban areas have incorporated these collectors into their street plans as they have expanded.
3. It is with reference to the last point that we find the anthropic causes of floods—the most important nowadays—related to the intense invasion of fluvial plains and even watercourses for agricultural and urban-tourist uses. The major increase in cultivated land and urban growth experienced since the 1960s has not always taken the characteristics of the environmental setting into account. It is surprising to see that the areas at risk from flooding in Valencia, instead of becoming fewer as a result of the lessons taught by previous events, have increased. This is even after the approval, since the mid-1980s, of legislation aimed at reducing said risk (water, land, environmental impact, civil defence). One consequence of the abandoning of traditional agricultural methods, cultivation

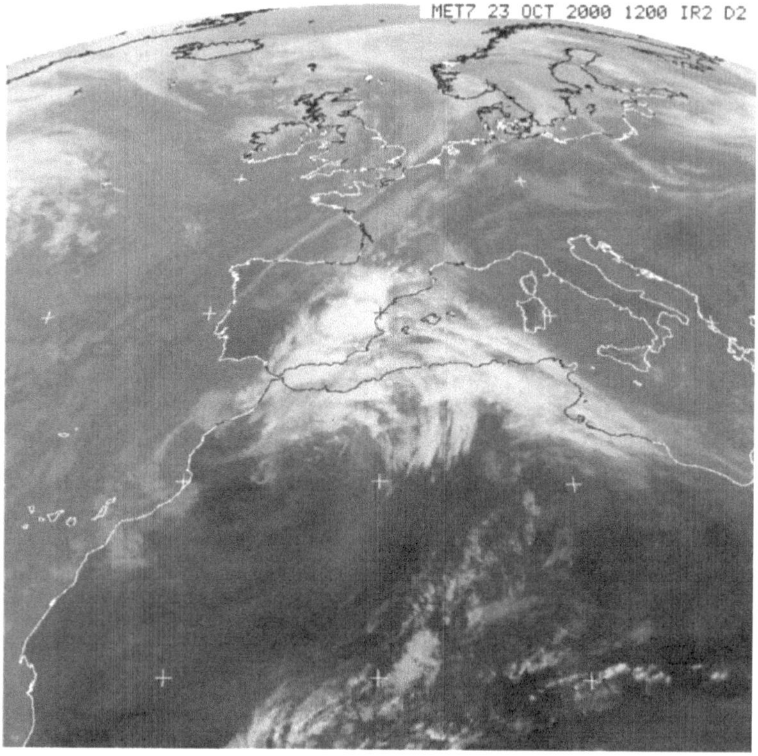

Figure 1. The installation of "cut-off lows" in the western Mediterranean, in autumns months, causes episodes of very intense rains in Valencian region. Episode of October 23–25, 2000. Source: EUMETSAT.

on terraces and irrigation with turbid water that were designed to use flood water by percolating it through the plots or distributing it quickly to the cultivated fields, has been the increase in the amount of solids carried by the waters of rivers, watercourses and ravines, as well as the increase in erosion on slopes where, with previous farming practices, there was almost none. Related to this, the increase of the burned area has locally increased soil erosion and therefore the amount of solids carried by floods, when forest fires have been immediately followed by heavy rainfall, something that is certainly not infrequent in the Region of Valencia. To all these factors we must add the incomprehensible permissiveness as regards compliance with legislation regarding the prohibition of occupying areas at risk and the deficient treatment of the problem of flooding in municipal planning documents, despite that laid down in the Valencian Land Laws (1994 and 1998) and the laws relating to territorial planning (1989 and 2004).

Late summer is the most dangerous time of the year for the development of torrential rain and is closely linked to the high seawater temperatures of these months. In fact, as shown in Table 1, the last twenty years have seen three events with torrential rain in February with very high maximum rainfall figures in a single day in a period that is not very given to their genesis. This fact confirms that cyclogenetic situations depend, above all, on the presence of favourable atmospheric conditions, combined with the remaining geographical factors mentioned above, to increase the amount of rain and the damage caused.

A study of torrential rain events in the Region of Valencia in the last fifty years allows us to make comments that should lead to future analyses aimed at improving our knowledge of the risk of fluvial overflows and floods. In the first place, the anarchic nature of heavy or torrential rain associated with mesoscale cloud formations. "Focussing" is a characteristic feature of rainfall along the peninsular Mediterranean coastline; the amounts recorded can vary by 50% or more between two observatories only a few kilometres apart. For example, on 2–3 October 1957 between the observatory of Xàbia (871 mm) and that of Cabo de San Antonio (409.7 mm) there was a difference of over 400 mm

Table 1. Exceptional events involving torrential rain and floods in the region of Valencia.

EVENT	EFFECTS
4 and 5 November 1864	Flooding of the Xúquer in La Ribera. Estimated losses of 70 "Reales de Vellón".
19 October 1879	"Flood of Santa Teresa" in the Segura basin. 777 víctims, 24,000 hectares of crops affected by flooding
21–24 April 1946	"The Good Friday Flood" in La Vega Baja del Segura. The most important recorded in this region of Alacant in the 20th century.
2 and 3 October 1957	Torrential rain and floods in the Marina Alta region of Alacant. Record rainfall in Xàbia (871 mm/24 hrs.)
14 and15 October 1957	Flooding of the River Túria in Valencia. 86 dead. Damage assessed at 4,000,000,000 pesetas (1957)
19 and 20 October 1972	Flood in La Vega Baja del Segura.
17–19 October 1973	Floods in Southeast Spain. 300 dead. La Vega Baja del Segura severely affected.
19–21 October 1982	Torrential rain and floods in Levante. "The Xúquer Flood" caused by the collapse of the Tous dam. Losses of 76,000,000,000 million pesetas. 38 dead.
19–21 February 1985	Bajo Segura. Floods in the municipality of Torrevieja. Damage to the salt harvest
14–15 November 1985	Floods in Levante. La Marina Alta de Alacant region affected. Damage assessed at 2.000 million pesetas.
27 September to 7 October 1986	Valleys of Alcoi, Marina Alta and Marina Baixa. Damage assessed at over 3,000,000.000 pesetas.
3–5 November1987	Floods in the catchment areas of the Rivers Xúquer and Segura. Damage assessed at 150,000 million pesetas. Urgent implementation of the Flood Defence Plans in the Segura and Xúquer catchment areas.
4–5 September 1989	Floods in Levante. Financial losses assessed at 100,000 million pesetas.
18–20 February 1992	La Marina Baja region affected. Damage to the loquat crops.
1–8 February 1993	Alto Vinalopó, Valles de Alcoi, Marina Baixa and Marina Alta. La Safor also affected. Destruction of roads, landslides at different points of Las Valles de Alcoi.
11–12 September 1996	Damage in La Safor, La Ribera and La Marina Alta regions. The towns of Tavernes de la Valldigna were severely affected by the River Vaca overflowing, as was Dénia. 400 people were evacuated from the Les Marines neighbourhood of Dénia. First estimates of damage rose to 6,000 million pesetas in different sectors.
30 September 1997	Torrential rains in the city of Alacant and surrounding areas. 3 dead. Plan against floods in the city of Alacant.
2 August 1998	Very heavy downpour in La Safor and La Marina Alta regions. 75 people evacuated from two campsites in Xeraco.
12–13 November 1999	Heavy downpour in La Safor and La Marina Alta. Floods in Oliva, Gandia, Xàbia, Dènia, Els Poblets and Calp. Over 300 litres in 24 hours.
23–25 October 2000	Torrential rain along the Mediterranean coast of the peninsula, from Almería to Tarragona. The provinces of València (Riberas Alta and Baixa, Foia de Buñol, L'Horta y Camp de Morvedre) and Castelló (Plana Baixa and Maestrat) were severely affected. Losses estimated at 100,000 million pesetas.

Source: Own data.

in barely five kilometres. We must also review our methods for calculating return periods and maximum probable rainfall, in order to establish more accurate risk thresholds. Maps of real maximum precipitation values are essential in order to demarcate sectors at risk from flooding. It is clear that the calculations of probability and return periods used in official prevention documents and plans offer results that are not very accurate as regards the pluviometric reality of the Spanish Mediterranean coast. This makes it necessary to focus our efforts on accurate studies of past events and on the extreme data really recorded for each of them. As we have already mentioned, we must consult historical documentation for a period of at least 100 years in order to identify the most damaging torrential rain events in each of the geographical areas in question. It is disheartening that, even today, infrastructures and the very safety of human settlements are based on the calculation of return periods. However, damage and destruction returns more frequently, leading also to loss of life.

The attached map shows the areas with high flood risk in the Region of Valencia, based on analysis of the events occurring throughout the 20th century. Despite the efforts made in recent years by government bodies, there are still many areas at risk from flooding. Here we should highlight the coastal strip, due to the concentration of conflictive points associated with the intensity of land occupation and the presence of the final stretches of rivers and ravines themselves. It is not rare to find the banks of watercourses or the surroundings of river mouths occupied by buildings (tourist housing developments) or recreational activities (campsites) (see Figure 2). Of special importance, due to the large area at risk, are the banks of the allochthonous rivers crossing the Region of Valencia (Xùquer and Segura). Indeed, La Ribera del Xúquer is one of the areas in Spain most affected by floods and not only because of the overflowing of this collector, but because of the confluence of other rivers and watercourses with torrential characteristics (Bolbaite, Sallent, Barxeta, Magro). La Vega Baja del Segura is the other main area at risk from flooding in the Region of Valencia. In the past it has suffered floods caused by this river, especially severe being those of October 1972 and 1973, and November 1987. The latter led to the execution of the Flood Prevention Plan in the River Segura catchment area that, among other works, included the excavation of a new channel from the city of Murcia to the mouth of the river at Guardamar. Despite these flood prevention works, the heavy rain of October 2000 underlined the need to act in those stretches of the watercourse that pass through Orihuela and Rojales, as it is unable to evacuate rainfall greater than 200 mm in 24 hours. The city and surrounding areas of Alacant make up another area at great risk in the Region of Valencia. The events of September 30th 1997 led to the implementation of a Flood Prevention Plan that included the channelling and burying of the watercourses that cross the city and surroundings of Alacant. Other areas with high flood risk, due to the high rainfall they receive in unstable atmospheric conditions in autumn, are the La Safor and La Marina Alta regions, where the towns of Gandia, Dènia and Xàbia have suffered frequent heavy flooding. Other areas especially at risk include L'Horta d'València—due to the overflowing of the Poyo and Carraixet watercourses,

Figure 2. Areas at risk from flooding in the region of Valencia. Own creation.

the Port de Sagunt—due to the overflowing of the Palancia, Benicàssim—with little drainage of the water stored there and Vinaròs—due to the overflowing of the Servol.

2 SPATIAL PLANNING AND FLOOD RISKS IN THE REGION OF VALENCIA

Different measures have been taken in the Region of Valencia to defend against natural hazards: preventive measures aimed at avoiding damage, and curative measures designed to repair the losses caused by extraordinary meteorological events. These have either been of sectorial nature—aimed at dealing with a single risk, or general—dealing with the whole range of risks. They have usually been carried out by central or autonomous government, without forgetting actions taken by provincial and municipal administrations. Many examples of private defence measures against natural hazards have also been implanted in both rural and urban areas.

Curative defence measures have been aimed at mitigating the problems associated with the excess or lack of water. Here we include, together with financial assistance and the declaration of areas affected by floods or droughts as disaster areas, through national and autonomous decrees, all the flood prevention work which has largely been carried out since the 1950s, as well as the work carried out to ensure agricultural and urban water supplies. The former includes channelling, diversion and the building of dykes, as well as the construction of large collectors in the urban stretches of watercourses, and even the relocation of urban nuclei in areas away from plains at risk from flooding. Among these, special mention should be made of the "Plan Sur", carried out on the watercourse of the River Túria through the city of Valencia after the devastating flood in October 1957. Another project with identical aims is the channelling of the River Segura between Murcia and Guardamar, triggered by the disaster that occurred in the Vega Baja in November 1987. While the former has shown its effectiveness in the city of Valencia when faced with later events, the new channel of the Segura through Orihuela and Rojales found to be ineffective and the Hydrographic Confederation of the Segura is now considering the construction of a by-pass to move the watercourse away from urban nuclei.

Extensive work has been carried out in the city and surroundings of Alacant since the flood in September 1997, based on the construction of large collectors that run beneath the city, and the channelling of the middle and lower stretches of certain watercourses (Juncaret, Orgegia). The Xúquer Catchment Area Flood Prevention Plan is to channel the middle and lower stretches of the river and has relocated the towns of Gavarda and Beneixida which were flooded as a result of the collapse of the Tous dam in 1982. Many measures have also been taken as regards the watercourses and ravines that cross urban and other areas with intense human activity.

However, defence against flooding based on rational planning of land use in the territory is still very recent. In the Region of Valencia, despite the early approval of the Spatial Planning Law (Law 6/1989, revised by Law 4/2004), no regional scale land management document has been drawn up (Territorial Planning Scheme, as stated in Law 6/1989 or Valencia Spatial Planning Scheme as per Law 4/2004). The Territorial Action Plan is the territorial planning process above municipality level that has been used and is defined as the document that develops the Planning scheme of the Territory of the Autonomous Community and which can have a) sectorial or, b) integrated nature. With reference to natural hazards, the autonomic government has recently drawn up a sectorial Territorial Action Plan for the prevention of flood risk (PATRICOVA, January, 2003), which aims to provide an integrated solution to this phenomenon, through the diagnosis of areas at risk and a set of structural and non-structural measures that include the scale of fluvial basin and land usage planning, to be applied at municipal level (PGOU).

It must be pointed out that the same Land Management law (Law 4/2004) gives preferential treatment to natural hazards and, among them, to those caused by flooding. Article 14 of this legal text is entirely devoted to the Prevention of natural or induced hazards, with specific measures described for flood risks. It thus states that all public riverbeds must be kept free of permanent uses and that urban development planning must direct future urban developments away from areas prone to flooding or towards those with a lesser risk of flooding.

Within the territorial planning framework laid down in Valencian land management law, January 2003 saw the approval of the PATRICOVA Plan. It is, therefore, a land management document of sectorial character and regional scope, whose legal obligations must be complied with by all municipalities in the region.

The PATRICOVA Plan is based on the analysis of potential flood areas by gathering the results of studies carried out by the Administration in coordination with experts from the Technical University of Valencia. They have a scale of 1:50.000, demarcate potential flood areas and are legally considered to be a Territorial Action Plan of sectorial nature, which implies the fulfilment of a series of functions foreseen in Art. 15 of Law 6/1989 concerning Land Management (see Figure 3).

Among the general functions laid down by the Law the PATRICOVA Plan must include: 1) The criteria and aims of territorial nature for sectorial performance, according to the directives of the Region of Valencia Territorial Planning scheme; 2) The planning, execution and management of the general supra-local and regional infrastructure, equipment and services systems; 3) The territorial priorities for public investment of sectorial policies; 4) The identification of projects and specific measures aimed at reaching the proposed aims; 5) Provide the Land Management Programme with the framework for the programming and "territorialisation" of sectorial resources.

In identical fashion, Art. 17 of Law 6/1989 of Land Management regulates the content of the PATRICOVA Plan: a) The area of application; b) Analysis and diagnosis of the situation of the sector to which it refers; c) Definition of the aims and territorial evaluation of the needs of the sector; d) Synchronisation with existing municipal planning; e) Justification and definition of the planned infrastructures, works, facilities and services; f) Technical

Figure 3. "Map demarcating potential flood zones in the region of Valencia. Scale 1: 200.000. This map was the basis of the final draft of the PATRICOVA Plan.

protection procedures that must be applied when carrying out the work planned; g) Identification and location of the projects and actions to be carried out; h) Cost of the projects and planned actions. All the above must be (Art. 19) included in due form in the following documents: a) A report accompanied by the analysis, diagnosis, aims, criteria and measurements of the actions carried out; b) A programme including the estimated cost of the work planned and the order of priority in which it is to be carried out; c) Procedure for carrying out the actions; d) Graphic Documentation, with information and planning maps; e) Any other item or content that may be required in relation to the type of Territorial Action Plan to be carried out.

The Plan, once approved, will be the first of its type in Spain, which will make the Region of Valencia the first Spanish region to have specific regulations to anticipate and manage flood risks. It must be realised that all dispositions laid down in the PATRICOVA Plan are binding and must be complied with by all territorial and urban development planning schemes drafted in the Region of Valencia.

As regards the demarcation of flood risks (Art. 15), the PATRICOVA plan states that this be decided on the basis of: a) The contents of the General Plans from each municipality; b) The Studies and Plans approved by the Regional Government or by the Hydrological River Catchment Area Bodies; c) The Studies of potential flood risks carried out for this purpose. The question receives special treatment in the regulation, since it is admitted (Art. 16) that as "the PATRICOVA regards itself to be a regional study realized in origin at a scale of 1:50.000 and it can be made specific, extended, and even modified by means of studies of potential flood risks". This article constitutes the explicit recognition of the profound difficulties involved in using the above mentioned scale to analyse hazards or risks of floods in major territorial detail, especially when it is a question of urban areas or local situations that require more detail, with more precise scales—5.000 or even 2.000.

To supplement the lack of territorial detail, the PATRICOVA Plan establishes the need to finalise potential flood risk studies, the scale of which shall depend on the defined flood risk. These studies must be accompanied by basic documentation (Art. 17) consisting of: a) a Geomorphological Study, indicating the potential extension of the flood; b) a Study of past floods, to support the results of the geomorphological study and other questions concerning hydrology and hydraulics; c) a Hydrological study to identify the levels of probable risk by means of statistical and hydro-meteorological models; d) a Hydraulic study to evaluate the capacities of watercourse outlets, overflow areas and the depth of the water, e) Maps of areas at risk, watercourses, flood areas and areas where water accumulates.

A paragraph that has deserved special interest in the PATRICOVA Plan is that relating to territorial and urban development planning in relation to flood risks, which has been included in Chapter III of the regulation. The most rigorous limitations of use for future urban development in the territory correspond to those municipalities that suffer high flood risks. Within this category we include those municipalities that have two thirds of their municipal area affected by high risk of flood and that have specific conditions of risk, which force the limitation of further land use development.

In the PATRICOVA Plan six distinct levels of risk are established, from high to low severity, on the basis of probability calculations (return periods) and of the maximum possible depth reached by the water in case of flooding. For example, Risk 1 includes areas with an upper probability of 0.04 (equivalent to a return period of less than 25 years) that in any one year there could be at least one flood with a water depth of over 80 centimetres. At the other extreme, a Risk of 6 includes potential flood areas with a probability of between 0.01 and 0.002 (equivalent to a return period of between 100 and 500 years) that in any one year there could be at least one flood with a maximum water depth of less than 80 centimetres (see Figures 4, 5 and 6).

It is important to note that a regulation has been established to take into account flood risks in urban development planning (Art. 20), which must contain minimum information in its informative section: a) The Public Hydraulic Authority, with its rights of way and police zones; b) The areas of flood risk identified in the PATRICOVA Plan itself, and by other studies carried out by catchment area bodies; c) All watercourses and water accumulation areas larger than 0.5 km^2 linked to catchment area slopes. In addition, the existence of flood risk would lead to the imposition of a series of limitations on land classification and on its possible urban development. For example, General Plans will have to classify the Public Hydraulic Authority and the flood areas demarcated in the PATRICOVA Plan with Risk level 1 as especially protected non-development land.

In identical fashion, limitations are also imposed on non-development land affected by other levels of flood risk. In the areas classified with Risk 2, 3 and 4 the building of housing, farms, hotel establishments and tourist camps, schools and health centres, fire stations, cemeteries, and other uses and activities is prohibited. On the other hand, we should note the greater flexibility and tolerance of uses granted to non-development land with Risk 5 and 6, where the building of housing and hotel establishments is allowed, provided that appropriate construction techniques are adopted.

With land classified as "development land", the existence of flood risk means that a study of potential flood risk to determine the urban development planning conditions, the defence works that it was necessary to complete, and the shape and regulations of the buildings that must be carried out. In addition, when the flood risk affects urban and development land with an approved programme, the PATRICOVA Plan grants the Town Halls significant flexibility in verifying the effect of the risk and in deciding adaptation measures for future buildings. In this case, the reduction of flood risk

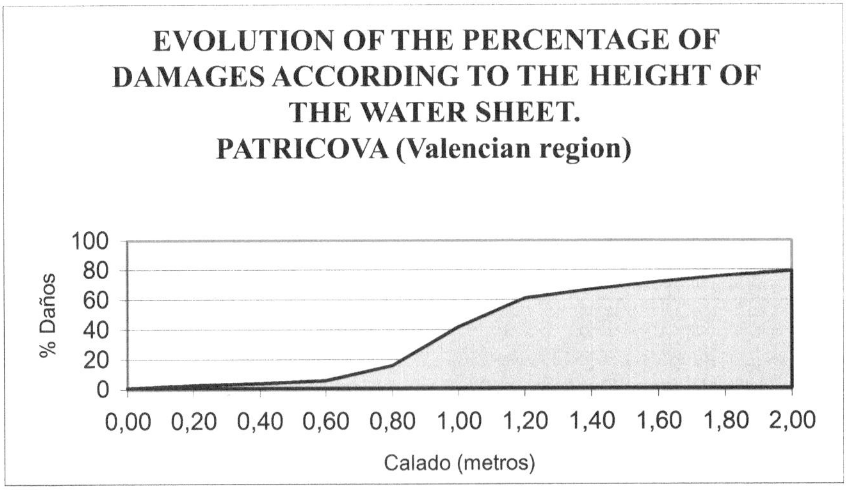

Figure 4. "Evolution of the percentage of damage according to water depth PÀTRICOVA (Region of Valencia)".

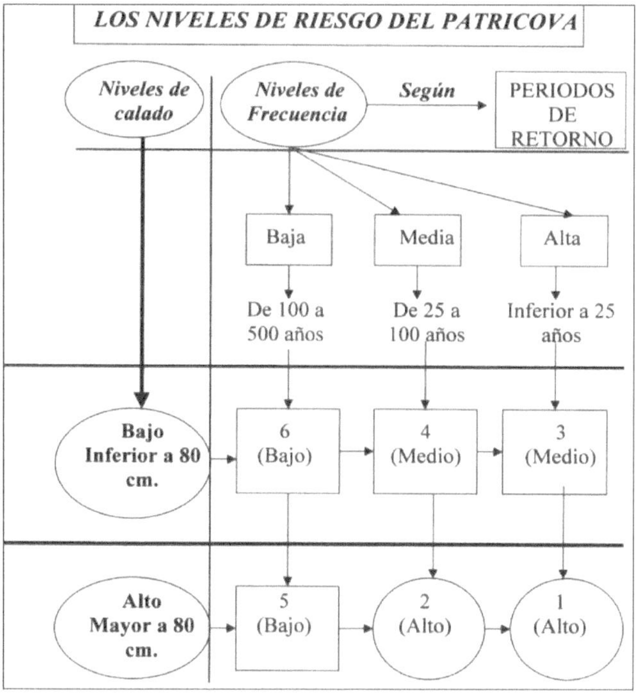

Figure 5. "Levels of flood risk established in the PATRICOVA plan (Region of Valencia) according to water depth and return period".

Key:

- Water depth: low (lower than 80 cm), high (higher than 80 cm)
- Frequency levels: Return periods: Low (from 100 to 500 years), medium (from 25 to 100 years), high (less than 25 years)
- Flood risk levels: 1 (high), 2 (high), 3 (medium), 4 (medium), 5 (low), 6 (low)

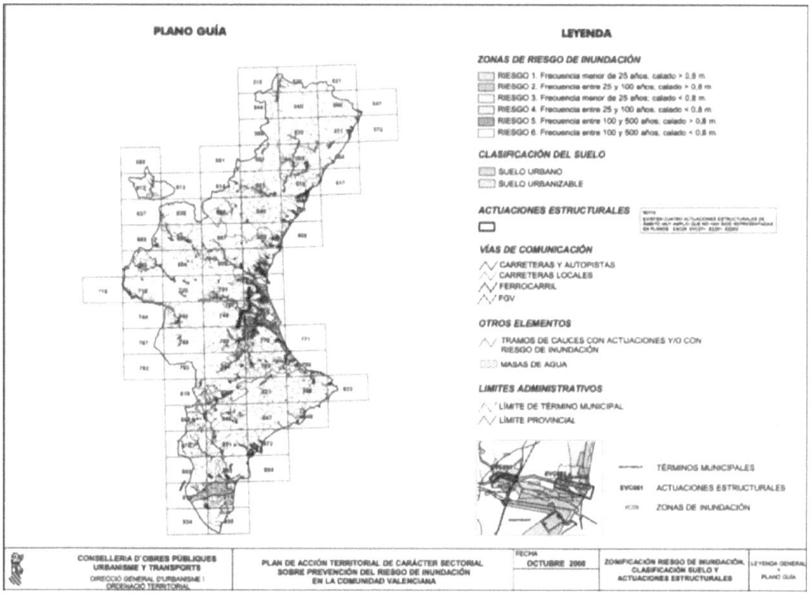

Figure 6. "Plan-guide to flood risk areas and risk levels. PATRICOVA. Region of Valencia. All maps to scale 1:80,000.

would depend on the building of defence works and the possible level of adaptation of buildings and infrastructures. In fact, Articles 26, 27 and 28 of the PATRICOVA Plan are dedicated to establishing the conditions of adaptation of infrastructures and buildings according to risk levels (see Tables 2 and 3).

The PATRICOVA Plan also includes a section (Chapter IV) containing regulations related to the defence works that should be included in the Municipal General Urban Plans that suffer from flood risks. In these cases, urban development would depend on the defence structures needed to reduce the risk, which must also withstand low probability floods, with return periods of at least 500 years. Article 33 also states the need to preserve what it calls the "areas of sacrifice and overflow routes" that correspond to natural flood plains and to fluvial palaeo-channels, avoiding the implantation of new uses that increase the risk in these areas.

We also welcome the regulation relating to rainwater drainage (Art. 34), since it states that urban areas of over 100 hectares in size must incorporate the following design conditions into their a rainwater drainage systems: a) They will be sized for rain return periods of at least 15 years; b) The pipes will have a minimum diameter of 400 mm; c) The drainage-holes and sinks will be 50 cm. Long at the kerb edge and have 1,250 cm^2 of surface horizontally.

The PATRICOVA Plan also includes the execution of hydrological-forest restoration schemes, some of them with high priority, in order to reduce erosion and run-off coefficients from the slopes of the catchment areas that create flood risks. Other, more general defence measures are included (Art. 36), including the active promotion of insurance policy programmes, the signposting of potential flood zones, the development of technological procedures for buildings and infrastructures in areas at risk and educational programmes for the general public.

According to the PATRICOVA Plan, the major flood risk zones detected in the Region of Valencia add up to 1.256 km^2 located in 278 areas and 393 municipalities, representing 5.4% of the surface area of the Region of Valencia (23.268 km^2), although we must realise that this area does not include the surface occupied by fluvial watercourses and reservoirs. Approximately 420.000 people live in this area suffering from high flood risk—over 10% of the total population of the region (see Figure 7 and Table 4).

Table 2. Characteristics of adaptation of buildings in flood risk areas (included in the general urban planning documents) Patricova Plan (Region of Valencia).

In areas subject to any flood risk (Art. 27)	a) The roofs of one-storey buildings must be accessible by stairs. b) The buildings will not be arranged crosswise to the direction of water flow. c) Residential, industrial and commercial uses are forbidden beneath street level.
Additional measurements in zones with risk 2, 3 and 4. (Art. 28)	a) No basements or semi-basements will be allowed, except for intensive residential use exclusively designed for car parking. A series of requirements must be complied with: 1) The access doors must be watertight 2) There will be a pumping system that is not connected to the sewer and powered by an electric generator unit. 3) The access ramp of access will be 10 cm lower than the pavement. b) Sewer accesses will not allow water to flow in the opposite direction. c) Doors, windows and front closures on the ground floor will be single units up to a depth of 1.5 meters.

Source: PATRICOVA Plan.

Table 3. Other defence measures that must be included in general urban planning scemes, Patricova Plan (Region of Valencia).

Art. 33. "Sacrifice zones" and the overflow routes	The General Urban Plans must preserve the "sacrifice zones" and overflow routes, that is to say, the flood plains and run-off channels
Art. 34. For the rainfall drainage system in urban areas of over 100 hectares in extent.	Rainwater drainage systems will comply with the following: a) They will be sized for rain return periods of at least 15 years; b) The pipes will have a minimum diameter of 400 mm; c) The drainage-holes and sinks will be 50 cm. long at the kerb edge and 1,250 cm^2 of surface area horizoantally

Source: PATRICOVA Plan.

Figure 7. Territory occupied by Medium and Low Júcar river banks form one of the major flood risk area in the Valencian region. Red color indicates risk level 1; violet color indicates risk level 2.
Source: PATRICOVA. Generalitat Valenciana. E: 1:160.000.

Table 4. Provincial distribution of floods risk areas.

Province	Area		
	Total (hectares)	Risk (hectares)	% Area affected
Alacant	581,515	42,645	7.3
Castellón	666,874	16,337	2.4
Valencia	1,078,390	66,670	6.2
TOTAL Region of Valencia	2,326,779	125,652	5.4

Source: PATRICOVA Plan.

The PATRICOVA Plan has also analysed 62 urban tourist resorts on the Valencian coast, with fairly worrying results, since 46 of the number mentioned above (62) have serious flood risk problems, due to their proximity to areas of marsh or large salt-water lagoons and to inadequate natural and artificial drainage systems. Apart from the areas with high flood risk, the PATRICOVA Plan states that there are 450 critical urban points with very serious localised flooding problems. 450 are mentioned although it is recognised that there may be very many more, as this is undoubtedly the case. In many cases, they correspond to problems of inadequate urban drainage due to multiple reasons, of which the following are most important: 1) Bad street design; 2) Absence of drain-holes, and 3) Insufficient capacity of the main collectors and sewer system.

To face to all these problems, the PATRICOVA Plan sets the reduction of the territorial impact of floods as its first aim, which implies significant coordination between the different administrations with territorial competences in this matter, such as the Hydrographic Confederations and the Ministry of the Environment, as well as Ministries and Town Halls. Finally, special mention should be made of a series of principles aimed at guiding land use planning processes at the local level. This level is the most important when it comes to finding solutions to reduce the exposure of the population to natural hazards (see Table 5).

3 CONCLUSION

The implementation of flood prevention practices based on rational territorial planning measures is a recent process in Spain and the Region of Valencia. This means that control of land use processes in relation to potential flood risk really refers to the last five years, after the obligation imposed by the Spanish Land Law of 1998. Article 9 of said Law

Table 5. Principles contained by the PATRICOVA plan for land management of urban land at risk from flooding.

PREVENTION WORK ON DEVELOPMENT LAND WITHOUT AN APPROVED PROGRAMME AND ON NON-DEVELOPMENT LAND	A) OBJECTIVE CONDITIONS MUST BE MET FOR THE DEVELOPMENT OF DEVELOPMENT LAND WITHOUT AN APPROVED SECTORIAL PROGRAMME. B) ITS RECLASSIFICATION AS DEVELOPMENT LAND MUST BE PREVENTED.
ON DEVELOPMENT LAND AND DEVELOPMENT LAND WITH AN APPROVED PROGRAMME	A) DEALS WITH STRUCTURAL MEASURES TO REDUCE RISK. B) SPECIFIC ACTIONS TO ADAPT BUILDINGS.
FLOOD RISK FOR LAND CLASSIFIED AS: • URBAN: 72,000,000 M^2 • DEVELOPMENT: 44,270,000 M^2	
FOR FUTURE URBAN DEVELOPMENT	A) THEY MUST BE DIRECTED TOWARDS AREAS NOT AFFECTED BY FLOOD RISK. B) BUT....ITS APPLICATION COULD BE MORE FLEXIBLE IN THOSE TOWNS WHICH DO NOT HAVE ALTERNATIVE GROWTH AREAS.

Source: PATRICOVA Plan.

states the need to declare all those areas with a recognised natural hazard as "non-development land" and the land use regulations of the Autonomous Communities have adopted the article mentioned above. However, the Region of Valencia has gone further than other Spanish regions by carrying out studies and drawing up regulations concerning the reduction of natural hazards (specifically for floods) through territorial planning, as we have explained above. The approval of the Territorial Action Plan for the prevention of flood risk in the Region of Valencia (PATRICOVA) is an important step forward in this sense. However, a number of questions raise doubts as regards the real effectiveness of this plan when it comes to reducing flood risks:

- Firstly we have the real situation of the intense occupation of the Region of Valencia that occurred before the PATRICOVA was approved in 2003. This means that the immense majority of the areas at risk from flooding were so before the Plan has been approved, meaning that they are not applicable. Flood risk reduction in these areas therefore involves carrying out hydraulic infrastructure work. These areas at risk that existed prior to the PATRICOVA are the most numerous in the Region of Valencia. We have not considered the possibility, as was laid down in the French Barnier Law in 1995, of public expropriation or exchange of land in those urban areas occupying the areas of greatest risk. In other words, mitigation of flood by territorial planning involves what is to be done once the plan comes into force.
- We must also consider the very method of demarcating areas at risk chosen by the PATRICOVA Plan. As we have already mentioned, two criteria are used: the potential depth that might be reached by floodwater from a river or watercourse and the return periods of an event. Using these two parameters and a digital model of the terrain, the areas at risk from flooding have been identified, specifying in each case the corresponding degree of risk from the 6 level scale contained in the Plan. The criteria are mathematical and probabilistic, complying with the legal requisites laid down in Spanish water legislation as regards the demarcation of areas subject to flooding (Art. Regulation of the Public Hydraulic Domain, 1986) and even exceed those requirements by listing different levels of risk. However, the resulting maps contain errors in the demarcation of areas and levels of risk, as a consequence of lacking fieldwork that would have allowed the results of the mathematical model to be qualified. In a territory as dynamic as the Region of Valencia, where property development has been one of the pillars of development in recent years, it is vital to check areas at risk on site. This is especially true when, due to the accelerated rate of house building, the territory has undergone very significant changes in very short periods of time (2–3 years).
- In this respect, we must highlight the fact that the Region of Valencia has led property development processes in Spain over the last decade (see attached Table 6) and this has often taken place without taking into account the characteristics of the natural environment where new housing developments have taken place. This has resulted in the occupation of areas at risk from flooding and the creation of new areas at risk. We can state that, over the last twenty years in the

Table 6. A comparison between estimated and real growth figures for resident population and housing in the region of Valencia (1981–2001).

	ESTIMATED GROWTH OF RESIDENT POPULATION (1981–2001)	REAL GROWTH (1981–2001)	ESTIMATED GROWTH OF HOUSING (1981–2001)	REAL GROWTH (1981–2001)
Alacant	+232,319	+312,744 (+224,275)	+359,577	+390,959 (+234,636)
Castellón	+18,724	+53,299 (+52,682)	+75,149	+98,748 (+82,406)
Valencia	+112,960	+150,581 (+147,853)	+256,148	+303,803 (+301,062)

Source: Censuses of population and housing (1981–2001). The estimates are from the study *Population and housing forecasts for planning in the Region of Valencia*. Ministry of Public Works, Town Planning and Transport, Valencian Regional Government. 1997 (ISBN: 84-482-1307-6).

Region of Valencia, flood risks have increased instead of diminishing, precisely because of the massive invasion of land caused by the significant growth in housing stocks.

If we study the figures and reports used by both central government and private bodies with regard to housing growth trends on the coastal strip of the region of Valencia for the next ten years, it seems that this trend will be maintained in coming years. By 2015, if we bear in mind the behaviour of the building sector in recent years, we should expect an increase of 1.8 million dwellings along the coastal strip (beachfront and second line) in the Region of Valencia (see Table 7).

This significant rate of growth must be closely monitored by the Government if new residential areas are to be prevented from invading areas at risk from flooding. A major role must be played at municipal level in this respect.

- And the latter is a vital aspect. When we analyse flood risks in Spain at a territorial level, we must not forget that it is at local level that the essential processes of land occupation and allocation of land uses in the territory take place. The physical environment is a basic factor in the configuration of urban areas and has been recognised as such by town and territorial planning legislation since the appearance of contemporary Spanish town planning in the mid-20th century. It is true that references to the need to include studies of the physical environment are very scant, and this is combined with the lack of interest and training of the teams drawing up territorial and municipal planning documents when it comes to analysing and assessing the importance of the physical environment with respect to the town or city. National legislation and, since the 1980s, autonomic regulations concerning territorial and land use planning have incorporated this question into successive legally binding documents. However, land use and territorial planning legislation, both national and autonomic, gives the local level a leading role. In fact, general urban planning schemes have become the key instrument for defining territorial policies. It should be realised that a correctly drafted municipal planning document can, as we have already mentioned, be an effective tool for preventing natural hazards.

However, this comes up against the fact, common throughout Spain, that little importance has been given both in the past and in the present to natural hazards by the municipal level PGOU's, despite legal obligations. The technical documents associ-

Table 7. Housing growth trends on the region of Valencia coast (perspectives for 2015). Own creation.

PROVINCE	TOTAL DWELLINGS TREND 2015	HABITUAL RESIDENCE TREND 2015	SECOND RESIDENCE TREND 2015
Alacant Coast	+665,077	+372,116	+292,922
Valencia Coast	+870.814	+565,248	+302,826
Castellón Coast	+258,925	+137,050	+121,868
TOTAL REGION OF VALENCIA COAST		+1,784,816	

ated with these plans hardly refer to the question and, what is even more serious, they do not attach maps showing areas at risk. It is thus necessary for Local Administrations to pay more attention to and monitor areas at risk from flooding in order to ensure that they are not occupied. This demands greater rigour when it comes to drafting and approving municipal territorial planning schemes.

The territorial planning laws and schemes implemented in different Spanish regions in recent years will play a major role in reducing flood risks through measures designed to ensure rational spatial planning. These measures must be accompanied by programmes for managing emergencies, as well as for educating and raising public awareness of the risks in order to develop integrated policies for reducing natural hazards. However, this will not have the desired effect if the Administration is not aware of the fact that natural hazards require this global vision if they are to be approached effectively, and that the local level is vital to carry out these actions. We must eliminate, for once and for all, the irresponsible attitude of blaming nature for actions carried out by man in a territory whose physical characteristics are very often ignored or underestimated. If not, our constitutional right to the defence of the safety of individuals contained in Art. 17.1 of the Spanish Constitution would be called into question. As it usually happens, we only will remember what McHarg (2000) said about the importance of "planning territory while taking nature into account" when a new disaster occurs.

ACKNOWLEDGEMENTS

I want to express my deep gratefulness to prof. Scira Menoni for having permitted to take part in SRA-Europe Meeting, Como 2005. For her and for Raffaella Cozza my gratefulness for all the attentions and nice treatment that always had with me in relation with all the organizational matters.

REFERENCES

Agencia Europea del Medio Ambiente 2004. *Mapping the impacts of recent natural disasters and technological accidents in Europe*. Environmental issue report n°35. Luxembourg, 47 pp.

Alberola Romá, A. 1999. *Catástrofe, economía y acción política en la Valencia del siglo XVIII*. Institució Alfons El Magnànim. Diputació de València, Valencia, 333 pp.

Ayala-Carcedo, F.J. & Olcina Cantos, J. 2002. *Riesgos Naturales*. Editorial Ariel. Col. Ciencia, Barcelona, 1512 pp.

Camarasa Belmonte, A.Mª 1995. *Génesis de crecidas en pequeñas cuencas semiáridas. Barranc de Carraixet y rambla del Poyo*. Madrid, Ministerio de Obras Públicas, Transportes y Medio Ambiente, Confederación Hidrográfica del Xúquer, 252 pp.

Consellería d'Obres Publiques, Urbanisme i Transports 1997. *Delimitación del riesgo de inundación a escala regional en la Comunidad Valenciana*. Serie Cartografía Temática n°1. Generalitat Valenciana, València, 56 pp + mapa.

Generalitat VAlenciana 2001. *Plan de Acción Territorial de carácter sectorial sobre prevención del riesgo de inundación en la Comunidad Valenciana*. Consellería de Obras Públicas, Urbanismo y Transportes. Documento de exposición pública.

Gil Olcina, A. & Morales Gil, A. (coord.) 1989. *Avenidas fluviales e inundaciones en la cuenca del Mediterráneo*. Instituto Universitario de Geografía. Caja de Ahorros del Mediterráneo. Alicante.

Gil Olcina, A., Olcina Cantos, J. & Rico Amorós, A.M. (Edits.) 2004. *Aguaceros, aguaduchos e inundaciones en áreas urbanas alicantinas*. Publicaciones de la Universidad de Alicante, Alicante, 735 pp.

McHarg, I.L. 2000. *Proyectar con la naturaleza*. Ed. Gustavo Gili, Barcelona, 197 pp.

Morales Gil, A., Olcina Cantos, J. & Rico Amorós, A. 2000. "Diferentes percepciones de la sequía en España: adaptación, catastrofismo e intentos de corrección" en *Investigaciones Geográficas* n°23, Instituto Universitario de Geografía, Universidad de Alicante, Alicante, pp. 5–46.

Olcina Cantos, J. 1995. "Riesgos climáticos en las tierras valencianas. Incidencia en la actividad agraria", en *Investigaciones Geográficas*, n°15. Instituto Universitario de Geografía. Universidad de Alicante. Alicante, pp. 99–143.

Olcina Cantos, J. 1995. *Riesgos climáticos en la península ibérica*. Acción Divulgativa, Madrid, 415 pp.

Olcina Cantos, J. 2004. "Riesgo de inundaciones y ordenación del territorio en la escala local. El papel del planeamiento urbano municipal", *Boletín de la Asociación de Geógrafos Españoles*, n°37 (monográfico "Agua y Ciudad"), pp. 49–84.

Olcina Cantos, J. 2004. "Riesgos fluviales y ordenación del territorio", en *Alteración de los regímenes fluviales peninsulares* (Gil Olcina, A. coord.), Fundación CajaMurcia, Murcia, pp. 45–97.

ONU 2004. *Living with Risk: A Global Review of Disaster Reduction Initiatives*, ISDR, Nairobi.

Pérez Cueva, A.J. (Coord.) 1994. *Atlas Climàtic de la Comunitat Valenciana*. Conselleria d' Obres Públiques, Urbanisme i Transports, Generalitat Valenciana, València, 205 pp.

Quereda Sala, J.J. & Montón Chiva, E. 1994. *Las lluvias torrenciales en la Comunidad Valenciana. Interacciones Atmósfera-Mar*. Excma. Diputació de Castelló. Colección Universitaria. Castelló de la Plana.

Smeder, M., Christou, M. & Besi, S. 1996. *Land Use Planning in the Context of Major Accident Hazards. An Analysis of Procedures and Criteria in Selected EU Members-States*. Ec Institute for Systems, Informatics and Safety Major Accident Hazards Bureau. Informe UER 16452 EN. Ispra. Italy.

Vera Rebollo, J.F. & Olcina Cantos, J. 2002. "Riesgos Naturales en la costa", en *Revista OP, Ingeniería y Territorio* n°61, vol. III (El litoral), pp. 50–59.

Risks Challenging Publics, Scientists and Governments – Menoni (ed)
© 2010 Taylor & Francis Group, London, ISBN 978-0-415-58072-4

From structural to systemic: Proposals to make French flood management more consistent and robust

P. Pigeon
Université de Savoie, France

ABSTRACT: This paper reveals contradictions and rapid evolutions in the French "natural" risk management system, and tries to understand them. Among other factors, the instability of the French legislative framework derives from the central role the French State plays in this matter, since 1982 and Tazieff's law implementation. The unwanted effects of this system, which is poorly risk prevention-oriented, lead to stress on the limits of corrective works, such as dykes, sills. It explains the recent will to strengthen civil security management, and local settlement resilience. But the effectiveness of these new tools supposes a clear acceptance of the impossibility to fully get rid of risks, and the necessity to adapt urbanization pattern to them. That would mean a political and social acceptance of the necessity and limits of political actions.

1 INTRODUCTION

This paper stresses on the contradictions and weaknesses that can be found in the French flood management system. Recent texts are contradictory by themselves, and maps displaying risks levels as well. Furthermore, the legal basis of this system evolves very rapidly, though claiming to promote sustainable development. How can we understand these contradictions and juridical weaknesses? What is done, and what can be done, in order to give this system more effectiveness—that's to say, trying to thwart the trend towards growing economic losses, while gaining more social acceptance of political decisions? Wishing to answer the first question requires to find back the historical construction of the French system, which is only poorly oriented towards prevention. Therefore, its economic or politic sustainability appears questionable. The third point of the paper shows measures the French State took in order to break the trend leading to the system disruption, and to strengthen its resilience. It stresses on the limits of structural measures, and, even more, on the fact that they may upgrade flood risks levels while acting on physical processes. But shifting the official point of view on structural measures has numerous implications. It should lead political actors to admit officially the impossibility of eliminating risk, and, therefore, the growing uncertainty of political action. Yet, at the same time, political actors have to achieve social compromises on risks levels displayed by land-use zoning, while taking numerous other social constraints into account. Field studies prove that these projects are about to become reality, but only in part. Systemic representations of the interactions between physical processes, risk management measures, and settlements evolutions may help giving flood risk management policies more effectiveness. They deal with the matter of responsibility, and prove the necessity of reactive, flexible politics.

2 WEAKNESSESS OF THE FRENCH FLOOD RISK MANAGEMENT SYSTEM

Three elements justify this assertion will be provided in the following paragraphs.

2.1 *Logical contradictions embedded in the juridical texts themselves*

Regulations such as the "prescription en matière d'urbanisme et d'aménagement en cas de risque d'inondation" (Direction Départementale de la Haute-Savoie, janvier 2004), display a contradictory information concerning hazards, and the so-called effectiveness of works managing them. We can read, at page 1, that "no protection device, such as dykes, flood contention basins, can modify natural hazards"[1]. But, in the next page, that any dyke or protection device not protecting existing settlement should be forbidden[2]. Can dykes modify, or not, natural hazards? Do they have a protective role? If they do not, what are they made for, according to the writer of this regulation?

But we can find this type of contradictions in the maps displaying risks as well.

2.2 *Discrepancies between PLU, SCOT and PPR zonings*

PLU (Plans Locaux d'Urbanisme) and SCOT (Schémas de COhérence Territoriale) try to manage urbanization by the way of land-use zoning, juridically binding, at two separate scales (Trapitzine, 2002). PPR (Plans de Prévention des Risques) come from Barnier's law (1995) and display risks zonings, which are juridically binding as well, and have to be taken into consideration by the PLU and SCOT zonings (Blanchi et al. 2003). Three zones are usually found: the red one does not allow any building, while the white one is supposed to be free of the type of risk dealt with. The blue one is opened to construction, but buildings and activities have to take into account the existence of some levels of ascendance risk. If they do not so, within financial limits -the cost of adaptative measures should not exceed ten per cent of the market value of the property-, insurers may react. In agreement with the Central Bureau of Tariffs, and according to the article L 125–6 of the insurer's code, they may decide additional premium to be paid by the landowners concerned, especially in case of frequent damages (Mission des Risques Naturels, 2004).

In spite of these clear juridical bases, discrepancies between the PLU-SCOT and PPR zonings are plentiful. Let us compare the PPR of Serrières-en-Chautagne (Savoie) and the SCOT Savoie-Métropole (Figures 1–2). In the first kind of plans, red and blue zones in the Rhône valley display a risk level which poorly fits future industrial development. The industrial zone of Motz-Serrières appears like a patch of blue amidst the red zone, giving therefore a clue on the local political constraints at hand when the final zoning was decided. No doubt that the risk level could be ranked higher. However, the SCOT Savoie-Métropole recognizes a major economic role to these same zones. They are depicted as "major zones of economic activity, with a capacity of 40 hectares, and further development of at least 4 hectares".[3]

These contradictions are not limited to the local scale. The instability of the juridical French system in that matter may be considered responsible for them, at least in part.

2.3 *A managing system which evolves very rapidly*

If we take only the major texts into consideration, we find no less than five laws since the Tazieff's law (13 of July 1982), which created the fundamentals of the PPR. They were named PER (plans d'Exposition aux Risques) at that time. The PER were modified by the 1987 law, and, mainly, by the Barnier's law, the 2 of February 1995. Since then, we had the Bachelot's law, on the 30 of July 2003, and the law regarding civil security (13 of August 2004). With its article 3, this last one states that it is time to give more effectiveness to the French natural risk management policy. All in all, the juridical French system acknowledges its own weaknesses. Let's stress on the fact that these laws are intended to enforce the principles of sustainable management.

How can we understand such weaknesses? Next paragraphs will be devoted to answering this question.

3 THE HISTORICAL BASIS OF THE FRENCH SYSTEM AT THE ROOTS OF ITS CONTRADICTIONS

This system is essentially driven by the French State, and it is supposed to match its institutional needs. The first necessity is to display risks by mapping. The second is to manage its financial constraints. And this point tends to question the sustainability of the French system. It enlightens the poor juridical stability of juridical management tools.

3.1 *Displaying risks by mapping*

At the beginning of the 1970's, two catastrophic events, a snow slide in Val d'Isère and a mudslide in the Plateau d'Assy, which amounted to 70 dead in each case, triggered the present system (Pigeon, 2005). Families were motivated to know if those disasters could not have been avoided, and if only natural processes were implicated, as claimed the other political actors concerned, a means to avoid human responsibility. Finally, about ten years later, the Conseil d'Etat[4] decided that the French State was responsible for risk mapping.

---

[1] L'aléa naturel n'est pas modifié par l'existence d'ouvrages de protection (digues, bassins de rétention etc.).

[2] "Les digues ou toute autre protection ne seront autorisées que si elles permettent de protéger des biens ou des lieux habités existants".

[3] "Le projet pour le développement urbain", page 58. In: Schéma de Cohérence Territoriale (SCOT) Savoie-Métropole. 2004. Rapport de présentation: 110 p.

[4] Conseil d'Etat, arrêt monsieur Blanc contre ministre de l'Equipement (27 juillet 1979): "considérant que le préfet s'est abstenu jusqu'en 1970 de mettre en œuvre la procédure de délimitation des zones exposées aux risques naturels, ce retard est constitutif d'une faute qui n'a pas mis l'administration en mesure d'apprécier la réalité et l'importance du risque lorsqu'à été délivré en 1969 le permis de monsieur Blanc", dans un couloir d'avalanche, à Tignes.

Should such events occur again, the absence of map would mean direct recognition of the French State's responsibility: damages could be avoided, but not in case of lacking information. 25 years later, the PPR still tries to fill the gap, as this policy does not necessarily matches the needs of the citizens or the municipalities (Blanchi et al. 2003).

3.2 *Managing the financial constraints of risk policies*

The Tazieff's law, on the 13 of July 1982, decided a broad mutualisation of these financial constraints. As soon as you have an insurance concerning a house or a car, natural risk is automatically taken into account. The French citizen has, however, to pay more for his insurances, an additional premium. But, basically, it means that policyholders not living in flood areas pay the same premium that policyholders who do. It also means that, without the existence of PPR, policyholders or municipalities are not financially refrained from living or building in flood areas. Therefore, this policy is poorly oriented to prevention (Mission Risques Naturels, 2002).

3.3 *The sustainability of this system is questionable*

Urbanisation in the flood areas can explain, in part, the increasing costs related to natural risk management. According to a "circulaire interministérielle" (30 of April 2002) insurers had to pay 7.3 billions euros for damages related to natural risks, between 1982 and 1999, and 1.2 billion in 1999. But the "Mission des Risques Naturels" (2004) amounts to 3 billions euros the cost of damages for the insurances related to September 2002 and December 2003 flooding. This trend has to be considered with a critical mind, yet, it puts the French system under stress, and questions its sustainability.

4 HOW TO GIVE MORE EFFECTIVENESS TO THE FRENCH MANAGEMENT SYSTEM?

Contradictions and rapid evolutions of the French system may, at least in part, reveal its resilience, its capacity to adapt. Contradictions, at first sight, may appear as weaknesses, but they could reveal adaptation abilities as well. Three points immediately below tend to underpin this statement. But there is still a lot to do: the last two points intend to prove it.

4.1 *Recognition of unwelcome side effects of any policy*

That's to admit the limits of any structural measure, whatever it may be: dykes, contention basins, dams, for example. In the Arve Valley, dyking up of the mountain rivers increased their specific power far over the value of 35 watts/m2 (Piegay, 1997). When the bed-rocks allowed it, rivers carved down their beds: "locally, rates of incision exceeding 50 cm/year were frequent during the period 1950–1970", according to Peiry et al. (1994)[5]. It contributed, among other factors, to a reduction of the flood frequencies, allowing the municipalities to think that this type of risk had disappeared, whereas damages shifted to dykes and bridges collapses. Consequently, municipal zonings poorly displayed flood risk, and allowed constructions without taking this type of risk into account. The French State, before 1995 (Barnier's law, and the PPR), poorly counteracted this trend. At present, official management texts in France, such as Bachelot's law, acknowledge policies side effects. No matter what the policy, it won't eliminate the targeted risk, but transform it, rather. It's easy to find contradictions in the juridical texts, therefore.

4.2 *Recognition that structural measure may upgrade the risk level*

It's the next step and it concerns the evolving zonings of the PPR. For example, the PPR Chautagne (2004) had to integrate a new red zoning: "the new plan foresees a new security areas next to the dykes. This security strip fits the type of works and protection level they are likely to afford.... This zone Rd matches the dyke risk". Therefore, this risk level may be higher in the vicinity of the dykes than it could have been without dykes. At a broader scale, dyking up the rivers Rhône or Arve leaded to the false opinion that flood hazards had disappeared. The Farmer's curve proves that the intensity of risk has never been higher during the history of the settlement, while the frequency of flooding has become extremely scarce (Figure 3). In the Arve valley, the last major event dates back to 1968.

4.3 *Positive co evolution between risk management and urbanisation should be reduced*

Dykes should not be used to downgrade risks zonings of the PPR or to urbanize large tracts of

[5] Peiry, J.L et alii. 1994. L'incision des rivières dans les Alpes du nord : état de la question. Rev. Géogr. Lyon, 69(1) : 49–56.

lands in flood-prone areas, relying on the so-called effectiveness of these structural works. In France, protective works are frequently used in order to reach that goal. We find in the regulation of the PPR de Ville-La-Grand (1998), page 16: "*aucune implantation nouvelle ne pourra être effectuée avant la réalisation de travaux collectifs réduisant le risque en supprimant les débordements à l'origine de ce risque jusqu'à la crue centennale*". A report concerning the PLU does go farther, at page 15: "*on précisera qu'un bassin écrêteur des crues est prévu au lieu-dit Marsaz. Ce bassin devrait permettre de réduire l'étendue de la zone bleue en limitant le risque d'inondation*" (Mairie de Ville-La-Grand, 2004). Protective works do help alleviating political conflicts related with PPR zonings as well. Let us remind ourselves that red and even blue zones downgrade land market values. But, for the French State, it's a means to counter the trend of economic losses, and to adapt the present system of natural risk management. Most of the contradictions we mentioned at the beginning of this paper come from these very fast shifting points of view concerning the effectiveness of structural measures. It leads automatically to the next point.

4.4 *Political acceptance of residual or transformed risks*

How to admit the necessarily incomplete effectiveness of structural measures while not constraining too much local development policies, or the economic interests of landowners? That would lead to politically unsustainable land use plans as the PPR and, therefore, the PLU. This goal seems hard to reach. Yet, we can find at least an example of this type of agreement with the municipality of Magland, in the Arve valley. A dyke, partially built, is intended to shield buildings from rapid flooding coming from upstream, but not from slower flooding. The political acceptance of this type of works allows a blue zone in this quarter of the municipality, instead of a red one, and a type of urbanisation which is, in part, adapted to the risk level officially admitted. However, there is still poor acknowledgment that this dyke would upgrade the risk level in case of a flood that would be higher than centennial. It's clearly written in the study reports (SOGREAH, 1995, inter alia), but the perception of this fact by the citizens or the municipal councils appears still wanting. The last point of this paper stresses this aspect of the problem.

4.5 *Finding ways to strengthen risk acceptance and to broaden responsibilities*

The example of Magland may not be unique, yet it should be extremely rare. The main problem at hand is set by the political acceptance of the impossibility to eliminate risk, and its implications for land use. It becomes harder as a more responsible flood management will take very rare events into account, floods over the centennial frequencies, for example. One of the more sensitive and critical topic will be the integration of these events in a local management plan. At present, the French State tries to perform it with the PPR concerning snow slides, and for the municipality of Chamonix. Any solution would need to admit that the growing uncertainties related to risk management, and yet the impossibility to eliminate it, leads to consciously and collectively accepted levels of risks. It would broaden responsibilities, at the scale of communities. At present, the Plans Communaux de Sauvegarde, which are very adaptable and try to involve numerous actors in the decision process, prove that this trend is about to become reality in France, at least, in part. They come from the aforementioned 2004 law on the civil security.

5 CONCLUDING REMARKS

To make natural risk management more effective vogues:

- Mapping hazards and risks in spite of uncertainties related to natural and social processes evolving very rapidly. Therefore, a systemic way of thinking better meets political needs, as it depicts the limits of structural measures, and the impossibility of risk elimination. Instead, it states that risk, risk management and settlement mutations tend to co evolve in a way that makes the French system poorly sustainable in the long run, in spite of the efforts to strengthen its resilience.
- Not forgetting that these maps have to take numerous constraints, and type of risks, into account: for example, responsibilities of the political actors implicated, economic constraints at every scale.

Trying to find social agreement regarding decisions concerning levels of admitted risks. Which is a new way, pragmatic and progressive, to share responsibilities, and to recognize that managing risk is essential to human development.

REFERENCES

Blanchi, R., Robert, F. & Guarnieri, F. 2003. Première analyse du contenu et de la qualité des Plans de prévention des risques naturels (PPR). *Annales des Mines*, mai 2003: 61–70.

Circulaire du 30 avril 2002 relative à la politique de l'Etat en matière de risques naturels prévisibles. Ministère de l'aménagement du territoire et de l'environnement, ministère de l'équipement, des transports et du logement. In http://www2.equipement.gouv.fr/bulletinofficiel/fiches/bo200219/a0190045.htm

DDE Haute-Savoie. 2004. *Prescriptions en matière d'urbanisme et d'aménagement en cas de risques d'inondation*. Mairie de Ville-La-Grand.

Mairie de Ville-La-Grand. 2004. Révision du PLU, diagnostic. Bureau Latitude, archives municipales.

Mission Risques Naturels. 2004. *Catastrophes naturelles, prévention et assurance*. Paris: Fédération française des sociétés d'assurance.

Peiry, J.L., Salvador, P.G. & Nouguier, F. 1994. L'incision des rivières dans les Alpes du nord: état de la question. *Rev. Géogr. Lyon* 69(1): 47–56.

Piegay, H. 1997. Principes nouveaux en matière de gestion des marges de rivières en tresses. *Revue de Géographie Alpine* 85(2): 33–48.

Pigeon, P. 2005. *Géographie critique des risques*. Paris: Economica-Anthropos.

PPR de Chautagne. 2004. Communes de la plaine de Chautagne. Présentation de la nouvelle procédure de consultations. Mairie de Serrières-en-Chautagne.

PPR de Ville-La-Grand. 1998. Rapport de présentation. Mairie de Ville-La-Grand.

SOGREAH 1995. Commune de Magland. Etude hydraulique, protection de Magland contre les crues de l'Arve. Mairie de Magland.

Trapitzine, R. 2002. Comment passer de l'ancien POS au nouveau PLU? *Etudes foncières*, janvier –février, 95: 18–23.

Risks Challenging Publics, Scientists and Governments – Menoni (ed)
© 2010 Taylor & Francis Group, London, ISBN 978-0-415-58072-4

Environmental liability as a prevention tool: A GIS application

C. Simeoni & S. Bellagamba
Department for Production Plants interaction with the environment, ISPESL, Italy

ABSTRACT: The European parliament and Council directive on environmental liability with regard to the prevention and reclamation of environmental damage, defines future guide-lines for an environmental liability damages system. The proposal's aim is to apply the principle according to which "the polluter pays". A correct achievement of strict environmental liability gives good incentives for avoiding damage, increasing awareness behaviour regarding on ecologically oriented by managers operating in major hazard facilities. The aim of this work is to verify the actual availability of sources of information on environmental data and on major hazard facilities in order to develop a methodology to allow the characterisation of environmental hazard indicators and to standardise hazard identification procedures. Proposed methodology has been applied in a Central Italy site (Arno river). Available data on environmental, industrial and social aspect have been integrated and overlaid through GIS technique to obtain land sensitivity maps.

1 INTRODUCTION

By 30 April 2007 U.E member states must abide by the European directive 2004/35/CE on prevention and reclamation on environmental damage.

Site contamination represents a threat to human health and the environment. In recent times there has been an increased loss of biodiversity as a result of releases of contaminants to ground or surface waters. Estimates published by the European Environment Agency (EEA, June 2000) specify that in Europe there are about 1,5 million of potentially contaminated sites. Only few Country have already been identified as definitely or potentially contaminated quantified in 300,000 sites and the costs associated with its cleanup give a sense of the significance of the problem: just for some Member States or regions and some sites costs are estimated between 55 and 106 billion €.

In such cases, there is clearly a need to ensure that the damage environmental assets are restored, however a better solution would be that damage did not occur. Damage prevention is, of course, the directive's primary objective. When environmental damage nevertheless occurs the question is "who should pay". So far, in most Member States cleanup costs associated with contaminated sites have been paid with public money.

2 THE DIRECTIVE

The Directive's objective of is to establish a legislation framework to ensure, through an environmental liability system, prevention and remedying of environmental damage 'environmental damage' means: damage to protected species and natural habitats, water and land damage. With a few exceptions, the basic concept is that an operator causing environmental damage or creating an imminent threat of such damage should, bear the cost of the necessary preventive or remedial measures. This Directive shall apply to environmental damage caused by any of the hazard activities listed in pertaining community legislation, and to any imminent threat of such damage occurring by reason of any of those activities; It is also applied to damage to protected species and natural habitats caused by any occupational activities other than those listed in community legislation like hazardous activities and to any imminent threat of such damage occurring from those activities whenever the operator has been at fault or negligent.

Finality of liability for environmental damage is to introduce duty to refund and restore caused damage. Until now payment burdened on State funds.

A correct achievement of strict environmental liability, could help in fulfilment of two goals:

- As an incentive for avoiding damage, through more awareness towards ecology from managers operating in major hazard facilities
- to recover the cleanup costs.

To bring about a liability, negative environmental effects must be linked with acts or failure to act of certain individuals. This is possible when we can:

- identify legal person of damage
- identify and quantify damage

A relevant role is played by insurance companies to prevent further contamination and awaken the industrial sector to environmental problems.

As a matter of fact, insurance premiums for environmental damages is linked to the risk level and presence/absence of industrial health and safety processes and environment protection

2.1 *Problems related to the directive application*

Premiums represent the price paid from contracting parties to take out insurance coverage. To fix a premium it is necessary to identify and quantify risks. In insurance field risk is defined as "probability that an event will take place and the amount of possible loss". Therefore the premium is a result of the estimate number (frequency) of damages related to contract and their medium value. The other two are a stochastic variables that can be determined through statistical observation. Availability of reliable data is important to estimate frequency of an event and it's related clean-up costs.

Data type to use on a characterisation of site risk are:

- emission sources
- type and quantity of hazardous materials in place
- type of productive process
- set-up of the general environmental management organisation and procedures
- type of air and water pollution control
- waste management procedure

characteristics of target area:

- density of population
- soil use
- infrastructures

state of migration matrix (air, water, soil).

To carry out an efficient system for environmental liability damage we must dispose of "information" on company, surrounding area and reclamation costs.

USA model Super found has produced historical series on number of contaminated sites, on clean-up costs for site type, on site distribution for type of company, on number and type of accidents and related restoration costs.

European data is available only for some member States, restoration costs are partial and aren't referred to any specific site. The limited data on accidents and related restoration costs are not subdivided on site types and the type of industry that has caused contamination. In consideration it's very hard to make a true esteem of risks.

A tool for European companies and other organisations, to evaluate, report and improve their environmental performance is the Eco-Management and Audit Scheme (EMAS). Since 2001 Emas has been open to all economic sectors, participation is voluntary and extend to organisations operating in the European Union and the EEA. Up to now the distribution of Emas organisations are not homogeneous in Europe and Emas organisations in Italy are few and concentrated in large sized enterprises.

3 METHODOLOGIES

Different typologies of data set are collected and managed through GIS methodologies to locate areas with homogeneous land sensibility.

GIS systems are able to manage different georeferenced data (raster, vector, alpha numerical) through dedicated spatial operators. In fact the GIS system allows us to acquire, store, process, view and manage available information, monitor variations and update data.

The area of study, the Arno river (Central Italy) is characterised by the presence of major hazard industries. Other information is also available for this area.

The following informative layer has been used:

- Hidrogeological risk

The layers, elaborated from Italian civil Protection Department, define the risk situation in observance of art.2 L. 365/2000. The risk area, also, support the drawing-up of provincial and communal land-use planning and external emergency plan. Related to landslide risk the area of study shows null values.

- Seismic risk

Elaborated from National Group Major Hazard and realised on communal base according to classification as provided in DM 16/1/96 (S = 6,9,12).

- Localisation of major hazard industries

In Italy major hazard industrial plants are subject to D.Lgs 334/99, an implementation of European directive 96/82/EC.

The area of study present plants classify as art. 6 and art. 8 of D. Lgs. 334/99. Classification is based on presence of specific substances or compounds which are reported in risk's and quantity's categories included in annexes of ib. D.Lgs. Particularly the industries utilised substances classified as toxic for ingestion and inhalation and hazard for the environment. Short and long time Risk is also related to aquatic flora and fauna.

Information on activities and substances are filed at MATT (ministry of environment—Territorial Protection).

- Land-use map

"Corine Land Cover" was used to scale 1:100.000, with minimal area of 25 ha.

- Census

Last Census of Italian population (data of 1991) elaborated from ISTAT (Italian institute of Statistics). Census sections has been taken as reference level for GIS elaboration.

Software utilised for analysis is ArcGIS 9.0 with extension 3D Analyst, Spatial Analyst and Geostatistic.

Applied techniques:

- With spatial overlay of different layers have been obtained, rather than a visual overlay, also information on relating objects attribute.
- Matrix analysis to transform vectorial layer in raster layer for particular models as hidrogeological modelling, terrain elevation, gradient and exposure calculation.
- Query to extract necessary data to spatial analysis.

Each layer has been reclassified to apply matrix analysis between thematisms. Reclassification connects increasing coding value to increasing risks.

Digital cartography Corine Land Cover has been reclassified to single out macro areas homogeneous compared to use reach destinations (Table1).

Table 1. Reclassification of macro areas.

Class	Recoding	Territorial element
Urban	1	Continuous urban fabric Discontinuous urban fabric Road and railnetworks and associates land Airport and port areas Construction sites Green urban areas Sport and leisure facilities
Industrial	2	Industrial and commercial units Dump sites
Natural system	3	Arable land Permanent crops Pastures Heterogeneous agricultural areas Forest and seminatural areas Scrub and/or herbaceous vegetation association Open space with little or no vegetation Wetlands
Extractive	4	Mineral extraction sites
Water bodies	5	Inland waters Marine waters

Classes related to population density has been computerised on base of 1991 ISTAT census; width class is based on quantile value.

New classification of seismic risk is consist of 4 zones: The first three correspond to high seismcity zone (S = 12), medium seismicity zone (S = 9), low seismicity zone (S = 6) agree to L 64/74. The fourth zone is a new entry; in this zone the duty to aseismic design is assign to regions resolution.

4 RESULTS AND CONCLUSIONS

The produced cartography allowed the evaluation of industrial activities interaction with environmental matrix. In this way it is possible to obtain a greater informative detail otherwise not highlighted out from aggregated data.

The map related to flooding risk is obtained from aggregation of density of population, land use and hydrogeological risk layers (Figure 1). In this way has been considered the events of slow release contamination in water resources and the value of their effects on surrounding population and soil.

Map in Figure 2 shows the overlay between density of population and land use to evidence the anthropic pressure near industrial sites.

Seismic sensibility (Figure 3) overlies seismic risk and density of population.

The localisation of Industrial sites subjected to the law 334/99 allows conclusion to be drawn of industrial impact on population and territory.

Particularly produced sensibility maps show that industrial plants are localised area with high alluvial sensibility, high anthropic pressure and medium seismic sensibility.

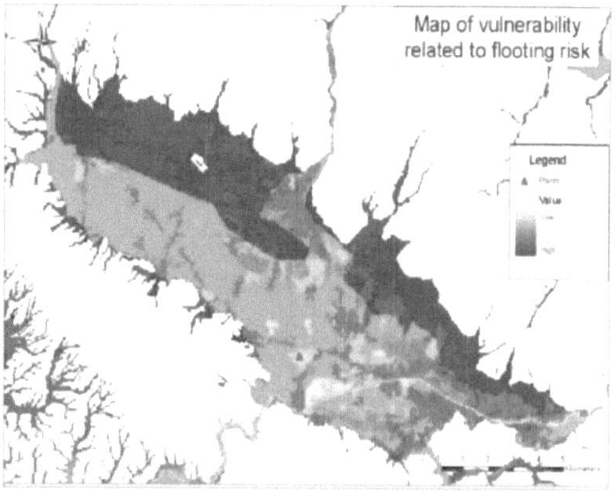

Figure 1. Map of vulnerability related to flooding risk.

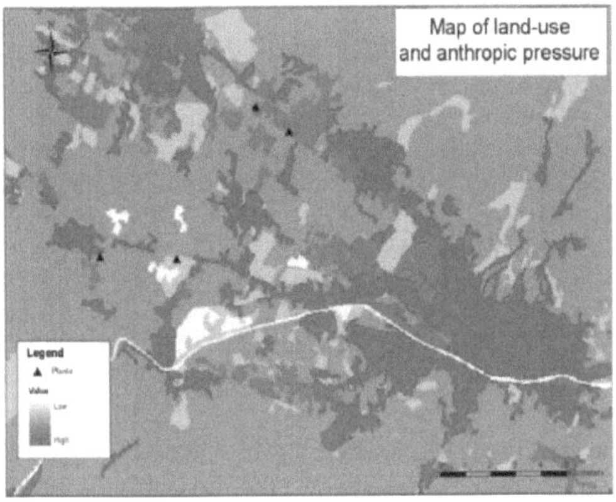

Figure 2. Map of Land-use and anthropic pressure.

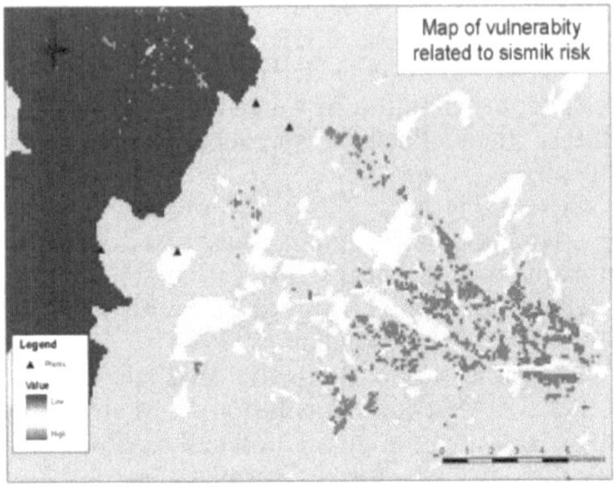

Figure 3. Map of vulnerability related to seismic risk.

For development of a future environmental liability damages system, this method is able to value interaction between single environmental aspects and associates to them a class of risk. Further development will be represented by inputting in GIS data base quantitative data and other information not yet available on natural risk and industrial hazard (localisation of industry, type and quantities of dangerous substances stored and/or manipulated, frequency and typologies of damages to human health and natural resources and related cost, etc.). For this it is important that governmental institutions develop homogeneous alphanumeric and cartographic data bases, constantly updating, for each European regions. So, we will be able to carry out specific analysis and quantitative risk assessment.

In this context a relevant role is played by small and medium-sized enterprises (SMEs), that represents Italian industrial structure. SMEs are frequently causes of environmental damage of major size than you might expect from their dimension.

REFERENCES

Burrough, P.A. 1986. *Principles of geographical information system for land resources assessment*. Oxford: Clasedon Press.

Environmental insurance for enterprises, Swiss Re, 1999.

Environment in Europe, EEA, 2000.

Maguire, D.J., Goodchild, M.F. & Rhind, D.W. 1991. *Geographical Information System: Principles and Applications*. New York: Longman Scientific & Techical.

Musmeci, L. 2001. *Modalità di applicazione dell'analisi del rischio nel contesto del D.M. 471/99*, Atti del Convegno "Bonifica dei siti contaminati: aspetti giuridici e amministrativi", Milano.

Risks Challenging Publics, Scientists and Governments – Menoni (ed)
© 2010 Taylor & Francis Group, London, ISBN 978-0-415-58072-4

Nation's critical infrastructures: Inherent vulnerabilities and economic fall backs of their unavailability

M. Martelli
AGORA' 2000, Rome, Italy

P. Spagnesi
Technip Italy, Rome, Italy

ABSTRACT: This paper aims to provide an operative method for identifying particular systems and assets 'at risk', analysing their specific characteristics in terms of vulnerability and mutual dependency relations. The threat which a technological infrastructure is exposed to and the possible direct and indirect effects on the potential receivers has to be the starting point for an identification analysis of infrastructure's criticality. The second part of the paper concerns the presentation of the possible economic effects arising from the unavailability of systems. In this respect, for making a choice regarding investments aimed to protect "critical infrastructures", it is necessary to refer to economic analysis instruments and methods. The main result of the work is represented by an operational methodology aimed at the definition of the specificities, which make essential a national technological infrastructure, considering the effects and technical-economic falls back derived from the weak non-operational elements—or subsystems—which constitute it.

1 INTRODUCTION

In the last years, the major western democracies, as well as several states of the area going from North Africa to South East Asia through Middle East, were exposed to terrorists' attacks, aiming at their destabilisation.

Also in countries where *post-conflict management* operations are today in action, the vulnerability of systems supporting civil and economic reconstruction, as well as infrastructures supporting everyday life, has to be reduced, through an adequate analysis and planning action of prevention and protection strategies. However targets (as well as direct and indirect consequences) of such attacks, seem to be more and more diversified and imponderable. Consequently, the management strategies, aiming at reducing relevant risks, have to take into consideration the application of proper risk analysis techniques, to preliminarily assess all possible interdependencies.

To this concern it is worth to remind as the concept itself of 'critical infrastructure' has expanded over time, starting from the vital systems for the economic and national security, as well as for the government continuity, up to include those assets where complete loss or partial damage can determine a negative impact on the 'nation's morale'.

The definition does not take into consideration only civil, military or religious tangible assets, but even other kind of assets, the so-called "intangibles". Therefore, at present, there is an increasing need to rationalise this issue, providing a methodological approach usable not only by government agencies, but also by the private industry, which has to face—at a reduced scale—the concern related to the criticality of its own infrastructures in different contexts among those considered until now.

In such context, the study on interdependencies among critical infrastructures becomes extremely important, because the analysis of the before and after '*domino effects*' type potential costs and effects, which can appear in case of unavailability of an apparently not essential system (or part of it) can cause instead the inclusion of such system among those which must absolutely be guaranteed. Once critical infrastructures are characterised in the light of such aspects, they will be classified attributing them different vulnerability levels, on the basis of the identifiable management and structural "weaknesses".

2 BASIC ASPECTS OF CRITICAL INFRASTRUCTURE

In this paper we consider an infrastructure and Critical Infrastructure (CI) adopting the EU following definitions [1]:

- An infrastructure is the framework of interdependent networks and systems comprising identifiable industries, institutions (including people and procedures), and distribution capabilities that provide a reliable flow of products and services, the smooth functioning of governments at all levels, and society as a whole.
- Critical infrastructures are those physical resources, services, and information technology facilities, networks and assets which, if disrupted or destroyed, would have a serious impact on the health, safety, security or economic well-being of Europeans or the effective functioning of the EU or its Member States governments.

At the European Union (EU) level the CI sectors are the following [1]:

- Energy installations and networks (e.g. electrical power, oil and gas production, storage facilities and refineries, transmission and distribution system)
- Communications and Information Technology (e.g. telecommunications, broadcasting systems, software, hardware and networks including the Internet)
- Finance (e.g. banking, securities and investment)
- Health Care (e.g. hospitals, health care and blood supply facilities, laboratories and pharmaceuticals, search and rescue, emergency services)
- Food (e.g. safety, production means, wholesale distribution and food industry)
- Water (e.g. dams, storage, treatment and networks)
- Transport (e.g. airports, ports, intermodal facilities, railway and mass transit networks, traffic control systems)
- Production, storage and transport of dangerous goods (e.g. chemical, biological, radiological and nuclear materials)
- Government (e.g. critical services, facilities, information networks, assets and key national sites and monuments).

When analysing a CI, mainly in industrialised countries, it is worth considering the following issues:

- Critical infrastructures are Complex Adaptive Systems (CAS), highly connected and highly interdependent. Interconnectedness and interdependence make these infrastructures more vulnerable to disruption or destruction.
- Critical infrastructures have become more dependent on common information technologies, including the Internet and space-based radio-navigation and communication.
- Within each CI Sectors, infrastructures are owned or operated by both the public and the private sector.
- Physical well-being of people, democracy's values, economic freedom and culture represent themselves inherent vulnerabilities for countries which accept to live in a global world.
- Identifying and prioritizing which assets of an infrastructure are most essential to its function, is necessary for developing an effective protection strategy. "The assets, functions, and systems within each critical infrastructure sector are not equally important. The transportation sector is vital, but not every bridge is critical to the Nation as a whole [2]".

3 THE PROTECTION OF CRITICAL INFRASTRUCTURE—THE EU APPROACH

The criteria for determining the factors that make a particular infrastructure or element of an infrastructure critical need to be studied. These selection criteria should also be based on a sectoral and collective expertise. Three factors might be suggested for identifying potential critical infrastructure [1]:

- Scope—The loss of a critical infrastructure element is rated by the extent of the geographic area which could be affected by its loss or unavailability—international, national, provincial/territorial or local;
- Magnitude—The degree of the impact or loss can be assessed as None, Minimal, Moderate or Major. Among the criteria which could be used to assess potential magnitude are i) Public impact (amount of population affected, loss of life, medical illness, serious injury, evacuation); ii) Economic (effect on GDP, significance of economic loss and/or degradation of products or services); iii) Environmental (impact on the public and surrounding location); iv) Interdependency (between other critical infrastructure elements), and v) Political (confidence in the ability of government);
- Effects of time—This criteria ascertains at what point the loss of an element could have a serious impact (i.e. immediate, 24–48 hours, one week, other). However, in many cases, psychological effects may escalate otherwise minor events.

Therefore the main recommendations for CI Protection (CIP) of a EU Member States are the following:

- Develop a National CIP Programme;
- Set minimum requirements for health, safety, security, economic well-being to be guaranteed to get a resilient nation;
- Unify the CIP effort in a single accountable official, coordinating the activities of state departments and agencies that have highly specialized expertise and long-standing relationships with industry;
- Identify the state departments and agencies responsible for CI sectors, as well as relevant private sector representatives;
- Enable effective partnership and to share information, data and expertise with other Member States, international community and private sector;
- Contribute to the creation of harmonized EU method and European risk-analysis system, harnessing expertise of local government and private sector developed in similar contexts (e.g., Seveso II Directive, Business Continuity, Disaster Recovery, etc.).

4 THE PROTECTION OF CRITICAL INFRASTRUCTURE—THE USA APPROACH

Since July 1996, the Government of the USA formally started to deal, in a structured and organized manner, with the issue of CI and their protection. The methodological and legal production has increased following the events of September 2001: after that, the Government established a centralised body for the management of national security issues, the *Department of Homeland Security*. The main laws and relevant documents issued after 1996, are the following:

- Executive Order 13010—*Critical Infrastructure Protection*. Federal Register, July 17, 1996
- President's Commission on Critical Infrastructure Protection, *Critical foundations: Protecting America's Infrastructure*, October, 1997
- *The Clinton Administration's Policy on Critical Infrastructure Protection: Presidential Decision directive N0. 63*, White paper, May 22, 1998
- Executive Order 13228—*Establishing the Office of Homeland Security and the Homeland Security Council*. Federal Register, October 8, 2001
- U.S. Office of Homeland Security. *The National Strategy for Homeland Security*. July 16, 2002 [2]
- Office of the President. *The National Strategy for the Physical Protection of Critical Infrastructure and Key Assets*. February, 2003 [3]
- Homeland Security Presidential Directive/ HSPD-7—*Critical Infrastructure Identification Prioritization, and Protection*. December 17, 2003

In the following graph is presented the USA approach for the interrelations between Infrastructures vs. USA National Functions.

5 A PROPOSAL FOR ASSESSING THE CRITICALLY OF AN INFRASTRUCTURE

5.1 *The identification of the CI*

In order to assess the criticality of a selected infrastructure a logical de-construction of the infrastructure is needed in order to list and order all the *assets* and *links* which constitute the CI and make it functioning. A relevant function is associated to each *assets/links* in order to assess its continuity and then influence other *asset/link* of the same infrastructure, or other sectors.

The hierarchy of CI elements [6] has to be considered as follows.

- Part: smallest component of a system that can be identified in an analysis
- Unit: a functionally related collection of parts (e.g., a steam generator)
- Subsystem: an array of units (e.g., a secondary cooling system)
- System: a grouping of subsystems (e.g., a nuclear power plant)
- Infrastructure: a complete collection of like systems (e.g., the electric power infrastructure)
- Interdependent Infrastructures: the interconnected web of infrastructures and environment.

All the single elements are linked together on a one or multiple basis (see Rinaldi and al. [6]).

In many models/methodologies that have been developed, threats, vulnerabilities, and risks are integrated and then used to inform the allocation of resources to reduce those risks.

These methodologies consist of the following elements, performed in the suggested order:

- identify assets and identify which are most critical
- identify, characterize, and assess threats
- assess the vulnerability of critical assets to specific threats
- determine the risk (i.e. the expected consequences of specific types of attacks on specific assets)
- identify ways to reduce those risks
- prioritize risk reduction measures based on a strategy.

In order to assess the criticality of the infrastructure we need to go through the following steps:

1. Select pertinent sector
2. Identify assets & links

National Factors → / Sectors ↓	*National Defence*	*Eco-nomic Security*	*Public Health & Safety*	*National Morale*
Telecommunication information networks	X	X		
Energy	X	X		
Banking/Finance		X		
Transportation	X	X		
Water			X	
Emergency services			X	
Government			X	
Health services			X	
National defence	X			
Foreign intelligence	X			
Law enforcement			X	
Foreign affairs	X			
Nuclear facilities, in addition to power plants			X	
Special events				X
Food/Agriculture			X	
Manufacturing		X		
Chemical			X	
Defence industry	X			
Postal/Shipping			X	
National monuments Icons				X

3. Identify functions
4. Identify internal & external interdependency elements
5. Specify undesired events
6. Evaluate consequences
7. Evaluate existing countermeasures.

Then, we propose the following worksheet to identify and assess the criticality of an infrastructure.

SECTOR	
SUB-SECTOR	
LOCATION	
NAME INFRASTRUCTURE	

ASSET/LINK	FUNCTION	INTERDEPEND. Internal/ External	UNDESIRED EVENT	IMPACT			COUNTERMEASURES	CRITICALITY RANKING C1 ÷ C5
				Economical	Geographic	Duration		

5.3 *CI: The economic assessment*

In order to come out with a '*criticality ranking*' of a CI, we need to assess, in economic terms, the impact (or Consequences, References 8), connected to loss or damage of a CI.

In principle, losses (or, better, disadvantages) have to be estimated in monetary terms and might include political, morale, operational effectiveness, or other impacts.

In terms of unavailability of a CI, we need to assess Private Cost or what any one person gives up due to this unavailability and Social Cost or what the society gives up due to this unavailability.

The purpose of the analyst is to estimate the Net Social Cost which is Net Private Cost + 3rd Party Cost—3rd Party Gains.

The main issues when considering economic assessment and CI are the following:

- Baseline. The economic assessment is an incremental analysis that compares an action to be taken with a baseline. Assessing impacts of an action requires that we first identify what would happen if we did nothing (the counterfactual). We need a clear statement of the baseline: What will happen in future if no action is taken? This is often not easy, because the consequences of the current situation are often not clear.
- Opportunity cost. It is what we give up to get something or to do something. It means the cost/disadvantage of something in terms of an opportunity foregone (and the benefits that could be received from that opportunity), or the most valuable foregone alternative. The application of the concept of opportunity cost looks for the hidden cost of any and every individual economic decision.
- Irreversible change. The usual standard methods do not account for irreversible impacts. Some protection to the interests of future generations can be offered through the imposition of the safe minimum standards decision rule. The safe minimum standards decision rule recommends that conservation be adopted when an activity that has an irreversible impact (i.e., on the environment) threatens to breach an irreversible threshold (unless the costs of foregoing the development are regarded as *'unacceptably large'*). It is based on a modified principle of minimizing the maximum possible loss. Therefore, it differs from routine trade-offs, which are based on maximizing expected gains, e.g. cost-benefit and risk analysis. However, activities that result in potential irreversible change are not rejected if the associated costs are regarded as intolerably high. A critical aspect in the application of the safe minimum standards decision rule is specification of the threshold for unacceptable costs associated to foregoing development.

5.3.1 *Cost-benefit Analysis (CBA)*

CBA is a policy assessment method that quantifies in monetary terms the value of all policy consequences to all member of the society. The net social benefits measure the value of the policy (NSB = B − C). The underlying CBA logic: resources are scarce and the decision maker has to address them where they are supposed to maximize the benefit for the society as a whole.

In the analysis of the CI (the baseline), costs and benefits are assessed with the point of view of the community (through the eyes of the decision maker) utilizing, if necessary, a pricing system which might be very different from the market's one and may incorporate the development objective of the community. We will consider then direct and indirect effects and might include political, morale, operational effectiveness, or other impacts.

- CBA includes all members of society, and measures the project's positive and negative impacts in terms of willingness to pay for units of increased consumption, and to accept compensation for foregone units of consumption.
- Willingness to pay and willingness to accept compensation are used rather than prices actually paid or received because: i) many of the project impacts to be included in the economic analysis will be non-market (human life, environment, political, morale) ii) other project impacts will be bought and sold in markets where prices are distorted by various government interventions, by macroeconomic policies, or by imperfect competition.
- WtP might be estimated from people i) directly (through contingent valuations) or ii) indirectly (based on the concept of "revealed preferences, i.e. market analogy).

It's then worth mentioning that:

- CBA introduces a time dimension in the analysis: the baseline is presenting different temporal patterns of costs and benefits, or different in their duration. Discounting provides a common matrix that enables comparison of costs and benefits that occur at different points in time. Use of discounting is integral to cost-benefit analysis and cost-effectiveness analysis (only costs without benefits).
- Discounting converts the stream of costs and benefits over time into a stream of 'present' values. The difference between the value of the discounted benefits and costs is referred to as the NPV.
- The choice of discount rate

$$Discount\ factor = \frac{1}{(1+i)^t}$$

can have a significant effect on economic analysis. It signals the rate at which future consumption is to be traded against consumption in the present. Use of a high discount rate discriminates against the future.

The basic strengths of CBS as a tool for the economic assessment are: i) the model is quite simple (B > C) and ii) can be used effectively by non specialists in a pragmatic way. The main weaknesses are connected to the information limit (if the analysis is "feasible" it is not credible and viceversa), the possibility of missing distribution effects and the fact that CBS is good for for 'little' projects but has to be considered carefully at the moment we are trying to assess effects on the GDP (National and cross-border).

At present CBA is the most considered methodology at EU level to assess Impact (see, European commission—IA guidelines plus annexes, June 2005).

5.3.2 *Multicriteria analysis*

According to this method:

- The option under examination is chosen according to an explicit set of objectives (expressed in measurable terms). Multiple criteria are explicitly defined to determine the compliance rate with the objectives.
- For each objective a weight is fixed so as to establish its relative importance. Weights are assigned reflecting the relative importance given to the objectives
- Weights and scores are combined in a performance matrix. For each option a final impact is estimated.
- Welfare maximization is not the only criterion. The rule B > C does not apply.

5.4 *Matrix for criticality ranking*

On the basis of the above, we propose and evaluation grid which, at the same time, takes into consideration different criteria such as the economic impact, the duration and the geographical extension connected to the unavailability of a CI (Table 1).

Table 1. Impact evaluation grid.

	Impact			
	A1. Economical* (MEuro)	A.2. Geographical (Territory)	A.3. Duration (Days)	Score $[S_{Ai}]$
Weight $[W_i]$	40%	30%	30%	
	0–25	Local	1–2 days	1
	25–50	Province	1 week	2
	50–100	Regional	1 month	3
	100–200	National	6–12 months	4
	>200	*Cross-border*	>12 months	5

*Damages estimate, assessed with the economic analysis, 'baseline scenario'.

6 CONCLUSION

Starting from the results provided in Table 1, through the following formula

$$\Sigma (S_{ai} * W_i) = C_j \text{ with } j = 1 \text{ [min]} \div 5 \text{ [max]}$$
$$i = 1 \div 3$$

it is possible to assess, in synthesis, utilising all relevant issues which the decision maker considers useful and important, the criticality level for every *asset/link* under examination and then, for the whole CI. Whenever dealing with CI owned, used and/or managed on a private basis the weight level might be different.

REFERENCES

[1] Critical infrastructure protection in the fight against terrorism, com 2004. 702 Final—Commission of the European Communities, Communication from the Commission to the Council and the European Parliament, Brussels 20.10.2004.
[2] The national strategy for homeland security—U.S. Office of Homeland Security, July 16, 2002.
[3] The national strategy for the physical protection of critical infrastructure and key assets—Office of the President, February 2003.
[4] Security vulnerability assessment methodology for the petroleum and petrochemical industries—American Petroleum Institute and the National Petrochemical & Refiners Association, 2nd Edition, October 2004.
[5] Report with a proposal for a European Parliament recommendation to the European Council and to the Council on the protection of critical infrastructures in the framework of the fight against terrorism (2005/2044(INI))—Committee on Civil Liberties, Justice and Home Affairs—Rapporteur: Stavros Lambrinidis, May 30, 2005. Final A6-0161/2005.
[6] Identifying understanding, and analyzing critical infrastructure interdependencies—Steven M. Rinaldi,

James P. Peerenboom, and Terrence K. Kelly, IEEE Control Systems Magazine. December 2001.
[7] Site security guidelines for the U.S. chemical industry—American Chemistry Council, the Chlorine Institute, and the Synthetic Organic Chemical Manufacturers Association, October 2001.
[8] Valutare l'incertezza. L'analisi costi benefici nel XXI secolo-G. Pennisi-P.L. Scandizzo, Giappichelli, 2003.
[9] Cost-benefit analysis—concepts and practice (second edition)-A.E. Boardman, D. Greenberg, A. Vining e D. Weimer, Prentice Hall, 2001.
[10] Determinazione dei prezzi e giudizi di valore nell'abc—M. Martelli, in S. Momigliano e F. Nuti Giovanetti (a cura di), La valutazione dei costi e dei benefici nell'analisi dell'impatto della regolazione, Rubettino, 2001.

Risks Challenging Publics, Scientists and Governments – Menoni (ed)

 ISBN 978-0-415-58072-4

Safety information systems related to transport of dangerous goods in Norway

O. Njå
University of Stavanger/SEROS, Norway

J. Vatn
NTNU, Norway

ABSTRACT: In Norway transport of dangerous goods is mainly carried out on roads and at sea. Most of the onshore transport is on road, only 3% by rail. The transport is concentrated to the most populated areas, especially in the south-eastern region around the capital Oslo. This paper presents results from a study aiming to establish a methodology to evaluate the low- and high-level risk pictures related to dangerous goods transport. Low-level risk pictures relate to single carriers, either as responsible for single vessel/vehicle or a transport company. High-level risk picture relates to the national point of view. The paper discusses safety information systems that are employed in the Norwegian transport sector. The major concern is the actual use of data in these systems, and particularly how the different actors are linking data in a feed forward strategy. Subsequently the quality and the relevance of the obtained data sets are discussed.

1 INTRODUCTION

The term 'dangerous goods' includes a lot of different meanings. In this paper we relate dangerous goods to the definitions given in the ADR/RID-regulations (ADR/RID, 2005) and the IMDG-regulations (IMO, 2004), which contain the substances, materials and articles covered by current regulations. Thus, the framework of the nine classes with respective subgroups is what we mean by dangerous goods in transport.

In Norway approximately 12 million tons of dangerous goods are transported domestically each year, of this over 85% are flammable liquids. The two dominant classes besides flammable liquids are corrosive substances (Class 8) and gases (Class 2), cf. Madsen, Steffensen & Grue (2004). During the last ten years the average number of reported accidents/incidents in Norway is 59, related to domestic transport of dangerous goods per year. This number includes all non-normal situations connected to loading/unloading, handling at terminals and during transport on roads or rail. Last year (2004) 61 events were registered, none of them with serious consequences. Approximately 40% of the incidents lead to minor leakages, with an estimated total leakage of 2165 litres; a very small amount compared to previous years.

In recent years two major incidents have occurred. The Lillestrøm train accident occurred 5 April 2000 (NOU, 2000). A train carrying two propane tanks lost braking effects at the Lillestrøm station and collided with another waiting train. The propane tanks were damaged and propane leaked out. After a short time, the propane ignited. The situation was critical and came very close to BLEVE that would probably have killed a large number of people and laid the central areas of Lillestrøm in ruins. No persons were killed or physically injured. The other serious accident occurred at Sjursøya, just south of the centre of Oslo. A truck loaded with 17.500 litres gasoline and 19.000 litres diesel oil collided with a shunting train. The truck and train were set on fire but no persons were injured (AIBN, 2004).

This paper is based on a preliminary study of the Norwegian land based transport of dangerous goods. Norway is a country with long distances between the cities, a long coastline with many fjords and lots of mountains. These facts imply that the goods has to be transported inter modally, for example by trucks loaded on ro-ro ships over fjords, and further on roads requiring enhanced driving skills with respect to narrow roads, long tunnels and "aggressive" road-users.

The safety information systems' main function is to establish feedback control. The issue of this study has been to investigate how the feedback loops work within the different organisations in the transport sector, thus how the information transfer is carried out horizontally and vertically within the frame of the organisation. The intention has been to clarify the need for further information in order to develop risk methodologies and risk models, and investigate the different actors adaptation to different frame conditions, such as regulations, transport restrictions, risk of control/punishment, rewards/incitements, etc. Even though the safety information systems have not been analysed in detail with respect to contents and applications, several tendencies have become clear throughout this preliminary study.

Some of our findings are based on an analysis of two previously published case studies; one carried out in the sea transport (ferries—Bøen, 2005), and the second carried out in the railway sector (Åsland, 2005). Both studies investigated differences between the formal structures of the dangerous goods handling, and the observed behaviour and expressed attitudes amongst employees in the two transport companies. Furthermore, statistics, schemes, reports, a survey (Norddal, 2004) and interviews with key employees working in different levels of the transport systems bases this preliminary study.

The paper is organised as follows: Firstly, the Norwegian structure of the actors involved with dangerous goods is presented. Secondly, the theoretical premises for safety information systems are depicted in the context of the dangerous goods transport. Thirdly, the practical use of the safety information systems within and across different actors is discussed. The paper concludes on the status of safety information systems and recommends clarifications needed to assess the potential of safety information systems in the dangerous goods transport.

2 THE NORWEGIAN STRUCTURE OF DANGEROUS GOODS TRANSPORT

The actors involved in the transport of dangerous goods are numerous. Consider a transport from A to B; the products to be transported are often manufactured, packed and prepared for transport by the consignor. Upon request a carrier, normally a truck driver picks up the products. The goods are often reloaded onto trains at specific terminals or in harbours from ships to trucks and vice versa. The dangerous goods are finally delivered at the consignee's location. During the transport the regulations puts forward different requirements to the handling, the packing, the documentation, the actors' competence, the storage, and so on. At least two major regulations, ADR/RID (road and rail) and IMDG (sea), introduce comparative requirements between the different transport modes, which diverge in contents and rigidity. Currently we observe that there exists a lot of frustration amongst the actors in the transport sector.

The structure of organisations with responsibilities related to handling of dangerous goods transport from consignor to consignees is quite complex. It is characterised by the many parties involved, each one taking care of a limited set of activities and with unclear responsibilities. Thus, it can be claimed that the supervisory bodies have fragmented areas of expertise, and that their focus is on sector responsibilities. In order to achieve a functional and legitimate transport, the ADR and RID-agreements require that every enterprise packing and carrying dangerous goods must have a nominated approved safety advisor. The safety advisor shall plan transports and ensure that the employees involved follow up the mandatory requirements. This task entails need for an effective communication with the authorities at different levels. Table 1 depicts the structure spanning from bodies responsible for the international regulations at the *blunt end*, i.e. distant from the physical hazard, to the single truck drivers at the *sharp end*, close to the physical hazard.

Table 1 illustrates the large number of institutions involved, especially at authority level; and even some of them are not listed here. For example the municipal health authorities are by legislation empowered to require specific health care analyses with respect to dangerous goods transport. As far as we know, no environmental health analysis has ever been carried out in Norway, focusing specifically upon transport of dangerous goods.

3 SAFETY INFORMATION SYSTEMS

Safety information systems have become an important part of modern safety management. The idea is based upon the philosophy of high reliability organisations (HRO), developed by researchers at the University of California in the second half of the nineteen eighties (Marone & Woodhouse, 1986; Roberts, 1989, 1990; La Porte & Consolini, 1991; Wildavsky, 1988). The HRO philosophy puts weight onto organisational learning, in which learning from accidents, near accidents and deficiencies are important. An organisation that has safety as a superior goal will be prone to obtain knowledge about conditions that threatens the safety achievements and then develop countermeasures. Thus, the systems available to provide such knowledge are substantial to the high reliability organisations.

The safety information systems' main function is to establish feedback control. This principle requires that the prevailing situation continuously be monitored (on a day to day basis) in order to

Table 1. The structure of the Norwegian transport of dangerous goods.

Actor level (Rasmussen and Svedung 2000)	Rail	Road	Sea (ferries)
Regulatory and supervisory responsibilities regarding dangerous goods transport – blunt end			
United Nations - ECOSOC (International regulation/recommendations) – expert committees	Regional commissions, including the European Commission (ADR/RID (2005) is adopted Norwegian regulations)		International Maritime Organisation (IMO), IMDG-Code (IMO 2004)
National – political management	Government – Ministry of Transport and Communications, Ministry of the Environment, Ministry of Justice and the Police. The Storting (Parliament) and the Standing Committees, for example, on Transport and Communications		
National – administrative managements, general regulatory authorities	The Norwegian Railway Inspectorate	Norwegian Public Roads Administration	Norwegian Public Roads Administration
National - Special responsibility for dangerous goods and accident prevention	Directorate for Civil Protection and Emergency Planning (aided with supervisory services from the Police, the Norwegian Public Roads Administration, the Norwegian Railway Inspectorate and Norwegian Customs and Excise)		Norwegian Maritime Directorate
Regional authorities	19 counties (co-ordinating responsibility of traffic safety on roads), and 14 counties involving ferry transport		
Local Authorities	433 municipalities responsible for local administration of the ground and land use (they are also responsible for the local fire brigades, and environmental health issues).		
Special responsibility for emergency response and pollution prevention –sharp and blunt ends	Joint Rescue Coordination Centres, Local Rescue Coordination Centres, The Norwegian Coastal Administration, Inter-municipal resources to combat acute pollution, The ICE-network (International Chemical Environment – in Norway denoted RVK), operational rescue resources (police, fire brigades, emergency health services, etc.)		
Direct transport involvement – sharp end			
Infrastructure	Norwegian National Rail Administration	Norwegian Public Roads Administration, the counties and municipalities	Norwegian Public Roads Administration The Norwegian Coastal Administration (local harbour authorities)
Vehicles/carriers	Eight railway companies, from which CargoNet is the major carrier of dangerous goods. 12 terminals, Borregaard/Falconbridge (Chlorine), Sjursøya/Gardermoen (fuel)	The Norwegian Haulier's Association, and approx. 20.000 with ADR-competence related to their drivers licenses.	The Federation of Norwegian Coastal Shipping, approx. 30 ship owners and approx. 250 sections
Other actors related to dangerous goods	Consignors/consignees, such as members of The Federation of Norwegian Process Industries (PIL), Norwegian Industrial Gas Association, The Norwegian Petroleum Industry Association and other agents, and six approved schools for safety advisors		

improve the organisation to better achieve predefined goals. This is achieved by reflexive reporting within the production system. In our case, the system is the entire transport chain of dangerous goods. Typologies of events and weaknesses to be reported need to be agreed upon. A common understanding of concepts, categories and descriptions is vital. The data is collected, stored and analysed in order to develop safety indexes. The information is supposed spread to the users, the decision makers, which initiate work to develop measures to improve the situation. This completes the management loop, which is depicted to the right side of Figure 1.

Kjellén (2000) outlines theoretical foundations and prerequisites for safety, health and environment (SHE) information systems. Kjellén's background as safety advisor in the oil and gas industry has been important for his work, which has lead to the set of requirements to an effective safety information system:

1. The *data collection* must be reliable, the facts about accident risks must be accurate, and adequate coverage of circumstantial factors must be ensured.
2. The *distribution and presentation of information* must be relevant in relation to the decision making context, it must be comprehensible albeit easy to survey for managers and other parties, the time from registration to corrective actions are taken must be minimised, and the availability of information is important for easy access in hindsight.
3. The *safety information system must in general* be easily understood and acceptable for all parties involved, the system must promote involvement

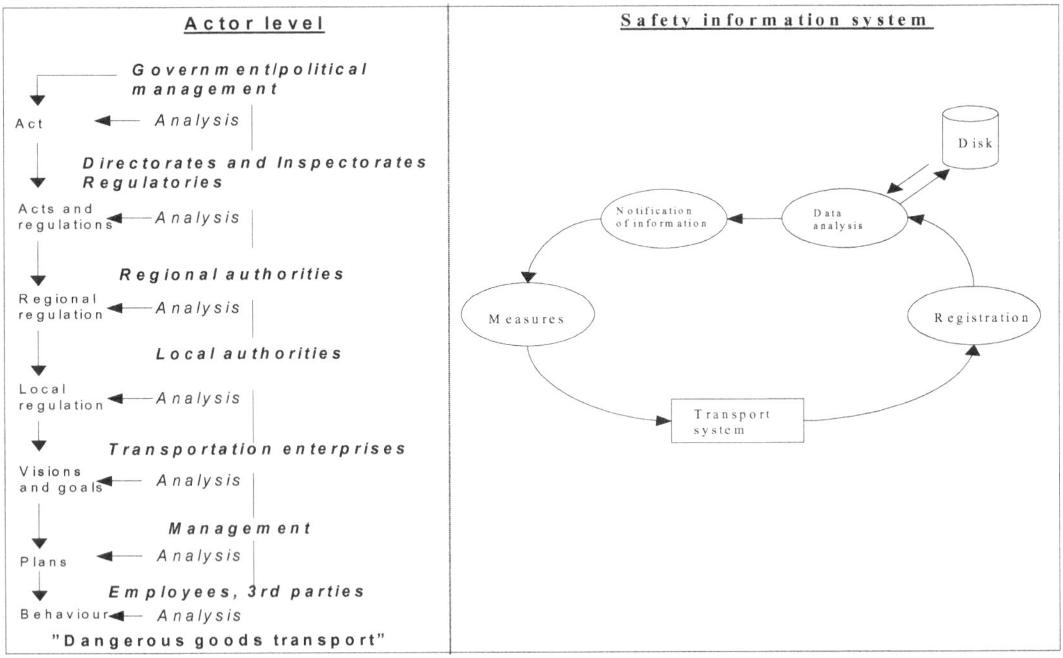

Figure 1. Safety information systems at different actor levels, based on Kjellén (2002), Rasmussen & Svedung (2000).

of managers as well as employees, and it must be cost-efficient with respect to operational costs assessed against the systems' ability to promote safety improvements.

Kjellén discusses the trucking industry in Sweden in light of which data that is commonly available, such as police reports and statistics from insurance companies. He envisages traffic-conflict techniques as a potential for improving the safety information systems, but no evaluations of systematic use within the transport sector have yet been conducted. Nicolet-Monnier & Gheorge (1996) list generic databases concerning dangerous goods and health hazards. The databases are general, nationwide and supposed used in quantitative risk assessments. Nicolet-Monnier and Gheorge do not evaluate the practical implementation of the databases as safety information systems or the related use of the systems within or across organisations.

Figure 1 also presents the hierarchical structure and different levels of the actors involved in dangerous goods transport, as seen in Table 1. The idea of this model is to investigate how the safety information systems work with respect to dangerous goods transport, and how the different actors approach safety management, both within their respective organisations and the interaction between the different actors at different levels.

4 SAFETY INFORMATION SYSTEMS, DANGEROUS GOODS TRANSPORT AND RISK

This section summarises the data material collected in the project. At the sharp end one major actor in each of the three transport modes and one consignor have been contacted with respect to their safety information systems;

- The ship owner is one of the largest in Norway on domestic ferry transport, and has over 800 employees, 27 ferries serving 17 sections.
- The railway company is also large in a Norwegian context, having 750 employees and serving a network of 12 terminals in Norway.
- The Norwegian Haulier's Association represents approximately 4000 goods transport companies and approximately 9000 lorries/road-trains from all over Norway.
- The consignor is part of a multinational corporate, which is a manufacturer of chemical substances. The company of this study is responsible for the gathering and analysis of accident/incident data from seven countries.

At the blunt end, the most central authorities' reporting systems, being the Directorate for Civil Protection and Emergency Planning (DSB), and the Norwegian Maritime Directorate (NMD) have been studied.

4.1 *Goals*

The purposes of the systems vary with organisational levels. The actors in the sharp end have designed their safety information systems to aid the safety management. However, only the consignor had prepared a wide definition of the purposes: "The purpose of this procedure is to ensure that all incidents and unsafe conditions, which have or would under different circumstances have caused damage to people, property or environment in short or long term shall be reported, and actions taken to avoid reoccurrence of similar incidents/conditions. The knowledge of the root causes (lack of management control) of these incidents/conditions shall be used to prevent unwanted incidents to happen and to improve the SHE (Safety, Health and Environment) consciousness".

At the directorate level the purpose of the reporting systems seems unclear. Besides being informed about specific events, the directorates do not analyse events or they do not have a systematic approach to the treatment of the data material. The DSB is a member of an international expert committee on dangerous goods and benefits from the overview given by the reports. Furthermore DSB publishes annual statistics on various accidents/incidents data, and dangerous good transport is one of the sections. NMD has its own database (DAMA), and the NMD forwards data to IMO in accordance with the agreements. NMD provide safety bulletins to the operative actors in the shipping industry.

4.2 *Current registration praxis*

In general the companies treat dangerous goods events as one small contribution to a variety of events within their organisations. For most actors, dangerous goods events are a minor contributor to the overall number of events.

After screening the incidents reports registered by the ship owner for the last five years, only two or three dangerous goods events were recorded in the period. One event was a lorry loaded with diesel oil that tipped over on board a ferry, and 1000 litres leaked onto the deck and washed to sea. One incident was due to erroneous marking, and the last was a report from a captain specifying information about how to handle a specific product in accordance with the IMDG-code.

Since the Lillestrøm and Åsta train accidents in 2000, the railway sector has gone through major changes with respect to implementing systems for safety management. The safety information system used in this company has recently been comprehensively developed, and efforts have been made to make the system function as intended. The rail company presents many reports, most of them concerned with erroneous marking, lack of documentation, and classification. Also structural flaws in the carriages are frequently reported.

Since 2001 the consignor has reported seventeen accidents/incidents in Norway with respect to cylinder or bulk transport of dangerous goods. The company has a reporting system that covers accidents and incident in seven countries involving the company employees as well as employees in all companies working for them. The safety advisor within the company analyses the incidents and issues recommendations to safety measures and corrective actions.

The hauliers' umbrella organisation was only acquainted with the DSB-system. The DSB-system has not a major role in their safety management work. Special data had been requested, but the relevance for their safety work was minor. The respondent, who was working with safety issues, was aware of the fact that "perhaps as much as 40 carrier companies (all kinds of transport) have systems that deals with non-conformances". However, the respondent had no knowledge about the practical use of such systems.

Both the DSB and the NMD have developed reporting systems. The local fire departments are the main contributors to the DSB registration. When they have been out combating an accident/incident involving dangerous goods they issue written reports, which to be sent to the DSB. Also the parties responsible for the vehicles, equipment and plants involved in the incident are required to send reports. At sea, the captain and the ship owner report to the NMD, in accordance with the procedure (NMD, 2003).

4.3 *Factors capable of influencing the systems*

The intentions with safety information system are attractive to all who favour safety improvements. However, there are a lot of factors that influence all the elements described in Figure 1, which could reduce the effect of the system, and even make things worse (Aven et. al. 2004; Kjellén, 2000; Aase & Ringstad, 2002). These psychological and organisational factors could be; attribution, incentives, filtering, local knowledge, and organisational frame conditions.

- *Attribution* refers to psychological mechanisms that influence how we understand an event. Those who are involved in the event are much more prone to explain the causes with technological failures than those standing outside the accident. A tendency seen in the railway sector is that events reported are rooted in third party activity (marking, documentation) and events which could be blamed on employees working in other terminals. The report written by the ferry captain, addresses the truck driver's request to enter the ferry in spite of bad weather as a major cause of the lorry tipping over.

- *Incentives* are related to the expected rewards and penalties from the reporting of events. The incentives could be socio-culturally conditioned without any observable reaction being paid. Negative incentives often imply lack of reporting, which in fact is the situation for all systems being mentioned. The impression, both in rail and at sea, was that the employees without management responsibilities did not prioritise reporting. The fire department has also special considerations. For them it is the number of reports that document their activity level, not the categories or the classifications. The dangerous goods incident schemes are regarded more comprehensive to fill in, thus, other schemes could be prioritised.
- *Filtering* refers to mechanisms that decide which information will flow through different stages in the safety information system. To reveal different mechanisms for filtering a deeper investigation into the specific safety information system is required. However, the statistics presented in the annual reports from the DSB are limited and regarded not applicable for the single companies or even sectors as background for safety measures. All respondents emphasised lack of feedback and systematic experience transfer within their organisations.
- *Local knowledge* relates to the different actors in the organisation that are confronted with the safety reports or recommended safety measures. The person's confidence to the system is dependant on how the feedback fits to his/her own understanding of the phenomena. Change of procedures is often the solution to the problems revealed by the safety information systems. Such measures are often perceived as cheap and the easy way out of problems that employees regard more comprehensive. This was indicated by a respondent in the railway sector, which could explain the lack of confidence to their system and the indifferent attitudes amongst the employees to report incidents.
- *Organisational frame conditions* include procedures for registration, categorising, quality assurance of data etc. For example introducing new systems, which are seen very bureaucratic, will meet resistance in the organisation, and thus neglect to report incidents. The need for sufficient competence amongst all participants in the system is often disregarded.

4.4 *Risk and dangerous goods transport*

The major frustration amongst the dangerous goods carriers is the regulations related to the ferry transport. The regulations imply that trucks loaded with dangerous goods may risk missing one and more trips at the ferry sections because of special restrictions. The confrontation seems to involve the NMD on the one side and the dangerous goods transport industry on the other side. The industry has complained through letters, press releases, report (Norddal, 2004) and seminars. Most parties claim that the restrictions set on the ferry transport do not enhance safety.

The project reported in this paper is partly sponsored by the industry. The industry's intention is to develop a model to identify the overall risk picture related to the dangerous goods transport in Norway. One objective of such a risk model is to use it to support harmonisation of the regulations, which is also a major goal of the National Transport Plan (NTP, 2004). The industry wants to change the regulation philosophy from detailed requirements regarding solutions to function based requirements where the safety performance to be achieved is more important than which solutions to be chosen for the ferry transport. In this perspective risk analysis, risk acceptance and risk management is making sense.

Modelling risk and carrying out risk analysis require in depth knowledge about the dangerous transport systems and the actors involved. Thus, the safety information systems become vital in a feed forward strategy, based on risk informed decisions. However, the present state of the design and use of the safety information systems covered by this study does not provide a credible support. It could be questioned whether it is possible to obtain reliable data on accidents, incidents or deficiencies with the existing systems at all.

The Norwegian Public Roads Administration provide reports for the European Commission concerning infringements and penalties from road checks. Over 35% of the ADR-vehicles checked from 2000–2003 have been reported to the police. A respondent from the Public Roads Administration in the South Region claimed that there were no signs of improvements in the industry. The South Region had a rate of police reported vehicles of over 50%, but this was due to their method to pick vehicles for road checks. He was also very critical to the practical use of the information systems related to the road checks, because they did never get feedback that could be used in their own planning or prioritising work of checking ADR-vehicles.

5 DISCUSSION

Kjellén (2000) deploys organisation theories to recommend use of safety information systems in road transport, which is a societal system. The hypothesis that, for example the theories of high reliability organisations are applicable also for safety management of dangerous goods transport within society needs to be challenged. In real life, the members of the actors in dangerous goods transport could have various preferences to the dangerous goods handling. And of course, a truck driver responsible for his/her own truck will probably have another view towards safety measures than a directorate responsible for the regulation and supervision.

In the following we will discuss different characteristics of the safety information systems at the various levels. The aim is to give a principal discussion on what is possible to achieve with safety information systems. This would of course be in contrast to current practice, and we would expect gaps in the possibilities and the current utilisation of such systems. The dangerous goods industry and several associated authorities demand that the regulations become function based and adapted to proactive safety management. Risk assessments play a major part in such systems. We discuss aspects to the interpretation of risk, when such systems are employed.

5.1 *Characteristics of the safety information systems at the various levels*

At the sharp end we have the companies and undertakings carrying out the dangerous goods transportation or loading. At this level there are differences in the organisation between the various transportation modes. For the railroad transportation there is primarily one carrier. In this case the safety information system could integrate both "ordinary incidents" and "dangerous goods incidents". The *potential* for organisational learning is high because a holistic approach is possible, and there could be resources to support the learning process.

When it comes to dangerous goods transportation on the roads, the situation is more challenging due to the fragmentation of private hauliers, often owning a single truck. Despite the fact that safety advisors should be available according to the ADR and RID requirements, it is far more challenging to ensure the systematic organisational learning. A functional safety management system is difficult to establish in such small organisations with one or only a few employees. Umbrella organisations, such as the Norwegian Haulier's Association, could involve as a co-ordination unit, but these ideas needs to be sufficiently investigated. Another possibility would be that the mandatory safety advisors, or nominated safety information analysts, could take an active part utilising the information to improve the hauliers and dangerous goods transport. With respect to dangerous goods transportation on ferries, the situation is something between the railroad and the road transportation with approx. 30 ship owners. For all parties mentioned in this paper the safety information systems have not been efficient. If they shall succeed enthusiasts are needed and general support in the organisations being provided. The safety advisors can play this role. However, the follow up practises regarding the exploitation of safety advisors both form the authority or the company perspectives seems not promising.

Emergency response and crisis management is conducted at the municipality and county levels. Accidents involving dangerous goods seldom occur within the district of the distributed fire departments. When such an accident happens, the local fire brigade often has insufficient knowledge on how to combat the various dangerous substances in a critical situation. Furthermore, the respondent from the chemical manufacturer listed examples that fire fighters had ignored truck drivers' knowledge at the accident scenes. We think that there is a need to improve the interactions between involved actors in the acute crisis handling. Johnsen et. al. (2005) have developed a method where real incidents are analysed by a scenario approach. Focus is put on the interaction between actors from different organisational units.

Today, the fire brigade being involved in the situation carries out most of the reporting from dangerous goods incident. Thus, the reporting "upwards" in the system starts at this level. However, the reporting form being used is mainly focusing on what happened before the incident was a fact. Little information regards those factors that influence the final outcome, i.e. the acute crisis handling. At the municipality and county level the current reporting system of dangerous goods incidents is rather irrelevant for fostering organisational learning. Methods representing more comprehensive investigations could be promising if the aspects of attribution, incentives, local knowledge, filtering and organisational frame conditions become considered.

The categorisation of dangerous goods events is difficult. As in all accident registration there is a general tendency that the more serious accident the higher probability that it will be recorded. At the national level DSB/NMD have a special responsibility to regulate the dangerous goods handling and transport in order to prevent accidents. Both DSB and NMD have put an accident reporting system into operation, where it is mandatory for the fire brigades or other actors involved to report incidents or accidents. It is important to recognise the limitations in such reporting systems especially with respect to organisational learning. By analysing accident and incident reports centrally at DSB/NMD, it is possible to reveal trends, important factors affecting the likelihood and consequence of undesired events. But it will be almost impossible to reveal the organisational setting in which the incident or accident occurred. Obvious important cues to what the problems are could be found, but further *investigation* at the sharp end is required to really understand the organisational aspect, and foster learning. The realistic learning from the safety information system at the DSB/NMD level is in terms of more *generic* measures. These could be related to rules, regulations, restriction in operation etc. It is very unlikely that DSB/NMD will reveal problems and challenges with e.g. the safety culture, or the pattern of interaction between actors at the sharp end. However, the reporting system could become more actively used as a starting point of experience transfer and communication between the industry and the central authorities, in order to achieve organisational learning in all parties. Today, this kind of communication is minimised. Westrum (1993) has introduced five level of safety culture, the pathological, the reactive, the calculative (rule based), the proactive and the generative level. A challenge with a safety information system where the operator report upwards in the system is that it could become difficult to advance in 's level of safety culture. When one passes the calculative or rule based level the "one way information flow" from the sharp end to the blunt end could stop the proactive and generative initiatives.

5.2 *Safety management based on risk assessments*

Modern safety management, as described in many textbooks, has an inherent interpretation of risk being an objective property of the activity or system being studied, and it is heavily based on historical numbers. This implies a sharp distinction between what is the true risk and what is perceived risk. This is a very invidious approach to risk. Particularly, due to the fact that many "experts" claim to know the truth and the experts' attitude is that lay people and others are driven by feelings and irrational behaviour. This is an old and positivistic perspective, which is widespread amongst many environments working with safety and emergency management. The explanation is simple. This perspective maintains a pattern of power with an utter authority to the experts. Nobody wants to give up one's authority and position. From a sceptic's point of view, it may seem that the aim is merely to create a stronger dependence to or demands for the experts' services. Norddal (2004) concludes that most of the special restrictions on the ferry transport can be removed, and the transport carried out in compliance with the ADR-recommendations. Norddals has a classical approach to risk, and he computes the societal cost benefit of the changes in the regulations. The background data is highly questionable suffering from limited datasets, and Norddal does not address uncertainty measures neither in the data collection nor the risk modelling.

However, we will claim that a goal-oriented management using high-level goals, generation of alternatives, analysis and evaluation of these is more appropriate. What is acceptable from a safety perspective should be sought solved with the use of holistic evaluations involving all relevant attributes. But, the rush to provide risk analyses, and the overall systems analyses to obtain the Norwegian risk picture on dangerous goods needs to be carefully considered. It is important to emphasise that a risk model, and the subsequent findings is to be used in a reflective decision support context. It will not be fruitful to use such findings to "prove" that we are right, and the others are wrong.

Understanding risk and probability theory is very difficult for employees in general. When the transport sector adopts risk based management as a fundament for managing safety, the sector must also be responsible for raising the competence. Risk and risk analysis should not be regarded as a "black box" unquestionable for other than the analysts themselves. This has to change, and risk in risk analysis settings should follow the principles of the predictive, epistemic uncertainty approach (Aven, 2003). This theory does not highlight a fictitious risk quantity, but quantities of the "world", such as the number of fatalities, the occurrence of a railway accident, accidental loads, etc. and risk analysis should be used as a tool for expressing uncertainty related to these quantities. This is in our view a much better foundation for involving decision makers and other parties in the risk analysis processes, with its models, data and assessments.

At the sharp end the risk analyses are very often of qualitative or semi quantitative nature. It could be job safety analyses, rapid risk ranking, and hazard identification where data from the safety information systems could be used in a heuristic manner to support the analysis. On a higher level, e.g. on a national level or international level where the aim is to establish risk based rules, the data from safety information systems could also be used to estimate parameters or assess probabilities for observable quantities in the risk model. Such quantitative risk models could also be used by companies that want to optimise the use of different transportation modes under constraints on minimum risk and maximum profit. This challenges a very structured manner to the data collection. In the nuclear industry and oil & gas industry reporting systems have been in place for several decades, and one have obtained some trust in the validity of the reported data. However, it is an open question whether it is realistic to establish safety information systems that could give the data required for quantitative risk models in the area of dangerous goods. Undoubtedly there will be a need for improving the safety information systems both within organisations and across organisations. The research project could be a starting point for introducing such changes, but the process must be driven by the industry itself.

Even though effort is paid to increase the knowledge of dangerous goods transport within the sector, the society needs to clarify what is acceptable or not. The different transport modes must agree upon how to measure risk and how to obtain acceptable risk levels. Risk assessments are not excluded from politics and use of power. Researchers in the risk and safety area seem to have a naive belief that risk analyses are objective and the use of risk assessments follows ideal rational decision processes. This is wrong. We are certainly in an area of power. Decision processes should be given more attention in future work. In this work the use of risk acceptance criteria should be critically examined. As of today, risk acceptance criteria transfer decisions to the risk analysts. That is an unacceptable situation.

6 CONCLUSION

We conclude that the safety information systems employed in the safety management of the dangerous goods transport in Norway have not played a major role so far. This is in accordance with the comparative study conducted by Lim and Nédélec (2006) of two French databases. However, it seems that large enterprises responsible for limited activities, such as railroad transport and consignors responsible for a limited number of dangerous substances do benefit within their respective organisations from safety information systems. These systems could be used as guidance for the other actors in the industry, which differ significantly with respect to frame conditions and incentives. In the preliminary phases of our work to establish risk models, to develop risk management approaches and to assess different actors' adaptations to frame condition in the transport of dangerous goods, we conclude that there are many challenges to be further investigated:

- Based on an intention to implement risk management, what is necessary provision of safety information?
- Existing safety information systems have major weaknesses. How can they become sufficiently improved, and is it at all realistic to establish such systems?
- The consignor's safety information system covers many countries and enterprises, which could serve as a pilot to other sectors, for example road transport actors.
- How could safety information become arranged across organisations/authorities as part of the societal safety management?
- How can we employ the risk concept? The concept must be clarified if it is to be used to manage safety
- How could the actors approach acceptable risks, from the single sharp end haulier to the society at large?

ACKNOWLEDGEMENT

The research reported in this paper is part of the research program; Risk and Safety in the Transport Sector (RISIT). The authors gratefully acknowledge the financial support from the Norwegian Research Council. We also thank Stine Bøen and Christin Berg (Rogaland Research) and Geir Sverre Braut (Norwegian Board of Health) for their comments to an earlier version of this paper.

REFERENCES

Aase, K. & Ringstad, A.J. 2002. Experience transfer and corrective measures. In Wilpert, B. and Fahlbruch, B. (ed.). *System Safety: Challenges and pitfalls of intervention*. Pergamon, Elsevier Science.

ADR/RID, 2005. *Regulations of November 11th, 2002, regarding transport of dangerous goods by road and rail, with guidelines.* Mandatory from January 1st 2005. Directorate for Civil Protection and Emergency Planning, Tønsberg, Norway (in Norwegian).

Agerup, N.H., Steffensen, S.M. & Andersen, R.M. 2000. *Dangerous goods handbook*. Elanders Publishing (in Norwegian).

AIBN, 2004. *Report on the collision between lorry and shunting train at Sjursøya oil terminal 17.02.2003, JB RAP.: 10/2004*, Accident Investigation Board Norway, Lillestrøm (in Norwegian).

Åsland, G.A. 2005. *A case study to shed light on the treatment and organisation of dangerous goods transport amongst different actors in the Norwegian rail sector.* Master thesis, University of Stavanger (in Norwegian).

Aven, T. 2003. *Foundations of Risk Analysis*. New York: John Wiley & Sons Ltd.

Aven, T., Boyesen, M., Njå, O., Olsen, K.H. & Sandve, K. 2004. *Societal Safety.* Oslo: Universitetsforlaget (in Norwegian).

Bøen, S. 2005. *Transport of dangerous goods on ferries in domestic waters.* Master thesis, University of Stavanger (in Norwegian).

IMO, 2004. *International Maritime Dangerous Goods, IMDG Code.* International Maritime Organisation.

Johnsen, S., Vatn, J. & Rosness, R. 2005. Cross border railway operations: Building safety at cultural interfaces, To be published in John Wilson, et al. 2005. *Rail Human Factors - Supporting the Integrated Railway*. Ashgate.

Kjellén, U. 2000. *Prevention of Accidents Through Experience Feedback*. London: Taylor & Francis.

Kjellén, U. 2002. Transfer of experience from the users to design to improve safety in offshore oil and gas production, Chapter in B. Wilpert & B. Fahlbruch, (eds.). *Systems safety: challenges and pitfalls of intervention.* Amsterdam: Pergamon.

La Porte, T. & Consolini, P.M. 1991. Working in Practice but Not in Theory: Theoretical Challenges of High Reliability Organizations. *Journal of Public Administration Research and Theory* 1(1): 19–47.

Lim, S. & Nédélec, B. 2006. *Comparative study of databases on accidents involving dangerous goods in the storage and transport to improve a crossover learning from experience.* In Guedes Soares and Zio (ed.). *Safety and Reliability for Managing Risk*. London: Taylor & Francis Group.

Madslien, A., Larsen, I.K. & Grue, B. 2004. *Hazardous goods in the Norwegian road and rail network. TØI report* 700/2004, Institute of Transport Economics, Oslo (in Norwegian).

Marone, J.G. & Woodhouse, E.J. 1986. *Averting Catastrophe: Strategies for Regulating Risky Technologies.* Berkeley: University of California Press.

Nicolet-Monnier, M. & Gheorghe, A.G. 1996. *Quantitative Risk Assessment of Hazardous Materials Transport Systems.* Dordrecht, the Netherlands: Kluwer Academic Publishers.

NMD, 2003. *Report from sea accident. KS-0197B (2003.12 SDir).* Norwegian Maritime Directorate.

Norddal, T. 2004. *Dangerous goods on ferries in Norwegian domestic traffic. Quantities, regulations and risk.* Report nr. 730049. Trondheim: Rambøll (in Norwegian).

NOU, 2001. *The Lillestrøm-accident, April 5, 2000. NOU 2001: 9, Main report from the commission of inquiry,* appointed by Royal Decree of January 7, 2000 (Åsta-accident, with extended mandate April 7 to cover the Lillestrøm-accident). Handed over to the Ministry of Police and Justice, January 30, 2001.

NTP, 2004. *National Transport Plan 2006–2016. Report No. 24 to the Storting (2003–2004).* Recommendations from the Ministry of Transport og Communications dated 12 March 2004, approved by the Council of State. Oslo.

Rasmussen, J. & Svedung, I. 2000. *Proactive risk management in a dynamic society,* Swedish Rescue Services Agency, Karlstad, Sweden.

Roberts, K.H. 1989. New Challenges in Organization Research: High Reliability Organizations. *Industrial Crisis Quarterly* 3(2): 111–125.

Roberts, K.H. 1990. *Some Characteristics of One Type of High Reliability Organization. Organization Science* 1(2): 160–176.

Westrum, R. 1993. Cultures with Requisite Imagination. In Hopkin, V.D., Wise, J.A. & Stager, P. (eds.), *Verification and Validation of complex systems: Human Factors Issues*: 401–416. Berlin: Springer.

Wildavsky, A. 1988. *Searching for Safety.* New Brunswick, New Jersey: Transaction Books.

Risks Challenging Publics, Scientists and Governments – Menoni (ed)
© 2010 Taylor & Francis Group, London, ISBN 978-0-415-58072-4

Methodology based on indicators for comparison of risks results from diverse energy systems

A. Colli, C. Kirchsteiger & A.L. Vetere Arellano
EC DG Joint Research Centre, Institute for Energy, Petten, The Netherlands

B. Ale
TU Delft, Faculty of Technology, Policy and Management, Delft, The Netherlands

ABSTRACT: This paper describes a methodology, developed by the Institute for Energy of the European Commission's Joint Research Centre, to compare and map the available end results of any energy risk assessment into common metrics in order to allow fair results comparison and detect possible energy trends. Based on the careful investigation of different energy fuel/life cycles, the methodology leads, as a first step, to develop a general life cycle scheme for all energy technologies. The Energy Indicators, divided into risk and sustainability indicators, have the assignment to characterize the energy supply process in both its positive and negative aspects. After completion, the project is likely to generate significant interest in the risk assessment community and its results could become in the long run standard tools for the dialogue among different stakeholders about energy risks.

1 INTRODUCTION

The European Union (EU) faces the problem of escalating external energy dependence. This requires the need to consider issues related on the one hand to secure the provision of uninterrupted availability of energy products on the market, and on the other hand to aim at minimising the risks to human health and the environment resulting from the use of various technologies for the generation and distribution of energy.

Based on the conceptual background previously exposed, this paper describes a methodology that is the basis of the Energy Risks Monitor (ERMON) project launched by the JRC's Nuclear Safety Unit—Risk Assessment Sector—as a contribution to JRC's Sustainable Energy Technologies Reference and Information System (SETRIS) project coordinated by JRC's Clean Energy Unit. It proposes the development of a general scheme for fuel and life cycles and introduces the recently initiated research on risk indicators.

The hypotheses and the research questions we want to answer during the study with the development of ERMON are listed in Table 1.

Table 1. Hypotheses and research questions concerning the development of ERMON.

Hypotheses and research questions	
Hypothesis 1	It is possible to compare risks from different energy systems. True or False?
Sub-questions	What is intended as an energy system?
	Why to compare risks/ benefits from different energy sources?
	How may I compare risks/ benefits from different energy sources?
	What can I compare and at what level/scale?
	What are the advantages and problems in this comparison?
Hypothesis 2	It is possible to anticipate future trends in energy risks.
Sub-questions	Why do we need to anticipate energy risks?
	How may I use results from Hypothesis 1 Study to identify and assess trends?
	What are the advantages and problems?
Hypothesis 3	It is possible to develop an energy risks communication tool.
Sub-questions	Why should an energy risks communication tool be developed?
	How can I use results from Hypothesis 1 Study to develop an energy risks communication tool?
	What format should this energy risks communication tool be?
	What are the advantages and problems?

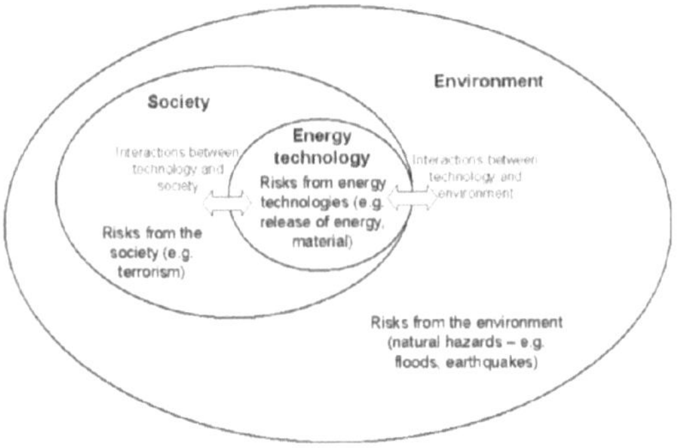

Figure 1. Interconnections and interactions between the different areas of interest for the development of ERMON.

Areas of interest for the study are environment, society and energy technology, as shown in Figure 1. The three areas involve different kinds of risks, and the interconnectedness and interdependence among each other part, bring to estimate a changing risk landscape.

During the development of ERMON particular attention is paid to study the interfaces and the exchanges (in terms of risks and benefits) among the different areas highlighted in Figure 1, with specific attention to the interactions energy technology vs. society, and energy technology vs. environment.

The relations among environment, society and energy technology focus on risks and sustainability issues. Aspects like effects of incidents/accidents on human health, property and environment are analysed, as well as all concerns affecting the economical, environmental, social and institutional dimensions of sustainability. The intention is to evaluate risk and sustainability aspects using a set of indicators, giving relevance to all different contact points to be studied and described in a overlap zone linking the two concepts.

2 METHODOLOGY

The project analysis and development is divided into two main phases described as follows. The first phase includes different tasks, starting with a detailed analysis of different energy systems and the investigation of all steps in their fuel/life cycles, with the purpose to develop a general fuel/life cycle scheme adaptable to all energy technologies. The following task in this first phase focuses on the development of a template of descriptors to characterise risk/benefit aspects.

Gray & Wiedemann, (1999) proposed to use risk management and sustainable development as frameworks for understanding and managing the environmental impacts of human actions, using indicators chosen on the basis of technical and communicative criteria. Following this proposal the descriptors in this project are divided into two main groups: those regarding sustainability and those regarding risk.

From a methodological point of view, the part related to the development of risk characterization descriptors will take into consideration the model of hazard causation.

This "causal model" has been developed by a group of researchers (C. Hohenemser, R.E. Kasperson & R.W. Kates) at CENTED (Center for Technology, Environment and Development) at Clark University in the eighties (Hohenemser et al. 1985). The initial aim of this model was to help in the comparison of different technological hazards.

The main objective is to get a set of measures for comparative hazard assessment avoiding arbitrariness. They propose a model of hazard causation anchored at one end in human activities and at the other by consequences, linked through a causal chain. Human activities produce changes in material fluxes, that are the origin of changes in valued environmental components; these changes induce some exposure that have some consequences to people and things that they value. Each link in this chain may be described by some characteristics and each characteristic may be described by a measurable descriptor (descriptors are normalised to some numerical scale). This methodology allows production of meaningful and easily understandable environmental hazard comparisons, in which not only final consequences are considered, but also ethical values also are explicitly taken into account (Kasperson, Kasperson, eds., 2001).

The causal classification clearly delineates the sequence of events that leads to a risk situation and constitutes a guide for discovering available points for intervention.

For each risk situation, specific descriptors are identified at different stages in the causal structure. The descriptors to be selected would be applicable to all technological hazards in a way to make comparison easier, could be expressed by common units or designations and must be comprehensible not only to experts.

With the view to complement risk indicators, sustainability descriptors are developed based on four individual dimensions of sustainability, influencing each other: economical, environmental, social and political. The choice of the indicators, according to (Gray & Wiedemann, 1999), should follow specific criteria: "they should be robust, problem- or action-oriented, chosen according to the particular management task involved and resonant".

An initial investigation of possible available sources of data has shown that the information is very different (data can be heterogeneous and/or comparative. e.g. costs), it can come from different sources and also vary in quantity and quality. In this context, an evaluation of the quality and the background of the information provided through ERMON is necessary.

Quality can be affected by two main aspects: process and uncertainty. The first consideration about quality of data takes into account the process of collecting data from failures, anomalies, incidents, accidents occurred on equipment in operation on industrial sites to join a database. Starting from raw data, the quantitative information passes through a process of validation, interpretation and quantification, until their use. During this process, the interpretation and the judgment of data by experts are relevant (ESReDA, 1999). Concerning uncertainty, two types are distinguished: random variability, i.e. statistical uncertainty related to a numerical statement, and non-random variability, i.e. refers to factors like model and knowledge uncertainty (Kirchsteiger, 2004).

The evaluation of risk qualification aspects of the project will be carried out based on ongoing work on different types of uncertainty considered in risk models, the way how they are included in the model and the quality (e.g. pedigree) of the information supporting it. An example could be the NUSAP (Numeral Unit Spread Assessment Pedigree) approach (Funtowicz & Ravetz, 1990). The NUSAP system addresses different types of uncertainty in a risk assessment, along with the quality of the information supporting the assessment. The NUSAP system, as reported in (Funtowicz & Ravetz, 1990), addresses three distinct levels of uncertainty, namely inexactness, unreliability and border with ignorance, providing a conceptual distinction among the technical, methodological and epistemological levels of uncertainty. A key concept in NUSAP is the pedigree of any piece of information included in a risk assessment. Pedigree accounts for the quality of the information. Assessment of pedigree involves the evaluation of several criteria about the scientific soundness of any piece of information. This is done through scoring the quality of information by expert judgement on a number of quality criteria, using semi quantitative scales defined by linguistic descriptors. NUSAP allows to address uncertainty and quality at different locations in a risk assessment, including input data, parameters, scenarios, model structure, model assumptions, indicators used, model system boundary, and problem definitions. NUSAP provides a systematic critical review of the available knowledge base for each of these components of a risk assessment and pinpoints specific weaknesses in the underlying knowledge base. It helps in assessing robustness of outcomes of a risk assessment in view of the uncertainties identified and in the setting of priorities for the improvement of the quality.

After development, the method will be validated with some case studies, according to the availability of data. Supported by positive results, the methodology will be implemented in a continuously updated web-based Information System on http://www.energyrisks.jrc.nl, under operational agreements with energy data and risks information owners.

This last part of the work includes development of different levels of access for different types of users. A preliminary identification of possible ERMON users and the investigation of their possible interests has been done and shown in Table 2. Six categories of stakeholders have been identified: government, NGO, research, company, citizen, and media. The listed stakeholders are not acting-alone entities, but there are mutual relations between every one of them and all the others. Most of all, among all stakeholders, the media play a fundamental role as they are pervasive of all other sectors (Mallon, 2006). The interests of the stakeholders are various even among the same category; they can vary in time and knowledge, according to the framing of the stakeholder itself, on the basis of personal conditions, experience, and the sustainable context in which a certain energy system in inserted into. It is then difficult to fix specific indicators to satisfy specific needs of a certain stakeholder, but all the possibilities has to be considered together in a larger view. The main advantage that ERMON will bring to all stakeholder is to reach a better knowledge and understanding of energy safety aspects, offering the instrument for harmonization of risk expressions allowing comparison.

Research investigation in the area of risk perception demonstrated that people rank risks not only on scientific studies of the probability of harm, but also on personal perception of how well the process is understood, its relation with cancer, the degree of catastrophe, the impartiality of danger distribution, the individual control level to exposure, and whether risk is voluntary or imposed. To investigate this process, Kasperson (Kasperson et al. 1988) proposed the concept of the social amplification of risk, suggesting that the actions of the media, government, and nongovernmental organizations, as well as disputes among scientists, can significantly increase or decrease public risk concerns.

This means that the right knowledge of every stakeholder is important, in a way that it can influence the thought of others. The point of view of the stakeholders in the energy environment can be fundamental to amplify or dampen perceptions of risk and, through this, create secondary effects

Table 2. The matrix of possible ERMON stakeholders, with identification of some example. Note that media are pervasive of all other sectors (Mallon, 2006).

	Who they are		What interest they have or represent	What ERMON can do for them
Stakeholders	Identification	Possible types	Possible interests	Added value from ERMON
Government	Government at all its level and government- related activities	Policy makers (councils, authorities, agencies, departments), bureaucracy	Decision making process (economically, environmentally, socially). Formulation of the agenda for energy-related issues, formulation and implementation of the policy program, and evaluation of actions	Better understanding of risk aspects to make political choices and their implementation
NGO	Commercial and non-commercial interest groups	NGOs for energy (different type related, pro and against), business, economy, environment, social and conservation, professional, development, and educational purposes	Energy knowledge and development and effects of energy related activities on business, economy, environment and society	Better knowledge of energy safety aspects, and better understanding to support political choices and their implementation
Research	Institutions for public and private research	Academic centres, research institutes and industrial research centres	Research departments of energy, epidemiology, safety, environment, sustainability	Easy accessibility of data for evaluations, better understanding of risk aspects from fair comparison
Company	Commercial stakeholders	Individual and massive transnational commercial entities (finance companies, insurance companies, institutional investors, manufacturers, suppliers, developers, generators, energy- intensive users, farmers)	Risk perception, safety, and security of supply	Risk perception facilitation, and better understanding of risk aspects from fair comparison
Citizen	Acting-alone individuals	People concerned or affected by human, environmental and economical impacts. Examples are site holders, individual investor, project contractors, employees	Energy availability, risk perception	Risk perception facilitation, and better understanding of risk aspects from fair comparison
Media	Conduit of information	Television, newspapers, magazines, internet	Survey and information delivery	Better understanding of risk aspects from fair comparison

such as stigmatization of technologies, economic losses or regulatory impacts. Both social and individual factors can act to determine the social perception of risk, and in this framework a good communication is of great importance (Pidgeon et al. 2003).

Representative data sets and case study results will be made part of the model. This paper mainly presents the results of the first phase of the proposed methodology.

3 DEVELOPMENT OF A GENERAL SCHEME FOR FUEL AND LIFE CYCLES

The transformation from energy sources into useful power, like electricity, is done through a process involving different consecutive conversion stages.

An energy system is a complex combination and interaction of many aspects (such as, human factors, technology, organization, policy, interactions with the environment, etc.) leading to the transformation of an energy source into useful power (that can be thermal, electrical or mechanical). Relevant in the energy chain are also human-related aspects, as people are involved in the energy system both as executors of the fuel transformation and conversion, but also as end-users.

The investigation has been done for the following ten most important and common energy systems: coal, natural gas, oil, nuclear, biomass, geothermal, hydro, solar, wind and hydrogen. For the purpose of the ERMON project, the investigation of different energy technologies takes into consideration only heat and electricity production (at different scale level) at the power plant site.

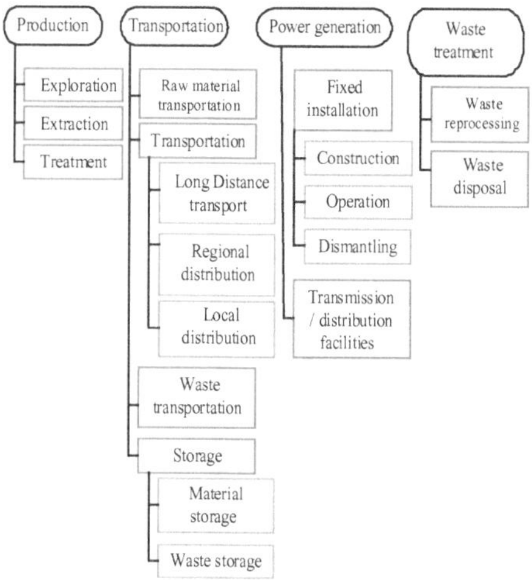

Figure 2. General fuel/life cycle scheme (Colli et al. 2005).

The general fuel/lifecycle scheme developed for ERMON (Figure 2) fits to some of the chains completely (e.g. nuclear), while some others (e.g. hydropower) leave some steps empty. The scheme has been developed focusing on fuel cycles and plant life cycles.

The main idea behind the development of the above-mentioned general scheme is that, however complicated a single chain can be, all its steps can be allocated within this general scheme.

The scheme is characterised by four main steps: production (related to all production operations); transportation (storage has been considered as a part of the transportation, as the stored material is waiting to be transferred to another intermediate place or to the place of use); power generation (power plant, including construction and dismantling operations); waste treatment (waste from the power plant as well as from other production activities. Waste can be treated or can be sent to a final disposal).

The application of the developed general scheme is made with the aspect of a matrix in which the different steps of the chain are filled with respect to the single fuel or life cycle; the filling level in the matrix is associated with the correspondent complexity of the chain into analysis.

Figure 3 is the resultant matrix that shows the application of the general scheme to fuel cycles (coal, natural gas, oil, nuclear, biomass, geothermal, hydro, solar, wind, hydrogen), as well as life cycles (solar modules and wind turbine production).

The investigation among different energy systems has been done also with the purpose to carry out a preliminary investigation of the different types of risks existing for each energy technology, in order to identify the most frequent risks, and prepare a knowledge base for the development of the indicators. The most risk-prone step in every energy chain has been identified, as shown in Table 3, and the related disaster scenario is described, according to available information, mainly found from historical events concerning the conventional energy sources (fossil, nuclear and hydro), and from safety studies in the case of new technologies.

To better understand the degree of risk of the different energy technologies into consideration, the evaluation of the electricity generation level for EU 25 in 2004, as available from (Eurostat, 2004), is also given in Table 3, together with their share to total, which has been calculated taking into account a total gross electricity generation of 3,178,603 GWh for EU 25 in the same year (Eurostat, 2004). This shows that the three main sources of electricity in the European Union are, in decreasing order, nuclear, coal, and natural gas. Values for renewable energies has been given only for hydroelectricity and wind, anyhow it is clear that the share of electricity

Stages of the life cycle			Fossil technologies			Nuclear technologies	Renewable technologies							Hydrogen technologies
			Coal	Natural Gas	Petroleum		Biomass	Geothermal	Hydro	Solar	Solar (PV modules)	Wind	Wind (turbines)	
Production	Exploration		Geological investigation	Geological investigation	Geological investigation	Geological investigation	N.A.	Geological, hydrogeological, geophysical, and geochemical investigation	Geographical / geological investigation	Geographical investigation	N.A.	Geographical investigation	N.A.	N.A.
	Extraction		Coal mining	Well drilling, pumping	Well drilling, pumping	Uranium mining	Biomass collection	Drilling	N.A.	N.A.	Feedstock and mineral collection	N.A.	Feedstock and mineral collection	Feedstock collection
	Treatment		Size reduction of mined coal; coal cleaning; coalification (bio-physical-chemical degradation); conversion into liquid or gas	Removal of hydrocarbon and non-hydrocarbon elements, liquefaction / compression	Refinery	Milling, transformation, conversion, enrichment and fuel fabrication	Residue processing, biofuels production	N.A.	N.A.	N.A.	Material processing, component manufacture	N.A.	Material processing, component manufacture	Hydrogen production from feedstock (with other auxiliary energy sources), liquefaction / compression
Transportation	Raw material transportation		Truck, railcar, barge	Upstream pipeline, tanker	Pipeline, ship tanker, short / long distance	Special container by ship, rail, truck	Residue transport	N.A.	N.A.	N.A.	Truck, rail	N.A.	Truck, rail	Pipeline, barge, railroad tank cars, rail, truck
	Transportation	Long distance transport	Railcar, barge	High pressure pipeline	Pipeline, barge, railroad tank cars	Special container by ship, rail, truck	Truck, ship	N.A.	N.A.	N.A.	Truck, rail, ship, plane	N.A.	Truck, rail	Pipeline, barge, air transportation
		Regional distribution	Railcar, barge	High / medium pressure pipeline	Pipeline, railroad tank cars, tank cars	Special container by rail, truck	Truck, ship	N.A.	N.A.	N.A.	Truck, rail	N.A.	Truck, rail	Pipeline, barge, truck
		Local distribution	Railcar	Medium / low pressure pipeline	Truck	Special container by truck	Truck	Pipeline	Water flow, canal, pipeline (also pressurised)	Solar concentrators	Truck, rail	N.A.	Truck, ship, helicopter	Local hydrogen grid, truck
	Waste transportation		Truck	N.A.	Tank cars	Special container by ship, rail, truck	Waste and ashes road / ship transport	Fluid reinjection	N.A.	N.A.	Truck	N.A.	Truck, ship, helicopter	N.A.
	Storage	Material storage	As necessary	As necessary	As necessary storage tank and local terminal	In special container and in the nuclear power plant	As necessary for energy consumption (close to plant)	N.A.	Dam construction and reservoir creation	N.A.	N.A.	N.A.	N.A.	As necessary pressure tank, insulated tank, solid compound
		Waste storage	After production and power generation stages	N.A.	Sludge' storage	Intermediate storage	N.A.	N.A.	N.A.	N.A.	N.A.	N.A.	N.A.	N.A.
Power generation	Fixed installation	Construction	Power plant construction	Power plant construction	Power plant construction	Power plant construction	Power plant construction	Power plant construction	Power plant construction	Solar system manufacturing and installation	Power plant construction	Turbine manufacturing and assembling	Power plant construction	Power plant construction
		Operation	Power plant operation	Power plant operation	Power plant operation	Power plant operation	Power plant operation	Power plant operation	Power plant operation	Power plant operation	Power plant operation	Power plant operation	Power plant operation	Power plant operation
		Dismantling	Dismantling procedures, material recycling and disposal	Dismantling procedures, material recycling and disposal	Dismantling procedures, material recycling and disposal	Dismantling procedures, material recycling and disposal	Dismantling procedures, material recycling and disposal	Dismantling procedures, material recycling and disposal	Dismantling procedures, material recycling and disposal	Dismantling procedures, material recycling and disposal	Dismantling procedures	Dismantling procedures, material recycling and disposal	Dismantling procedures	Dismantling procedures, material recycling and disposal
	Transmission / distribution facilities		Electricity / heat distribution facilities	Electricity / heat distribution facilities	Electricity / heat distribution facilities	Electricity / heat distribution facilities	Electricity / heat distribution facilities	Electricity / heat distribution facilities	Electricity distribution facilities	Electricity / heat distribution facilities	Electricity distribution facilities	Electricity distribution facilities	Electricity distribution facilities	Electricity / heat distribution facilities
Waste treatment	Waste reprocessing		Plant flue gas and ashes recycling	N.A.	Sludge' processing / incorporating	Reprocessing and conditioning	Ashes recycling	N.A.	N.A.	N.A.	Recycling	N.A.	Recycling	N.A.
	Waste disposal		Landfill	N.A.	Landfarm, combustion, underground injection	Final disposal / deposit, final geological repository	Landfill	N.A.	N.A.	N.A.	Landfill	N.A.	Landfill	N.A.

Figure 3. Resultant matrix that shows the application of the General scheme (Colli et al. 2005).

generation reserved to those technologies is still low if compared with fossil or nuclear. In fact, considering renewables in total, their ratio of electricity to gross national consumption in EU 25 is 13.7 in 2004 (Eurostat, 2004).

ERMON is a tool that wants to deal with risk expressions coming from safety studies, in which all the necessary information to evaluate risk are supposed to be already taken into account; the aim of ERMON is to offer a methodology to harmonise

Table 3. Identification of the most risk-prone step affecting the energy chains investigated in (Colli et al., 2005), description of the associated disaster scenario, and evaluation of the electricity generation level for EU 25 in 2004, as available from (a) = (Eurostat, 2004). The share to total has been calculated with a total gross electricity generation of 3,178,603 GWh for EU 25 in 2004 as given by (Eurostat, 2004).

Energy system	Most risky step	Scenario	EU 25 electricity generation 2004 [GWh] (a)	EU 25 share to total gross electricity generation 2004 [%] (a)
Coal	Extraction	Explosion in underground mines	647,685 (hardcoal)	20.4
Natural gas	Long distance transport	Gas release (with possible ignition and/or explosion) caused by external interference from ground work machinery impact.	602,280	18.9
Oil	Regional distribution and raw material transportation (to refinery)	Oil spills mainly due to collisions, groundings and hull failures. The predominant impact is environmental.	142,754	4.5
Nuclear	Operation (fixed installation-NPP)	Radioactivity release, explosion.	986,074	31
Biomass	Treatment (for occupational health) and transportation (for public health)	The main effect concerns air pollution. Release of biogas in material production stage with possible ignition and/or explosion. Consequences on drivers, pedestrians, cyclists and houses due to transportation incidents.	–	–
Geothermal	Extraction (drilling operation) and construction, operation and dismantling (fixed installation-power plant) + spontaneous natural phenomena (high temperature, gas emission, earthquake)	Burning, emissions of toxic gas, explosion. Damages to well and/or facility during earthquake.	–	–
Hydro	Construction and operation (fixed installation-dam)	Flood.	337,191	10.6
Solar (PV)	Treatment (module manufacturing)	Release of toxic substances, fire and/or explosion.	–	–
Wind	Construction, operation and dismantling (fixed installation-wind turbine, for mortality) and operation (fixed installation- wind turbine, for technical failure)	Loss of blades, burning, structural failures.	58,521	1.8
Hydrogen	Not identified energy chain associated to energy source involved in hydrogen extraction	Mechanical failure of containment device, chemical reaction, failure of safety device or operational error. Associate hazards can be chemical (ignition, deflagration, detonation), physiological (frostbite, respiratory problem) and physical (phase change, embrittlement).	–	–

risk expression and allow them to be compared. In this context, it is not necessary now to further investigate and evaluate risks from the different energy technologies considering additional data and information such as the localisation or the scarcity degree of the considered energy sources.

4 INDICATORS

As described earlier, the current approach is based on two main groups of indicators, related to two important concepts: sustainability and risk.

This division follows the idea in (Gray & Wiedemann, 1999), which focuses on the concepts of risk management and sustainable development as frameworks to understand and manage the environmental impacts of human actions.

Sustainable development, according to the Brundtland definition, "*means the capacity for an activity to be carried on indefinitely into the future*", while sustainable development is a "*development that meets the needs of the present without compromising the ability of future generations to meet their own needs*" [World Commission on Environment and Development].

In this sense a sustainable development looks deeply into the future, without forgetting the present, considering changes and improvements compatible with the limits of the society, in environmental, political, social and economical dimensions.

On the other side there is the risk management, a process that has its roots in the handling of economic risk in early capitalism, in the development of large-scale technologies after the second world war, and in natural hazards research. Risk management is referred to risk—including consequences and probabilities—and management—the process of controlling and making decisions; it is a process that deals with risk mainly in terms of probability of given undesirable outcomes, expressing a certain confidence that problems can be controlled in a rational way to protect the existing status quo.

In the development of the ERMON project, this approach is applied to study the energy supply process (energy supply: energy made available for future disposition; supply can be considered and measured from the point of view of the energy provider or the receiver), covering the whole of the energy system chain.

In this manner losses and benefits of the energy supply process are covered, using two different but complementary concepts, which will be discussed further below.

Moreover, using the general scheme for fuel/life cycles presented in the previous section and applying indicators, the intention is to be able to reflect not only the present hazards and risks, but also those to be expected in the decades to come. It is necessary to foresee there will be changes in technology. It will depend on the degree of change of the technology itself and based on these potential changes, one could expect possible positive or negative effects on the energy risk landscape, which will be reflected in the outcomes of the indicators.

4.1 *Risk indicators*

The aim of risk indicators is to provide comparable information for managing risks. Risk can have two different contexts: continuous risk, resulting from normal operation and activities, and accidental risk, involving incidents/accidents.

According to (Gray & Wiedemann, 1999), risk has various dimensions, including:

- The potential harm (consequences);
- Individual probability of this harm being experienced;
- Variability in risk within the population;
- Latency of harm;
- Catastrophic or group risk;
- Collective dimension (numbers of individuals exposed).

All these dimensions are affected by the cross-cutting dimension of uncertainty, which will be taken into consideration in the development of a qualification system for the indicators.

Dimensions are important and must be taken into account in the choice of indicators. This selection relies on both technical and communicative criteria. Technical criteria include robustness and problem orientation, together with specific factors depending on the individual dimensions of risk. Communicative criteria, mentioned in (Gray & Wiedemann, 1999) and based on the theory of the philosopher H. Grice (Grice, 1975), consists of truthfulness, informativeness, relevance and clarity. It is important to highlight that the selection of indicators is mainly a value-based activity, touching personal, social and political values, giving particular relevance to the process used for selecting indicators, which in turn is a process that does not represent the unique possible way.

The model and the set of twelve descriptors developed by C. Hohenemser, R.W. Kates and P. Slovic (Hohenemser et al. 2000) consider technological hazards as a sequence of causally connected events leading from human needs and wants, to the choice of the technology and to the consequences caused by the release and the exposure to energy and/or material. The twelve descriptors have relevant characteristics that are applicable to all types of technological hazards, comprehensible to non-experts and expressed in common units.

Descriptors are numerically quantified by a categorical distinction (intentionality, trans-generational, potential non-human mortality and experienced non-human mortality) and a logarithmic scale (spatial extent, concentration, persistence, recurrence, population at risk, delay of consequences, annual mortality and maximum potentially killed). The choice of logarithmic scale allows a practical representation, with the quality of matching human perception better than linear scales (like for decibel sound intensity scale or Richter earthquake intensity scale).

The original set of indicators has been slightly modified in this study with the introduction of new elements not reported in the original reference (e.g. aggregation level, economic loss) and, where necessary, more detailed classifications. Economical aspects needed to be added, as well as more detailed event classification and the distinction among different levels of population affected.

The revised risk indicators are (Figure 4):

1. *Aggregation level*: identifies the level of the event according to the system involved. A system can be divided into four levels (Perrow, 1999):
   a. *Part*: is considered as the smallest component of a system, for example a valve;
   b. *Unit*: is a functionally related collection of parts, for example those parts constituting a steam generator;
   c. *Subsystem*: is an array of units, for example the union of a steam generator and the water system;
   d. *System*: when different subsystems come together, then we have a system, like a nuclear power plant.

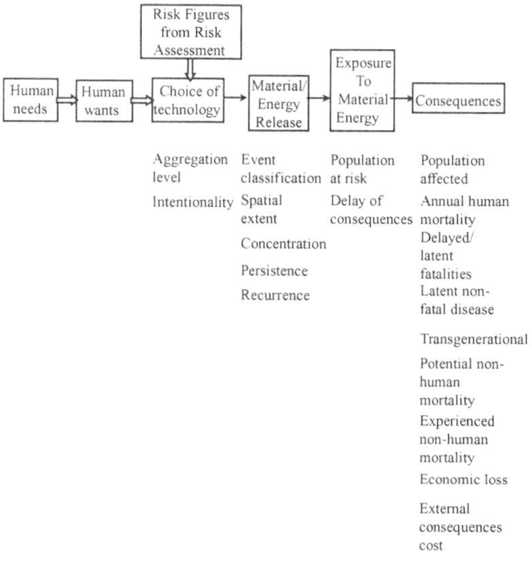

Figure 4. Causal sequence of technological hazards and selected indicators.

This classification wants to take mainly into account the technological systems and the way they work. Humans are considered as part of the system and they may constitute a part, a unit, as well as a subsystem.

2. *Event classification*: based on (Perrow, 1999):
   e. Incident: involves damages to parts or unit (whether the failure disrupts the system or not);
   f. Accident: is a failure in a subsystem or in the system as a whole;
   g. Component failure accident: involves one or more component failures (part, unit or subsystem) that are linked in an anticipated sequence;
   h. System accident: involves the unanticipated interaction of multiple failures.

The last two categories are distinguished on the basis of whether any interaction of two or more failures is anticipated, expected, or comprehensible to the persons who designed the system, and those who are adequately trained to operate it.

3. *Intentionality*: measures the degree to which technology is intended to harm.
4. *Spatial extent*: maximum distance over which a single event has significant impact.
5. *Concentration*: measures the concentration of released energy or materials to natural background.
6. *Persistence*: measures the time over which a release remains a significant threat to humans.
7. *Recurrence*: mean time interval between releases above a minimum significant level.
8. *Population at risk*: number of people potentially exposed to hazard. People are classified into first party (operators of the system), second party (associated with the system as non-operating personnel, suppliers or users, but without influence over it), third party (innocent bystanders) and fourth party (fetuses and future generation) (Perrow, 1999).
9. *Delay of consequence*: delay time between exposure to hazard release and occurrence of consequences.
10. *Population affected*: number of fatalities, injured and evacuees in a single event. People are divided according to the classification used for population at risk.
11. *Annual human mortality*: average annual deaths. People are divided according to the classification used for population at risk.
12. *Delayed/latent fatalities*: number of people affected by delayed or latent effects. People are divided according to the classification used for population at risk.

13. *Latent non-fatal disease*: number of people affected by non-fatal diseases originated from the event (cancer). People are divided according to the classification used for population at risk.
14. *Trans-generational*: number of future generations at risk.
15. *Potential non-human mortality*: maximum potential of non-human mortality.
16. *Experienced non-human mortality*: non-human mortality that have been experienced.
17. *Economic loss*: takes into account only property and rebuilding costs.
18. *External consequences cost*: distinguishes between environmental (impact on public and occupational health, agriculture, forests, biodiversity effects, aquatic impact, impact on materials, global impact) and non-environmental (impact on public infrastructure, security of supply, government actions) costs (Hirschberg et al. 1998).

The proposed set of indicators takes into consideration impacts on human health, environment and economical losses. In addition, the temporal dimension is also taken into account. The long-term effect is evident in indicators like persistence, trans-generational and, concerning human health effects, by the fourth party classification.

A first attempt to apply the indicators is presented in Table 4, which shows a test application to compare two disasters occurred in nuclear power plants. This application study compares risk expressions taken from two books, discussing the nuclear accidents of Chernobyl (Vargo ed., 2000) and Three Mile Island (Garwin & Charpak, 2002).

The information found in the two texts is in the form of exact value, and thus reported as given, or interpretation of written text to be adapted to the chosen scoring system for the indicators.

Two indicators (recurrence and population affected—evacuees) in the accident of Three Mile Island are given as a range between two values. This shows an uncertainty in the original risk expression. This uncertainty, together with other qualification aspects, will be evaluated with the application of the qualification methodology to be addressed in the next part of the development of ERMON.

The comparison was done taking into consideration indicators presenting information available for both events, which allow observing that the Chernobyl accident shows a higher degree of risk.

The work concerning risk indicators is not jet concluded, and other aspects need to be considered and introduced.

Following steps will be the consideration of security aspects like terrorism as a threat coming from the societal area, and natural hazards as threats coming from the environmental area (Figure 1).

Obviously, the set of indicators needs to be tested concerning its validity and the interdependence of the indicators. The final test will be done considering two events from energy systems of different nature (like, for example, nuclear and natural gas). The challenge in this case will be the development of a scale on which the comparison of the incidents/accidents can be based.

It is not possible to exclude that the future part of the work concerning indicators will not affect the present results and the indicators listed in this section.

4.2 *Sustainability indicators*

The sustainability descriptors have been developed according to the four key dimensions of sustainability, covering economical, environmental, social, and political/institutional aspects with a future-oriented perspective. Sustainability means the capacity for an activity to be carried on indefinitely in the future, which involves changes and improvements compatible with the highlighted dimensions in a present and long-term perspective.

Sustainability aspects do not have a stand-alone position, but interact with each other as shown in Figure 5. It is also important to stress that sustainability indicators are considered complementary and linked to risk indicators (Figure 6), two set of descriptors partly integrated into one another (overlap zone), with different aspects in common; all indicators from both parts are equally important and useful to each other in the evaluation of risk and benefits of energy systems.

Sustainability descriptors will be developed on the basis of a set of thirty Energy Indicators for Sustainable Development presented in (IAEA & others, 2005), which is the result of collaborative initiative between International Atomic Energy Agency (IAEA), United Nations Department of Economic and Social Affairs (UNDESA), International Energy Agency (IEA), Eurostat and European Environment Agency (EEA).

This set of indicators is presented as a staring point, mainly valid for applications at national level (cross-country comparison must evaluate the different condition—geographical, economical, etc. - of countries into analysis), for the development of a more comprehensive and relevant list of descriptors for sustainability.

Sustainable development indicators highlight:

- Sustainability of current energy use;
- Energy issue to interested stakeholders;
- Aspects or consequences of energy use;
- Links among the various dimensions;

Table 4. Example application study of risk expressions for the accidents at Chernobyl and Three Mile Island NPPs, using the proposed risk indicators.

Indicator		Chernobyl Value	Chernobyl Qualification	Three Mile Island Value	Three Mile Island Qualification
Aggregation level		System	Interpretation	Subsystem	Interpretation
Event classification		System accident	Interpretation	Component failure accident	Interpretation
Intentionality[1)]		3	Interpretation		
Spatial extent[2)]		4.5 (Exclusion Zone)	Exact value	There was little impact outside the US (Radio-active cloud over the Atlantic ocean)	Reference statement
Concentration[3)]		18	Exact value	16.6	Exact value
Persistence[4)]		7.2	Exact value	3.8	Exact value
Recurrence[5)]				3.5–3.6	Exact value
Population at risk[6)]	First party	5	Exact value		
	Second party				
	Third party	5	Exact value	Absence of harm to the citizenry	Reference statement
	Fourth party				
Delay of consequence[7)]					
Population affected[6)]	First party	1.3 (fatalities)	Exact value		
	Second party				
	Third party	5 (evacuees)	Exact value	4 – 4.9 (evacuees)	Interpretation
	Fourth party				
Annual human mortality[6)]	First party	1.5 (1986)	Exact value		
	Second party				
	Third party				
	Fourth party				
Delayed/latent fatalities[6)]	First party				
	Second party				
	Third party			0.3 (Estimation)	Exact value
	Fourth party				
Latent non-fatal disease[6)]	First party				
	Second party				
	Third party				
	Fourth party				
Transgenerational (non-human)[8)]		2	Interpretation	1	Interpretation
Potential non-human mortality[9)]				1	Interpretation
Experienced non-human mortality[9)]				1	Interpretation
Economic loss (property and rebuilding costs)[10)]		High	Exact value		
External consequences cost[10)]	Environmental	High	Exact value		
	Non-environmental	High	Exact value		

[1)] 1 = not intended to harm; 2 = intended to harm non-human organisms; 3 = intended to harm human.
[2)] Value expressed with Log10 scale. Base value in meters.
[3)] Value expressed with Log10 scale. Base value in Becquerel.
[4)] Value expressed with Log10 scale. Base value in minutes.
[5)] Value expressed with Log10 scale. Base value in years.
[6)] Value expressed with Log10 scale. Base value in number of people.
[7)] Value expressed with Log10 scale. Base value in minutes.
[8)] 1 = effect on exposed generation; 2 = effect on 1 future generation; 3 = effect on more than 1 future generation.
[9)] 1 = no mortality; 2 = mortality; 3 = species extinction.
[10)] High cost level is considered for more than 1 billion $.

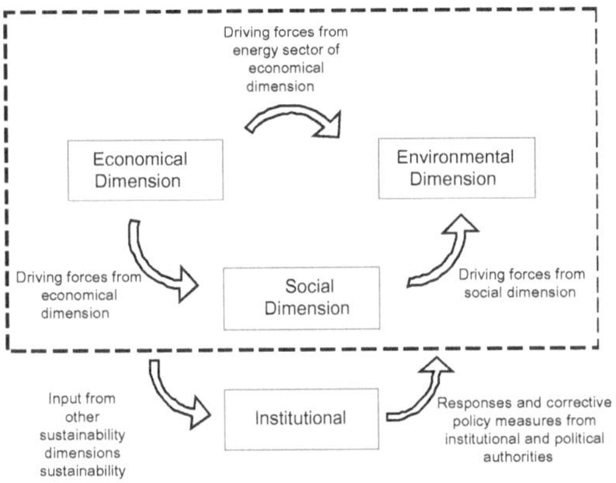

Figure 5. Interrelations among the different dimensions of sustainability, in relation to the energy sector (IAEA, IEA, 2001).

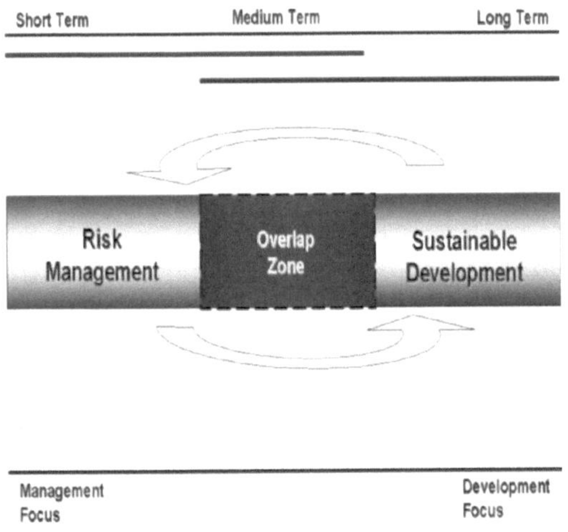

Figure 6. Risk Management vs. Sustainable Development.

- Longer-term implications of current decisions and behaviour;
- Progress or lack of progress in sustainable development.

The indicators for sustainable development, as from (IAEA & others, 2005), are classified according to different themes and sub-themes.

Considering the sustainability dimension they represent, indicators are characterised as follows:

1. *Economical dimension*: the economical indicators have two themes:
   - Use. Sub-themes are:
     - *Overall use*;
     - *Overall productivity*;
     - *Supply efficiency*;
     - *Production*;
     - *End use*;
     - *Diversification* (fuel mix);
     - *Prices*.
   - Production patterns, and security. Sub-themes are:
     - *Imports*;
     - *Strategic fuel stocks*.
2. *Environmental dimension*: the environmental indicators have three themes:
   - Atmosphere. Sub-themes are:
     - *Climate change*;
     - *Air quality*.
   - Water. The only sub-theme concerns *water quality*.
   - Land. Sub-themes are:
     - *Soil quality*;
     - *Forest*;
     - *Soil waste generation and management*.
3. *Social dimension*: the social indicators have two themes:
   - Equity. Sub-themes are:
     - *Accessibility*;
     - *Affordability*;
     - *Disparities*.
   - Health. The only sub-theme concerns *safety*.
4. *Institutional dimension*: the problem of institutional indicators has been left open in (IAEA & others, 2005), as there are no indicators developed to cover the institutional dimension. The reason is mainly that they are structural or policy responses to sustainable development needs, difficult to measure in quantitative terms (e.g. existence and effectiveness of a national sustainable energy development strategy or plan; adequacy and effectiveness of investments in capacity building, education or research and development; progress towards appropriate and effective legislative, regulatory and enforcement institutions for energy systems).

The institutional dimension, as shown in Figure 5, is considered separately, because it receives input from one or more indicators from other sustainability dimensions, and, in turn, gives feedback and response actions affecting one ore more target indicators, with impact on specific driving forces and affecting the sustainability of the whole energy system. The institutional dimension acts like a supervisor of sustainable development.

5 CONCLUSION

The results of this work serve as base for the ERMON project, which will help to compare the results of any existing risk study and incident/

accident statistics analysis for different energy systems across all steps in their specific fuel cycle or life cycle chains. The objective of ERMON is to compare the results in a consistent way. The comparison is based on the generic fuel/life cycle model presented in the paper.

The development of both the risk characterization and risk qualification template to create an energy risk compass supported by the creation of an energy risk knowledge base, has the purpose to improve the understanding and the communication of risks among all stakeholders (policy-makers, public, utilities, etc.), and to increase the acceptance and use of risk assessment approaches.

Such an energy risk knowledge base and tool present the benefit of providing users with a flexible methodology applicable to different energy systems in their fuel and life cycles. Furthermore, it allows different stakeholders to access and use the information according to their needs. Lastly, it provides a significant amount of and valid information, which can lead to a comprehensive evaluation of a specific energy-related hazard. This will also provide information about the quality-related elements of the energy risk assessment underlying the considered risk expression, in order to allow a comparison of different understandings of similar hazards and different risks from different energy systems, in order to judge their sustainability and to make decisions concerning policy-related issues, like environment, human safety, etc.

This work will be complementing and bringing forward work already started in the fields of life cycle analysis, accident and risk analysis, external costs and sustainability assessment, as well as qualification of different types of scientific/technical information (uncertainties mapping). Corresponding research co-operations are already underway. Further, co-operations with potential data suppliers, such as industry associations, have already been agreed or will be arranged in the near future.

REFERENCES

Colli, A., Vetere Arellano, A.L. & Kirchsteiger, C. 2005. *Development of a General Scheme for Fuel Cycles and Life Cycles from all Energy Technologies as a Basis for the European Energy Risk Monitor (ERMON)*, EUR 21735 EN, Publication of the European Commission Joint Research Centre, Petten, The Netherlands.

ESReDA Working Group Report, 1999. *Handbook on Quality of Reliability Data*, Det Norske Veritas, Eurostat,2004,Energydata,http://epp.eurostat.ec.europa.eu/portal/page?_pageid=1996,45323734&_dad=portal&_schema=PORTAL&screen=welcomeref&open=/&product=Yearlies_new_environment_energy&depth=4.

Funtowicz, S.O. & Ravetz, J.R. 1990. *Uncertainty and Quality in Science for Policy*. Kluwer Academic Publishers.

Garwin, R.L. & Charpak, G. 2002. *Megawatts + Megatons. The Future of Nuclear Power and Nuclear Weapons*. University of Chicago Press.

Gray, P.C.R. & Wiedemann, P.M. 1999. *Risk Management and Sustainable Development: Mutual Lessons from Approaches to the Use of Indicators*, Journal of Risk Research 2 (3): 201–218.

Grice, H. edited by Cole P. & Morgan, J.L. 1975. Logic and Conversation. *Speech Acts, Syntax and Semantics* 12: 41–58.

Hirschberg, S., Spiekerman, G. & Dones, R. 1998. *Project GaBE: Comprehensive Assessment of Energy Systems. Severe Accidents in the Energy Sector*, First edition, Paul Scherrer Institut, ISSN-1019–0643.

Hohenemser, C., Kasperson, R.E. & Kates, R.W. 1985. *Causal Structure in Perilous Progress: Managing the Hazards of Technology*. Boulder, CO: Westview Press.

Hohenemser, C., Kates, R.W. & Slovic P. 2000. *The Nature of Technological Hazards*, chapter 10 from the book: "The perception of Risk. Risk, Society and Policy", P. Slovic editor. Earthscan.

International Atomic Energy Agency, UN Department of Economic and Social Affairs, International Energy Agency, Eurostat and European Environment Agency 2005. *Energy Indicators for Sustainable Development: Guidelines and Methodologies*, IAEA, Vienna, http://www-pub.iaea.org/MTCD/publications/PDF/Pub1222_web.pdf.

IAEA, IEA, 2001. *Indicators for Sustainable Energy Development*, http://www.iea.org/textbase/papers/2001/csd-9.pdf.

Kasperson, J.X. & Kasperson, R.E., editors. 2001. *Global Environmental Risk*. United Nations University Press.

Kasperson, R.E., Renn, O.M., Slovic, P., Brown, H.S., Emel, J., Goble, R., Kasperson, J.X. & Ratlick S. 1988. The social amplification of risk: a conceptual framework. *Risk Analysis* 8(2): 177–187.

Kirchsteiger, C. 2004. Technical Communication on Status in Developing a Compass for Risk Assessment. *Safety Science*, 42(2): 159–165.

Mallon, K. Ed. 2006. *Renewable Energy Policy and Politics*, Earthscan, ISBN 1–84407–126-X.

Perrow, C. 1999. *Normal Accidents. Living with High-Risk Technologies*. Princeton University Press.

Pidgeon, N., Kasperson, R.E. & Slovic, P. 2003. *The Social Amplification of Risk*, Cambridge University Press, ISBN-10: 0521817285.

Vargo, G.J. (ed.) 2000. *The Chernobyl Accident. A Comprehensive Risk Assessment*. Battelle Press.

Risks Challenging Publics, Scientists and Governments – Menoni (ed)
© 2010 Taylor & Francis Group, London, ISBN 978-0-415-58072-4

Implementation of safety management systems in nuclear power plants in Germany and results of a German research project

B. Ernst & H.P. Berg
Federal Office for Radiation Protection (BfS), SalzgitterFederal Republic of Germany, Germany

H.P. Balfanz
TÜV Nord, Hamburg Federal Republic of Germany, Germany

ABSTRACT: The absence of any requirements for safety management in German nuclear regulations resulted in the need to develop regulatory guidance in which the requirements for safety management systems are systematically compiled according to a national standard. In June 2004, the Federal Ministry for the Environment, Nature Conversation and Nuclear Safety published a guidance document "Fundamentals of Safety Management Systems in Nuclear Power Plants". Furthermore, the licensing and regulatory authorities of some German Federal States made recommendations for the evaluation of a utility's safety culture self-assessment system. At first both strategies of nuclear safety management of power plants in Germany are compared. Finally a simple method will be presented called "Event-based Safety Culture Assessment". This method is based on the assumption that root cause analyses of incidents or near miss failures in nuclear power plants give indications of deficiencies in safety culture. An example of the application of this method will be shown.

1 INTRODUCTION

In the past, the German nuclear regulations did neither contain any explicit requirements for comprehensive safety management nor for structuring the utility's organisation under the aspect of nuclear safety that would allow the comprehensive planning and review of a safety management system. However, a number of sometimes rather general individual requirements to be fulfilled by a safety management system could be derived from existing German nuclear regulations. Examples are the Radiation Protection Ordinance and the Guideline on Technical Qualification of Nuclear Power Plant Personnel.

Owing to the reduction of cost as a result of liberalisation of the electricity markets, the importance of operational management is growing. Cost savings in the areas of personnel and organisation lead to a reduction in the number of personnel together with changes in the organisational structure and tighter working processes. Experience with accidents in other branches of industry shows the importance of safe operational management. Effective safety management is seen as one crucial element of safe operation of plants.

According to the German Atomic Energy Act and the Act on the Convention on Nuclear Safety, the licensee is responsible for the safety of the plant and has made great efforts to develop and implement safety management systems and safety culture self-assessment systems. One example is the safety culture assessment system /VGB-SBS/ developed by the VGB Power Tech, the association of energy utilities in Germany.

2 GERMAN APPROACHES OF SAFETY MANAGEMENT SYSTEMS AND SAFETY CULTURE ASSESSMENT

2.1 *Safety management systems*

The supervisory authorities shall ensure that the safety management systems meet acceptable quality criteria and that they are implemented correctly. For this purpose, the Federal Ministry for the Environment, Nature Conversation and Nuclear Safety published a guidance document 'Fundamentals of Safety Management Systems in Nuclear Power Plants", in which the requirements on safety management systems are systematically compiled according to a national standard /BMU 04/. Specific requirements will be put in more concrete terms in a further document at a later stage.

Existing rules from German nuclear laws and regulations, such as the requirements for organisation e.g. from the Radiation Protection Ordinance, the Guidelines on the Demonstration of the Technical Qualification of Nuclear Power Plant Personnel and the German Nuclear Safety Standard 1201 "Requirement for the Operating Manual", have to be taken into account for the development and implementation of a safety management system.

The essential requirements for comprehensive safety management as well as for the structures and elements necessary for the implementation of a comprehensive and systematic safety management system are listed in the following.

Requirements for safety management systems:

- General requirements
- Documentation requirements

Management responsibility:

- Management commitment
- Safety requirements
- Definition of safety policy
- Establishment of safety objectives
- Management by setting examples
- Definition of organisational structure
- Communication

Management of necessary resources:

- Provision of resources
- Human resources
- Infrastructure
- Work environment
- External organisations
- Specialist competence in key areas

Work realisation:

- Planning of work realisation
- Determination of requirements for process execution and process result
- Design and development
- Purchasing
- Materials management
- Execution of work
- Control of monitoring and measuring devices within the framework of work realisation

Measuring, analysis and improvement:

- General
- Monitoring of safety performance and processes
- Treatment of inadequate process results
- Internal audits
- External audits
- Improvements
- Corrective actions

For all these items conditions are determined that have to be strictly fulfilled. This approach is based on the historical fundamentals of the laws, which are focussed on responsibility in case of damage, and can be considered as compliance-oriented.

2.2 *Safety management assessment*

"Safety culture is that assembly of characteristics and attitudes in organisations and individuals which establish that, as an overriding priority, nuclear plant safety issues receive the attention warranted by their significance." /INSAG 4/

Depending on what a culture emphasizes, one can distinguish three categories of safety culture:

Compliance-oriented: Safety management is determined by regulations and rules. The licensee regards safety as an external requirement and assigns little importance to behavioural aspects. Accordingly, even for human performance issues solutions with procedural or behavioural orientation are not sufficiently pursued. Primarily technical solutions are sought.

Performance-oriented: An addition to compliance with regulations, good safety performance becomes an organisational goal and the licensee has flexibility selecting the methods to achieve these goals. Procedural solutions are applied in addition to technical solutions.

Process-oriented: Safety performance can always be improved. A characteristic of this category is a continuous learning. The organisation is a learning organisation. The level of awareness of behavioural and attitudinal issues is high and measures are being taken to improve behaviour.

These three categories are not mutually exclusive and, in fact, can coexist in a good safety culture. A direct quantitative assessment of safety culture is not feasible; therefore a combination of suitable safety culture indicators is used. These indicators should be periodically monitored.

The licensing regulatory authorities of some German Federal States emphasize performance-oriented and process-oriented elements of safety culture. In a process-oriented culture, feedback requires a rigorous self-assessment during which values for the safety culture indicators are determined. Safety indicators are primarily tools for use by the licensee to improve safety performance. An extensive self-assessment plays a key-role in monitoring the plant's safety culture. The authorities of these German Federal States made recommendations for the evaluation of a utility's safety culture self-assessment system and are interested in continuously using an efficient assessment system /ILK/. Such a system provides a timely warning of a declining safety culture and detects trends for those indicators providing an ongoing insight into the functioning of personnel and equipment with high safety-related significance.

For the evaluation of a utility's safety culture self-assessment system, the following ten recommendations are offered:

1. The regulatory authorities should ensure that the licensees have a self-assessment system that addresses organisational and personnel aspects. The authorities should review this system for appropriateness.
2. Licensees' planned actions and the underlying rationale resulting from the self-assessment should be discussed with the authorities.
3. The regulatory authorities should ensure that the self-assessment tools, e.g., questionnaires and work-study techniques, meet acceptable quality criteria.
4. The regulatory authorities should ensure that the self-assessment tools and the application manuals are implemented correctly, continuously used and improved.
5. The regulatory authorities should ensure that the self-assessment of safety culture also includes the consideration of a rigorous root-cause analysis of events. The root-cause analysis must include organisational aspects and human performance.
6. The regulatory authorities should request the licensees to keep, for at least five years, the source material relating to the self-assessment in order to assure continuity of the process.
7. The regulatory authorities and their technical support organisations should perform their own self-assessment regarding their supervisory activities and should develop appropriate action plans.
8. At intervals mutually agreed upon, senior members of the regulatory authorities should meet with senior managers of the licensees to discuss the overall results of the licensees' self-assessments and their actions plans.
9. The regulatory authorities should be particularly attentive to events which may entail the risk of a deteriorating safety culture.
10. In evaluating safety culture, the regulatory authorities should aim at supporting the activities of the licensee to improve its safety culture. The authorities' involvement should therefore not to be too prescriptive.

These recommendations intentionally allow flexibility in their practical implementation. Every regulatory authority needs to find its own path in this field and will be influenced by its past supervisory concept and licensee activities.

Correspondingly, the implementation of these recommendations will differ. Since the self-assessment of safety culture represents a relatively new field of regulatory activity, it is recommended that regulators adopt a stepwise approach in their efforts, taking care to pursue a dialogue with the licensee in the process.

An independent assessment of safety culture performed solely by the authority is not possible. Instead, the regulator's goal should be to encourage the licensees to address this topic in the necessary extent. But it is indispensable that the licensees bear the sole responsibility for the safety culture of their plants. The regulatory authority, in turn, should always ascertain that the licensees fully meet their high responsibilities.

3 RESULTS OF A GERMAN RESEARCH PROJECT

The BfS on behalf of the Federal Ministry for the Environment, Nature Conversation and Nuclear Safety has sponsored a project "Development of Practical Criteria for Safety Culture Assessment in German Nuclear Power Plants". The work started in November 2001 and was done by several experts, such as technical support organisations of the authority, research institutes and licensees.

The project was focussed on the following topics:

1. Evaluation of national and international approaches of safety culture assessment in the nuclear field as well as in other industries.
2. Development of evaluation criteria for integrating and monitoring safety culture.
3. Application of the developed set of evaluation criteria and verification in two nuclear power plants.

International publications (/Schein/ /KSA/) indicate comprehensive activities in the field of safety culture but there are still differences in interpretation and definition. The definition given in /INSAG 4/ and cited in section 2.2 is mostly referred to in publications.

Methods of evaluation which are in use can be divided into two main groups:

- The approach mainly based on interviews with plant management and operators to get direct information about the plant safety culture, called "direct assessment".
- The approach that uses information derived from the plant operation (e.g. incidents and accidents), called "indirect assessment"

The question appears how it is possible to identify good or weak safety management systems and how to distinguish one from the other? What are the indicators and mechanisms as early warning signs before an accident will happen to make poor safety management transparent for the responsible management and authority, e.g. what can be learned from safety-significant events such as the Davis-Besse incident and the Tokai-Mura accident.

An example of direct assessment is the "Safety culture evaluation system – VGB-SBS" /VGB-SBS/ which is created by the licensees. This system has been introduced in all German nuclear power plants and its main characteristics will shortly be described.

The core of this system comprises a comprehensive list of questions assigned to a set of 20 defined processes. The main issue of quantification is to gain indicators to see any development as positive and negative.

The system is focussed on self-assessment by the licensee to support the plant internal process of safety culture regarding the understanding about its meaning by the plant staff and the application of the system.

The role of the authority within an application process is under discussion. The aim is to find a balance between information about the current state of application, explicit results of safety culture assessment and the assurance of the confidence of the internal plant information process.

Furthermore, it has to be discussed whether the authority should have its own instrument assessing plant safety culture as it is in an approach of HSK in Swiss /HSK/.

To support a risk based decision process the American Petroleum Institute developed "Risk-based inspection" as a tool of safety management assessment /API/ under the focus of optimisation of plant processes. Additionally, the management factors based on the numerical results (weight factors) of a qualitative assessment of interviews are taken into consideration in the probabilistic safety assessment (PSA).

Both approaches based on interviews of plant managers and operators are called the "direct assessment" of safety culture.

An alternative way to get into the complex subject of safety culture is the root-cause analysis of incidents and accidents focussing on organisational and management factors. All known severe industrial accidents indicate a high impact of management and organisational factors caused by an eroding safety culture. /Reason/ as well as /Perrow/ indicate:

" Most we can learn about the impact of safety culture is from accidents".

The accident scenarios are described by multiple technical as well as organisational deficiencies, which were undetected - in the pre-accident-phase by safety management and inspectors. In many industrial accidents one can find a typical pattern of accident causes, such as sets of latent failures rooted by design, quality as well as organisational and management deficiencies. The incidents and accidents make them visible.

One can find many promising applications of root-cause analysis for identifying and improving safety culture, e.g., in the nuclear field as well as in other industries. This approach is called the "indirect assessment" of plant safety culture.

There are different types of root cause analysis available, such as

- "Safety through Organisational Learning" /SOL/,
- "Connectionism Assessment of Human Reliability" /CAHR/,
- "Event-based Safety Culture Assessment" /Balfanz/.

Within the project "Development of Practical Criteria for Safety Culture Assessment in German Nuclear Power Plants" a method was developed called "Event-Based Safety Culture Assessment". It is based on the assumption that root causes of incidents or occurrences in complex industries such as nuclear power plants give indications of deficiencies in safety culture.

For this method, a specific procedure of analysis and a coding system for identifying organisational and management deficiencies were created, taking into account the organisational and management characteristics of the other approaches, such as /VGB-SBS/ and /API/.

The analysis of the basic single causes of an event represents the core of the examination. One aims at identifying organisational and management factors concerning the failure causes. For that purpose "Cause complexes" are defined, covering different single causes.

These cause complexes will be classified (weighted) and quantified due to their frequencies.

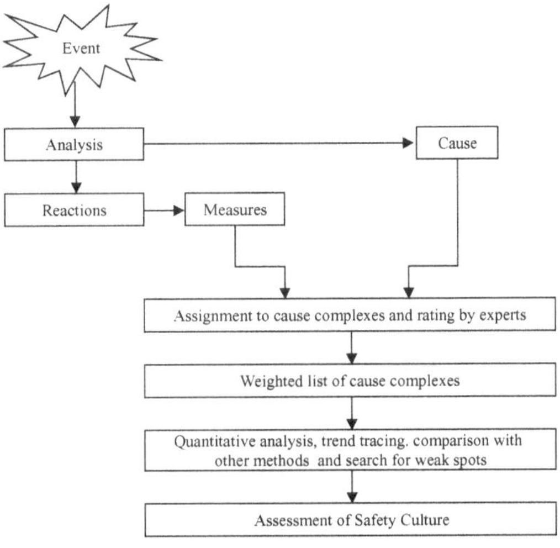

Figure 1. Event-oriented assessment of safety culture.

In association with the meaning of the assigned events, the valences of organisation and management factors are determinable with regard to possible weak points. The approach is subdivided into the following steps:

- Selection of events which should be investigated due to organisational and management factors, e.g. the set of safety significant events of a plant operational year.
- Obtaining and analysing available and accessible data, by the event reports
- Preparation of interviews with plant experts for further details and background information about the events and for clarification of the underlying causes.
- Assessment and classification of the identified causes with regard to the predefined cause complexes which should give indications about organisational and management impact.
- Expert rating about the relevance of events and the identified complexes.
- Summarising the type and frequencies of identified cause complexes which are based on the set of events to be investigated and drawing the resulting frequency distributions (see example below).
- Interpretation of the summary results such as indication of possible layers (cluster of cause complexes) or trends and indication of organisational and management deficiencies / safety culture.

The comparison of results of different sets of evaluation (e.g. trend analysis from several years of observation or from different plants) makes the interpretation of results more rational.

4 EXAMPLE

The example shows the event-oriented assessment of safety culture by a specific single-failure event:

Event description: During a reactor start-up after maintenance work, a failure in the turbine hydraulic control system caused an automatic turbine trip. The turbine protection system indicated this failure by a pressure difference at a hydraulic fluid filter.

Causes: The operators did not notice this pressure difference before the turbine was automatically tripped. The filter blocked due to normal increased oil pollution at restart of the unit. A check of the filter unit showed that one of the four filter cartridges was incorrectly installed by external staff.

Measure: Replacing of the filter cartridge and check for function.

Precaution against repetition: Specified regular filter controls during start-up process.

For each event, a numeric value representing a measure for the contribution of this category to the cause of the event is assigned to each cause-category (Table 1). The rating of the relevance of the identified cause complex to the event consists of three classes. The value 1.0 is valid for a direct cause that immediately leads to this event. Indirect causes that lead only in connection with other causes to an event, get the addition-value 0.5 and causes that have only a low contribution to the event the value 0.1.

The sum of all sequences of numbers of all events in a period yields a profile for the plant. For example, the profile (Figure 2) is the result of an analysis of 34 real events having happened in a plant. In the first picture, the assignment is done according to the categories human factors / technology / organisation from a characteristic catalogue.

The summary in groups of process planning (plan), implementation (do), supervision (check) and manage & support directly shows in a diagram the characteristics that weaken an organisation. In the displayed example the characteristics are the planning and implementation of maintenance.

The frequency distribution of cause complexes derived from a representative set of events of a nuclear power plant—as shown in Figures 2 and 3 – can be seen as a reflection of the safety culture of the plant and is an indicator of improvement potential. Based on the distribution main subjects of emphasis as well as trend analyses can be carried out. The higher the weight of a single cause complex, the higher is the need for corrective measure.

Table 1. Expert rating of cause complexes.

Identified causes	Coding of Cause Complex	Rating
Inadequate pressure indication	PD (d)	0.5
Known oil pollution at restart	OD (o)	0.1
Repeated occurrence	OD (o)	0.5
Quality control of external staff work	OD (o)	0.5
Incorrect operation procedure	PD (c)	0.5

P: Plan D: Do C: Check O: Organisation and Management
d: Design/Manufacturing o: Operation m: Maintenance c: change

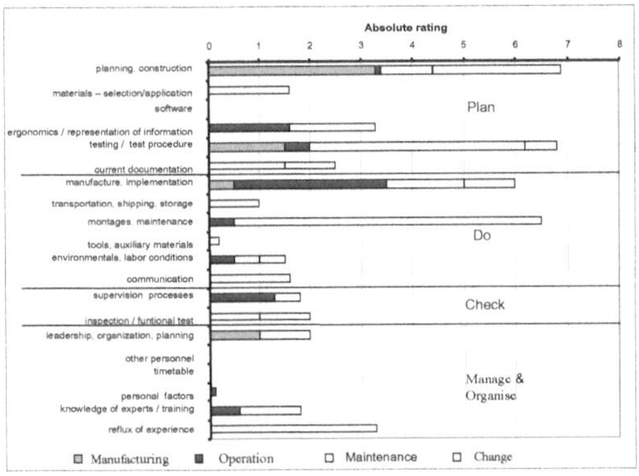

Figure 2. Frequency distribution of cause complexes derived from a representative set of events (34) of a NPP.

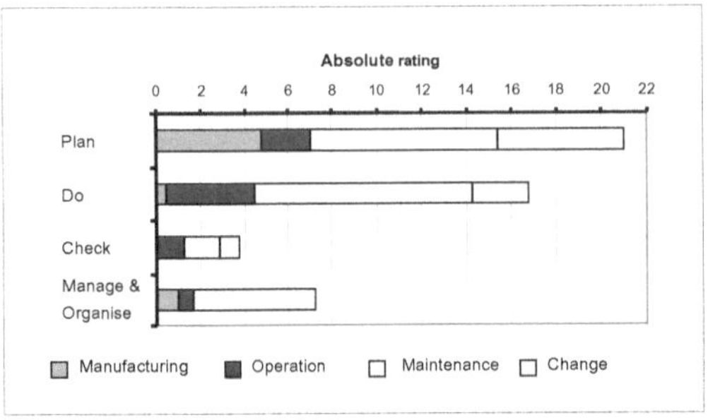

Figure 3. Summary of frequency distribution of groups of cause complexes.

The assessed frequencies and meaning of the events / causes reflect the characteristics of the risk assessment and give a qualitative picture of the risk relevance of organisation and management impact of events which have actually happened.

5 CONCLUSION

Indicators are integrated elements of a safety management system, e.g., to check the function of the processes and the attainment of the quality goals striven for.

With respect to the safety management system, supervision is performed by the introduction of suitable safety indicators as well as internal self assessment and external audits. Safety management expresses itself in a set of indications that can involve for example training of the personnel, sufficient staff, the quality assurance system, the technical performance of the plant, maintenance or ageing management.

However, also the non-compliance of data with safety relevant measures or an unusually high number of personnel from outside in key-processes can identify trends for a decrease of safety culture. In addition, there exist on the basis of that work—and organisation psychological research—so called soft indicators that have recognised the attitude of plant management and staff to safety goals, as approximately the operational contact with events without duty of notification, the work related status of employees' illness or the number of executed hours of overtime.

In compliance with its governmental supervision, the authority uses a set of instruments that is not normally identical with the measures of the operator to guarantee and survey safe plant operation. These are for example operational inspections, the recurrent examinations through experts or the extensive analysis of reportable events.

The current observation of indicators by which the effectiveness of the safety management can be measured as prospective as possible supplements this supervision concept.

However, experience has shown that is not reasonable for the authority to measure directly the proper function of the safety management system with indicators in order to underline that the licensee has the responsibility for the safe operation of the plant.

Those indicators are relevant for the authority which allow a permanent information on the function of especially significant safety systems or on the actual effects of a nuclear power plant regarding human factors and the environment.

Examples are unplanned reactor scrams, the number of reportable events, the unplanned non-availability of safety systems, the radiation exposure of the staff and the environment, the amount of accruing raw waste, the activity in the primary system, the exposure of radioactive materials over outlet air and sewage into the surroundings or the number of heavy work-accidents. These indicators describe not only technical situations but also processes that reflect the influence of human factors and/or organisational factors.

It is a disadvantage, however, that these indicators only allow a trend-statement developed from past to future plant-behaviour. Nevertheless, in order to be able to put these indicators into the sense of an early-warning, the respective indicator-value of an installation should be put down to a so-called "best-practice-value" as a reference value. Best-practice-values should be formed in such a way that they represent realistic goals in order to recognise changes in the plant by means of the present value at an early stage.

The procedure to achieve the best-practice-value is the following. For the first year to be viewed, the trend of the last 5 calendar years is used for the respective indicator through linear regression. With decreasing trend, the final value of the trend line, which has been reached, is used as best-practice-value for the following calendar year. The trend investigation is continued for the following years. If this trend continues the new lower value will be determined as the new best-practice-value.

In case the trend will not continue or even provides worse numbers the previous best-practice-value is retained. So it is excluded that a deterioration can affect the goals of the following years.

The practical application of this approach is still under discussion.

REFERENCES

API, 2000. American Petroleum Institute; Risk-Based-Inspection. Base Resource Document, API Publication 581.

Balfanz, H.P. & Schott, H. 2002. Entwicklung praxisgerechter Bewertungskriterien für die Sicherheitskultur in deutschen Kernkraftwerken. Stand der Arbeiten und erste Ergebnisse in TÜV Symposium "Sicherheitsmanagement in der Kerntechnik", München, (30./31.10.2002).

BMU, 2004. Federal Ministry for the Environment. Nature Conversation and Nuclear Safety: Fundamentals of Safety Management Systems in Nuclear Power Plants. *Bundesanzeiger*, 138: 16275 (27.07.2004).

CAHR, Sträter, O. 1996. Assessment of cognitive errors and organisational aspects based on evaluation of plant experience, in ANS. Proceedings of the International Topical Meeting on PSA 1966; Park City, Utah (USA); ISBN: 0-89448-621-7.

HSK, Humbel, C. & Frischknecht, A. 2002. Sicherheitsmanagement in den Schweizer Kernkraftwerken: Beschreibung, Erfassung und Bewertung durch die Sicherheitsbehörde, in TÜV Symposium "Sicherheitsmanagement in der Kerntechnik", München, (30./31.10.2002).

ILK, 2005. Statement on the Regulator's Management of the Licensee Self-Assessments of Safety Culture; ILK-19; Augsburg. http//.ilk-online.org.

INSAG, 1991. International Nuclear Safety Advisory Group; Safety Culture, INSAG Series No. 4, IAEA, Vienna.

KSA, 1997. Eidgenössische Kommission für die Sicherheit von Kernanlagen: "Sicherheitskultur in einer Kernanlage. Gedanken zur Bewertung und Förderung", Würenlingen (February 1997).

Miller, R. & Wilpert, B. 1997. SOL: Sicherheit durch Organisationales Lernen. Ereignisanalyse in der verfahrenstechnischen Industrie; 19. VDI-Tagung Technische Zuverlässigkeit (TTZ), Braunschweig.

Perrow, C. 1984. *Normal accidents: Living with high risk technologies*. USA: Basic Books.

Reason, J. 1997. *Managing the risks of organisational accidents*. Hamshire: Ashgate.

Schein, E.H. 2002. The Difficult Cultural Issues in Promoting Safety", IAEA – International Conference on Safety Culture in Nuclear Installations, Rio de Janeiro, Brazil.

VGB, 2001. VGB-SBS: Sicherheitsbewertungssystem, Kurzbeschreibung der Elemente, Essen.

Risks Challenging Publics, Scientists and Governments – Menoni (ed)

Determinants of investment risk in processes of transformation and modernization of polish heat energy sector

E. Sitek
Czestochowa University of Technology, Faculty of Management, Poland

ABSTRACT: Heat energy sector in Poland is characterized by high level of investment risk, which results from a variety of reasons. First reason for high level or risk results from the specific nature of heat energy sector that can be classified as an infrastructure sector. Another important reason for higher risk is political and economic transitions in Polish economy that have taken place after 1989. New Energy Law, legislated as late as in 1997, implemented market forms of business activity in heat energy sector. Third group of the reasons for high investment risk results from technical and ecological reasons. This paper aims to clarify these reasons behind the high level of investment risk profoundly and identify the sources of investment risk in heat energy sector.

1 INTRODUCTION

Heat energy sector in Poland is characterized by high level of investment risk, which results from a variety of reasons. First reason for high level or risk results from the specific nature of heat energy sector that can be classified as an infrastructure sector. Investment in heat energy sector, similarly to other forms of infrastructure investments, are characterized by their huge 'physical' size, increased financial expenditures, high capital intensity, high contribution of sunk costs and a number of other features that impact on the investment risk level (Ratajczak, 1999; Marcinek, 2000).

Another important reason for higher risk is political and economic transitions in Polish economy that have taken place after 1989. Before 1989, heat energy sector companies had operated as state-owned entities with particularly strategic importance to the economy. Access to production and distribution of heat energy on industrial scale was under public monopoly. New Energy Law, legislated as late as in 1997, implemented market forms of business activity in heat energy sector. The property of state-owned heat power companies was subject to communalization and managed by gminas (gmina is a principal unit (lowest level) of territorial division in Poland). Financial standing of gminas was then very difficult while their budgets were in significant debt. Political transitions in Polish economy, which preceded market principles of operation for heat power companies sparked increased level of risk in the sector. Some gminas sold their heat power companies to private entities since they were not able to finance their operation. It was not until accession of Poland to the EU that legal situation was stabilized in this sector, which had an effect on reduction in investment risk and rise in interest in investing in the sector.

Third group of the reasons for high investment risk results from technical and ecological reasons.

Heat energy sector requires technical reconversion since it is characterized by obsolete property, high level of equipment's wear and tear and huge heat loss throughout the grid (that ranges from 5% to 15%). High level of pollution emissions to the atmosphere is mainly caused by consumption of hard coal being a fundamental source of heat energy production (Figure 1). These conditions make it necessary to embark on serious reconversion in Polish heat energy sector, implemented in line with EU regulations.

2 IDENTIFICATION OF SOURCES OF INVESTMENT RISK IN HEAT ENERGY SECTOR IN POLAND

The goal of the investigations was to identify a profile of investment risk in Polish heat energy sector. The following categories can be distinguished:

- technical risk,
- market risk,
- financial risk,
- infrastructure risk,
- ecological risk,
- force majeure,
- political risk,
- legal risk.

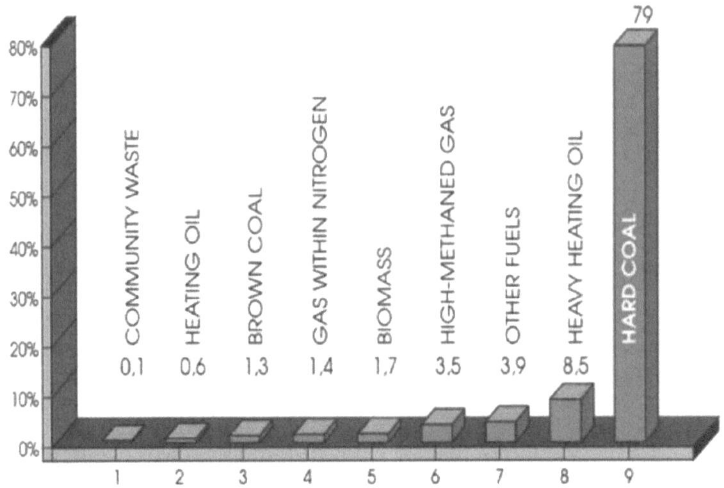

Figure 1. Structure of fuel consumption in Polish heat energy sector, Source: Buńczyk & Daniluk (2003).

It was accepted that these categories comprise a set of key variables recognized as domains that have impact on investment risk level and efficiency of risk limitation. The indexes that represent domains of each key variable were also distinguished. Then, they were presented to the respondents in the form of specific questions within questionnaire (Chapman, 1998).

The investigations were carried out by the Faculty of Management in Czestochowa University of Technology in 2003–2004. They encompassed 50% of the total number of heat power companies that employ *project finance* method. This group was predominated by the companies with domestic capital; only one entity owned foreign capital. The interviewed group comprised managers employed in special purpose vehicles (SPVs) that participate in investment projects. The goal of SPVs is to organize all stages in investment projects in terms of reduction of costs of production and transmission of heat energy. They also provide consultation, design, financing and execution of investment projects in terms of building new or modernization of the existing equipment and heat power systems.

Selection of this group of respondents was based on the assumption that the managers in these companies have proper know-how connected with the risk of investment decision-making.

3 RESULTS

The results of the investigations are presented in Figure 2 and Figure 3.

According to the respondents, the risk which is the most essential for investment processes in Polish energy sector is market risk (17.6%). Main component of this risk is changes in prices of energy resources used for production of heat energy. Other market risk factors include changes in demand for heat energy and changes that result from formation of heat energy prices.

Second, in terms of importance, risk (16.2%), according to the respondents, is political risk connected with government policy and policies in local governments. Lack of political stability combined with conflicts at both central and local level are often the reasons for changes in legal regulations and lack of conducive investment climate necessary for investments in the sector, which requires cooperation between public and private sector.

Another essential factor of investment risk is financial risk (14.9%). Its components include inflation risk, risk of interest rates and risk of exchange rates. A key role in Poland is also played by such factors as: access to cheaper sources of finance, security of conditions of investments' financing as well as ability to perform financial contracts. In *project finance* investments, legal risk is of big importance (13.6%). This results from the necessity of concluding of legal contracts between a number of investment project participants. Moreover, Polish law that regulates operation of businesses in heat energy sector, due to political transformations, has been subject to continuous changes. The Energy Law, for instance, has been amended several times. The changes are also caused by implementation, since 2004, EU legal regulations concerning energy market.

The infrastructure risk is connected with lack or underdevelopment of some infrastructural components, which might cause losses or necessity to bear

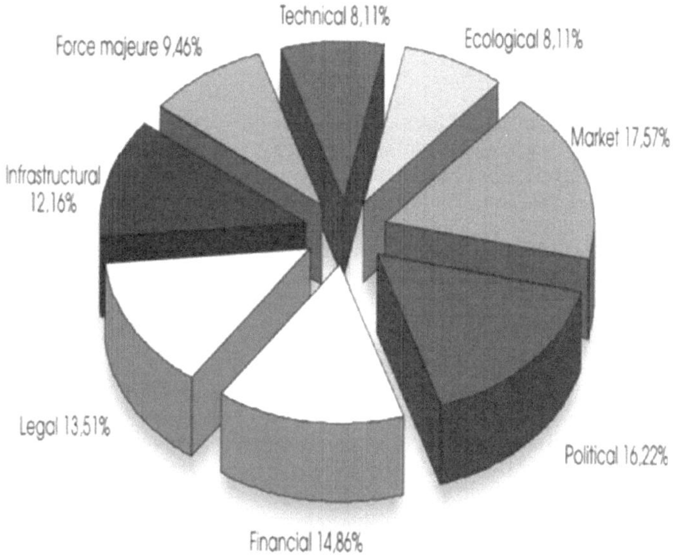

Figure 2. Factors that impact on the level of investment risk in Polish heat energy sector divided according to risk type. Source: Korombel, 2005.

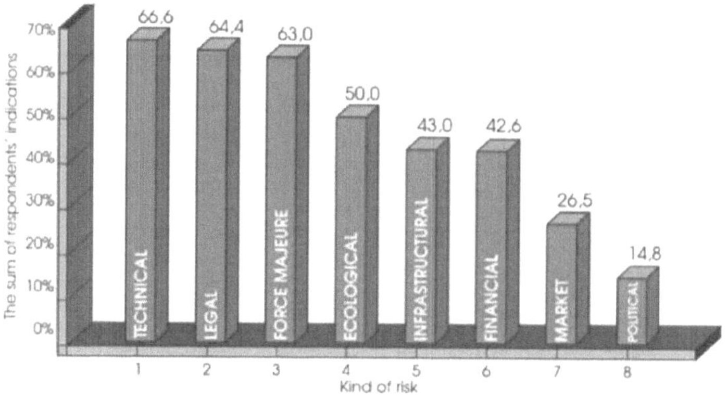

Figure 3. Opportunities of efficient limitation of investment risk in each risk group. Source: Korombel, 2005.

higher costs of investment project execution. This risk poses, according to the respondents, a risk of lower importance for execution of investment projects (12.1%).

Force majeure, understood as a possibility of unfavourable phenomena that occur during execution of the investment. Losses after floods, fires and other natural disasters can be, however, limited through insurances in specialized insurance agencies. Thus this risk was assessed by respondents as a risk of little importance (9.4%).

Technical and environmental (ecological) risks were found to be the risks of lowest importance to execution of investment projects (8.1%). This results from the fact that respondents have a range of methods to reduce technical risk, most of whom are characterized by high efficiency. The flexibility of methods here allows companies to be in control of technical risk.

Similarly, ecological risk also insignificantly impacts on execution of investment projects, despite adaptation of Polish regulations to EU legal requirements. This assessment might be caused by insufficient level of enforcing these regulations by the special public administration entities.

The respondents also expressed their opinions about the domains where efficient response and investment risk limitation is possible. The results are presented in Figure 3.

The biggest opportunities of efficient response and limitation of investment risk by the entities can be observed in the cases of technical and legal risk as well as *force majeure*.

The methods of technical risk limitation include: purchase of well-proven technologies, guarantees granted by contractors and suppliers of equipment or employing of highly-qualified staffs. These methods might ensure high level of control of technical risk and efficient limitation of the risk.

On the other hand, the risk of *force majeure* might be limited by adequate insurances from insurance agencies.

Legal risk might be limited by proper care during concluding of legal contracts with contractors that participate in *project finance* investment projects.

However, the most limited opportunities of efficient investment risk limitation appear within political and market risk. Low opportunities of market risk limitation in companies might result from the fact that market-related principles of operation in companies from heat energy sector are limited by the establishment of the Energy Regulatory Office (Urząd Regulacji Energetyki), which is entitled to e.g. grant licences for operation within this sector or control of the prices and fees in energy market.

During the investigations, the types of risk which were of little relevance or their occurrence is of little probability were neglected.

4 CONCLUSION

The issues of energy security have become a priority for Polish government policy. Therefore, political factors will significantly impact on the profile of investment risk in heat energy sector. The government will still influence, through the Energy Regulatory Office, on market risk and formation of energy prices in Poland.

Another key area of government intervention is to preserve natural environment and more rigorous enforcement of the respective EU regulations. This will make investors face major problems connected with ecological and technical risk. The importance of these factors to investment risk profile will be on the increase in Polish heat energy sector.

Polish heat energy sector requires execution of huge investment expenditures as it is characterized by large level of technical wear and tear and obsolete production property (Wilimowska, 1998). The level of risk is, however, still remains high.

In current conditions, investments in heat energy sector should be mainly executed as *project finance*. This method, compared to others, is more time-consuming and creates problems with investment projects management. It also requires involvement of a number of interrelated entities and intensified participation of *project finance* management.

However, it allows for reduction of investment risk in Polish heat energy sector through diversification of the risk between all project participants. It also permits, more than in the cases of other methods, bearing financial risk through banks an other entities that finance investments.

Project finance in Poland is mainly realized by ESCOs (Energy Service Companies), since it allows for generation of huge savings, particularly in the cases of modernization investment projects in heat energy sector.

REFERENCES

Buńczyk, A. & Daniluk, A. 2003. Energetyka cieplna w Polsce w 2002 roku (Polish Heat Energy Sector in 2002). *Biuletyn Urzędu Regulacji Energetyki*, No. 6.

Chapman, R.J. 1998. The Effectiveness of Working Group Risk Identification and Assessment Techniques. *Internal Journal of Project Management* 16, No. 6.

Korombel, A. 2007. *Ryzyko w finansowaniu działalności inwestycyjnej metodą Project Finance (Risk in Financing of Investments Using Project Finance Method)*. Difin, Warsaw.

Marcinek, K. 2001. *Ryzyko projektów inwestycyjnych (Risk in Investment Projects)*. Katowice: Akademia Ekonomiczna w Katowicach.

Ratajczak, M. 1999. *Infrastruktura w gospodarce rynkowej (Infrastructure in Market Economy)*. Poznań: Wydawnictwo Akademia Ekonomiczna w Poznaniu.

Wilimowska, Z. 1998. Ryzyko inwestowania (Investment Risk). *Ekonomika i Organizacja Przedsiębiorstwa*, No. 7.

Risks Challenging Publics, Scientists and Governments – Menoni (ed)
© 2010 Taylor & Francis Group, London, ISBN 978-0-415-58072-4

Environmental conflicts in Hungary—the case of the used battery reprocessing plants

Richard Szanto
Department of Decision Sciences, Institute of Business Economics, Corvinus University of Budapest, Budapest, Hungary

ABSTRACT: People usually do not think about their batteries, although these batteries are hidden in their cars, lap top computers, cell phones and in many other products. This ignorance does not change at the end of the life cycle of the product, albeit it is a very problematic process. The processing of waste batteries is not carried out in Hungary, although a lot of initiatives occurred in previous decades. The many failures drew attention to the management of environmental conflicts in the country and the social aspects of risk management, and brought on the need for effective tools for conflict resolution in this field. This paper explores the main decision traps during the siting decision-making process and gives some recommendations to avoid these pitfalls in the future. These decision traps relates to alternative seeking and selection of sites, to the different risk perception of companies and their stakeholders, in a world to the very complex environmental problem at stake.

1 INTRODUCTION

People usually do not think about their batteries, although these batteries are hidden in their cars, lap top computers, cell phones and in many other products. This ignorance does not change at the end of the life cycle of the product, albeit it is a very problematic process. Thanks to the important ingredient of batteries, lead, the waste management becomes hazardous; therefore siting decisions concerning waste management facilities often face public opposition in most countries all around the world. The reprocessing of waste batteries is not carried out in Hungary, although a lot of initiatives occurred in the previous decades. Many companies, with the support of the government, wanted to develop a solution to the problem, but their attempts in each case were miscarried. They could not set up a used battery reprocessing plant because of the protestation of the local community, various green activist groups and other local interest groups.

The many failures drew attention to the management of environmental conflicts in the country, and brought on the need for effective tools for conflict resolution in this field. Some case studies have been accomplished in recent years about the individual cases—but the author argues that the main problems cannot be understood by the investigation of individual investment setbacks, the whole process ought to be analyzed to comprehend the characteristics of these kinds of conflicts, and to give answers for the raised questions. This paper explores the main decision traps during the siting decision-making process, and gives some recommendations to avoid these pitfalls in the future. Although some argument is country-specific, the author states that most traps exist in western-countries too.

In the following section the research method is introduced. To understand the whole story of the battery reprocessing plants in Hungary some qualitative research methods were used. In the third section the almost two-decade-long story of the hazardous facilities is summarized. In the fourth section the main decision traps of the plant location process are discussed and some recommendations are given. In the last section the conclusion follows.

2 RESEARCH METHODS

During research qualitative research methods were applied. A case study about the individual cases was elaborated. As Yin (1994) proposes the author used several qualitative techniques to obtain a rich description of the cases. Ten research interviews were conducted during the study. The length of these interviews varied, but they were usually one hour long. Company representatives involved in the case studies were questioned, but also some important stakeholder group representatives were interviewed too (Table 1 shows the interviewees

Table 1. List of the interviews conducted during the research.

Interviewee	Year
Firm A—managing director	1999
Firm B—managing director	2001
Firm A—main engineer	1999
Wine Group—spoke person	1999
Grape Village—mayor	1999
Green Belt—consultant (two times)	1999, 2001
Green waste—group leader	1999
Environmental Regulatory Authority—head of division	1999
Small Village—village clerk	2003
Ministry for Environment—head of division	1999

and the year when the interview was conducted). Real names for persons and places are not used in this paper since the author wants to keep the anonymity of the main actors of the story.

Besides the interviews some company documents were also analyzed, such as corporate environmental policies or the reports of the environmental impact assessments. Stakeholders, like environmental activist groups, local interest groups also provided some documents relating to the observed cases, such as expert assessments, media announcements, etc. An intensive query of the national and local press was executed in the same time. Although the individual cases did not get the same recognition in the media, the local and the nationwide newspapers concerned with each cases. Some academic papers concerning this case were investigated too, and environmental conflict literature was reviewed.

3 THE BRIEF HISTORY OF THE BATTERY PROCESSING PLANTS IN HUNGARY

Many articles and books tried to summarize the situation of environmental protection in Central and Eastern-Europe after 1990, some of them focused particularly on the Hungarian case. These works usually agree that one of the most severe problems of environmental protection in Hungary (and in the whole region as well) is hazardous waste management. The country produces relatively high level of hazardous waste, whereas their utilization (recycling) is low (Kerekes & Bulla, 1994). A small portion of the hazardous waste comes from used batteries. Since the processing of these batteries is not carried out in Hungary, the country has to export approximately 22,000 tons of used batteries[1] to Slovenia and Austria which own large recycling capacities and let these neighbor countries make profit from the reprocessing of used battery. Hungary is a net lead importer, and trade deficit could be decreased (or eliminated) by the local utilization. Exporting hazardous waste raises some ethical questions too, since some international agreements do not favor transportation of hazardous waste across borders.

In this section the brief history of battery reprocessing plants is summarized (see the timeline in Figure 1 for overview). This section does not cover the full story; it focuses on several decisive moments highlighting the most important pitfalls and drawbacks.

3.1 *Early initiatives*

The story begun almost two decades ago when the State Development Agency decided to support the state-owned National Mining Company's idea to set up a used battery reprocessing plant in North-Eastern Hungary. The location of the plant would have been in Mine Town, where the Mining Company had a mine already closed down. The investment did not succeed mainly because of two reasons: (1) Experts criticized the plan vehemently because of its obsolete technology and inefficiency. (2) The local inhabitants did not trust the National Mining

Company because of its bad reputation in the area (people considered the firm as the polluter of a small creek next to the town, and they expected a similar lead pollution from the new plant). According to the mayor of the town, 70% of the people rejected the idea of the hazardous facility.

After 1990 the Mining Company was dismembered into smaller firms and these were privatized[2]. One of the successors Firm B decided to search for a new location after the bad experiences of the Mine Town case. The new location was in the same county, not far from the original placement. Surprisingly the village agreed to set up a processing plant in the territory of the municipality, and the Regional Environmental Agency also supported the investment. An Environmental Impact Assessment was elaborated, and the technology seemed more modern that it was some years before. However, protestations emerged; the surrounding municipalities rejected the plan. They

[1]These are mainly car batteries. For example cell phone batteries are only 0.03% of the total amount.

[2]Before 1990 Hungary was mainly a centralized economy—the most companies were owned by the State. During transition the majority of the firms went to private. The privatization revenues hit 2000 billion Hungarian forint (14,44 billion USD) in 2004 according to the Hungarian Privatization and State Holding Company (www.apvrt.hu).

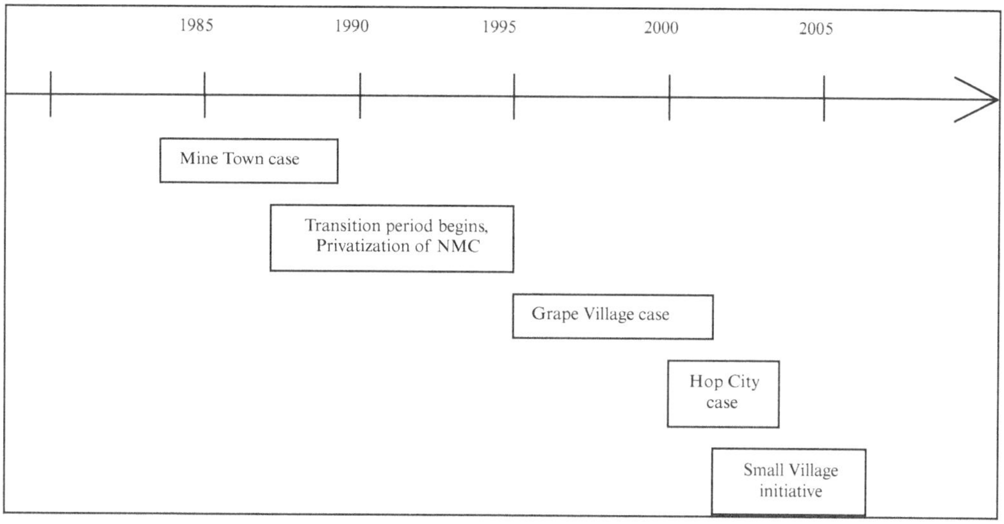

Figure 1. Timeline of the cases.

felt that the new plant would infect the over polluted area. They also asserted that although the plant would have been located within the official boundaries of the target municipality, the facility would have been closer to the houses of a neighbor village. Firm B and its investment plans did not succeed; the heavy protestation impeded the new plant.

3.2 *The Grape Village case*

In the late 90s a new company appeared on the scene: Firm A. Firm A is the largest battery-producer of the country, who foresighted a big opportunity in battery-reprocessing business. Lead extracted from used batteries can be reused in new devices and the state support would have meant a promising beginning[3]. Firm A decided to site the plant in another county close to Grape Village. In 1997 with its plan Firm A won the state support and began to persuade its stakeholders to accept the hazardous plant.

Grape Village is located at the gate of a very famous vine-region of Hungary. The vine has a grand reputation world-wide, the vines of this area take some of the biggest prizes every year, and provide the largest portion of the Hungarian vine export. When it turned out the Firm A wanted to site its battery reprocessing plant to Grape Village immediate protestation began. The case became a national issue; the protestation got high publicity in the country media.

The selection of the location in this case differed from the previous ones, since in this region no industrial facility was incumbent, no prior pollution occurred (it was a mainly agricultural land). However, the local farmers and vine producers protested against the new plant, mainly because they worried about the reputation of their vines. They usually did not argue the environmental qualities of the new facility; they simply rejected the plant because of its proximity to the vineyards. They claimed that customers (mainly foreigners who do not know the whole case) will not tend to buy their vine, if they realize that the vine is produced close to a hazardous facility.

The local inhabitants supported the investment, generally because they hoped new job opportunities in the plant (the unemployment rate was highly above the national average) and the municipality would have recorded increased tax revenues. However, some green activist groups opposed the planned investment since they did not find the technology appropriate (although the Regional Environmental Agency approved it), and they believed that with setting up the facility, heavy regional industrialization would start in the near future which would diminish the natural environment. Because of the enormous publicity that the case received the problem of the used battery reprocessing plants was put on to the national political agenda, politicians from both side (left wing government and right wing opposition) seized the opportunity to argue pro or contra in this case.

[3]After the failures the government decided to set up a state fund which would have given financial support for the private investment.

Finally the Firm A could not accomplish its plan because of the heavy public opposition. The protesters appealed to the Hungarian Environmental Protection Agency (HEPA) that found their arguments considerable, and the HEPA obliged Firm A to conduct a completely new impact assessment incorporating the new findings. Firm A did not do that and desisted from the investment.

3.3 *New developments*

The old-new actor Firm B did not give up its plan to build up a used battery reprocessing plant after loosing out in 1997. In 2000 when Firm A retreated they looked for another location in Southern-Hungary. In Hop City they obtained the support of the town, and the local authority approved the impact assessment. The surrounding smaller villages opposed the idea and enforced a referendum when only the inhabitants of Hop City voted with yes for the facility the others rejected the plan. The mayor of one of the surrounding municipalities claimed that they vision the region's future in a different way, they preferred to develop the local tourism, and a hazardous facility would have certainly decreased the attractiveness of the area. Most people did not forget the past failures of Firm B, and a significant distrust took place against the company. The national and regional politics again participated in the campaign, and the representatives of the Firm B considered the final rejection as "a political decision instead of a professional one".

The big actors had to withdraw, and until 2003 it seemed that Hungary would not have a used battery reprocessing plant in the near future. But in that year a new initiative emerged. A brand new Hungarian-Italian investor group announced that it would like to set up the plant in Small Village. The village has less than 200 inhabitants; there are no surrounding villages in that area. In the referendum the majority of the people supported the plan, and the mayor also welcomed the plant. Small village is located in an agricultural region like as Grape Village, but there are no historical vineyards in the area, and so far no interest group protested. The village seemed to be a good environment for the battery reprocessing plant. However, later it turned out that the Italian investor does not have enough funds, so the investment was delayed because of financial reasons.

4 DECISION TRAPS IN LOCATING THE SITE

In this section some decision traps are presented which might have caused the conflicts between companies and their stakeholders. The traps are grouped into three clusters. The first group relates to decision alternatives, the second to different risk perception of the companies and their stakeholders, and the last group explores the problem environment traps which relates to the historical and political embeddedness of the cases.

4.1 *Alternative seeking and communication*

4.1.1 *Creating alternatives*

Planning to locate the reprocessing plants firms tended to choose the obvious alternative; the site that they have already owned. Company managers pled cost cutting reasons when they were questioned why they chose those locations. To buy a new land for a reprocessing plant may be a costly investment but assessment of the total cost of the alternatives might have proven that social cost (for example convincing the local people) can outstrip the cost of the land investment. The quality of our decisions is often determined by the quality of our alternatives. Accepting the first (sometimes given) alternative and not creating others cannot be an efficient solution. Herbert Simon argues that decision makers tend to use the rules of thumb of bounded rationality accepting the first satisfactory alternative during a sequential decision making process. Kirkwood argues that both too many and too few alternatives can be problematic in the decision making process (Kirkwood, 1997). To strengthen the success of the decisions companies should create as many alternatives as they can produce, even some of them seem impossible. After accurate assessment it might turn out that the alternative regarded the best one at first sight has serious weaknesses.

4.1.2 *Shortcomings in communication and participation*

Using an accurate language during your communication can help to convince your partner. This strategy may work if your partner understands your language spoken. Some local people claimed that industry experts explained the companies' standpoint in the problematic issues with a very complex and professional vocabulary. They also stated (i.e. company representatives) their data and calculations are well-grounded and scientific, thus his argumentations are hardly disputable. Managers and company experts tried to force their solution on the local people and did not make the inhabitants and local farmers involved in alternative creation and selection. The local people's expertise may differentiate from the one of the company officials but certainly would provide some new aspects for assessment and decision-making. If the firms let the locals participate in decision making process and integrate (or at least try to do that) their opinion and interest into the final decision, and the acceptance of the decision would have been significantly higher. Of course, environmental laws call for public hearings as part of decision making process in Hungary[4]. However, these public hearings have many drawbacks: they happen usually at a late stage of the decision making process, and public authorities and companies have not provided wide notice of these events (Environment in the transition..., 1999).

4.2 *Risk perception and management*

4.2.1 *Technical, economic and psychological approaches to risk*

Both firms failed to assess the risk perception of their stakeholders. Laymen's and professionals' risk perception is generally different, the company experts considered environmental risk as a technical risk in this case. According to Covello (1989) risk is technically defined as the possibility of suffering harm from hazard. He claims that risk description of the technical approach includes a hazard that can cause the harm, the events creating the possibility of the harm and a statistical estimate of likelihood that harm will occur. According to the technical approach of risk management, the risk of the planned battery reprocessing plant was below the limits of the regulations and therefore should be acceptable by the public.

Some other aspects of risk perception, namely the economic and psychological approaches, state that risk may not be calculated by simply the technical formula presented above, but other considerations should be taken into account. The economic approach says subjective utilities should be calculated by interlocking the risks and the benefits of the planned facility. If the benefits overweigh the risk, the proposal should be accepted (Krimsky & Golding, 1991). The economic risk approach was represented for example by the local vine producers in the Grape Village case, since they assessed the economic risk (i.e. they cannot sell their product with a negative image) higher then the potential benefits (i.e. tax revenues or new job opportunities). Nevertheless the majority of the inhabitants of Grape Village were inclined to accept the risk, since they found more valuable the benefit side. The third approach to risk perception—the psychological one (see Slovic, 2001)—was represented mainly by the other local people, inhabitants of the surrounding villages. Laymen judge risk by different features then experts. They consider for example the frightfulness as a significant determinant of risk, but certain heuristics biases, like the accessibility and the proximity of the risk is also highly important in risk perception. Tversky and Kahneman's research (2000) showed that phrasing of a risk description can influence risk perception as well (framing effect).

4.2.2 *Not in my backyard*

Although most people agree that the battery problem somehow should be handled, almost everyone rejects that a battery processing plant is built close to their accommodation. This statement is an example of the well-known NIMBY (not in my backyard) principle, which states that people usually reject hazardous facilities in their direct environment claiming environmental injustice since the benefits of the certain facility are enjoyed by everyone in the country, but the negative effects affect only the local community.

4.2.3 *Inequitable distribution of risks*

Another important aspect of risk perception is whether the actors feel certain risks are equitable or not to them. Company managers often advocated their investment by stating it brings new tax revenues and job opportunities to the village or town where it is located. Local corporate tax is calculated by the revenue of the firm. For small villages this amount can be so huge that it completely changes the structure of the revenue side of their budget. Therefore the idea of the battery reprocessing plant was usually supported by the municipality where the companies planned to locate. Other local inhabitants rejected the idea saying they only can feel the negative effects since they live very close to the site, but cannot experience the benefits since they do not live exactly in the same place. For them the risk was not fair, and it was not considered by the applicant firms. They believed that the increasing tax revenues and the new job opportunities would be enough compensation for the locals since these villages were located in quite poor regions of the country. They did not take into account that some other local inhabitants will not percept the benefits but only the negative effects of the investment. Managers should have taken some measures compensating ones with non-equitable risk.

4.3 *Complex problem environment*

4.3.1 *Historically embedded Hungarian environmentalist movements*

Problems do not stand alone by themselves, but they usually have a very complex environment. The cases of individual companies constitute a long process and without looking at the whole process you can hardly understand the individual

[4]Environmental Laws were created just in those years when the cases had happened.

cases. The beginning of the protestations against the used battery reprocessing plants in the late 80's coincided with the transition in the country. After 1989 the socialist system collapsed and demonstrations against the environmentally irresponsible facilities proliferated in those years[5]. In the socialist era siting decisions were made by the central government with the exclusion of the local stakeholders. Public opposition against the central decisions were rare, since authorities usually retaliated against them by force, therefore people were mere spectators at decisions about hazardous facilities. Some researchers claim that the lack of the civil sector and the highly centralized political and economic structure can be blamed for the environmental degradation in the country during the communist decades (Szirmai, 1993). After 1989 the freedom to voice and protest strengthened the civil sector; environmental issues also became part of the national political agenda. According to Persanyi the environmental conflicts and demonstration in Hungary "reflect the grievous economic, political and cultural injustices that the society has suffered and that have made it distrustful of the center of power and its technocrats" (Persanyi, 1992, p.141). The investor companies usually did not take into account that their attempts to achieve their goals were happening in a historical moment, when people's thinking about politics, public decision making, local participation was going through a total change.

4.3.2 *Political influence*

Some environmentally hazardous facilities became political issues, the Grape Village case—as it was mentioned—was judged differently by the political parties, so to support or reject the idea of the battery reprocessing plant became a part of the election campaign. It did not change in the Hop City case either, mayors coming from different parties regarded the idea of the facility differently. Looking at the two-decade-long process, it can be stated that the government usually supported the initiatives (without reference to the political wing), and the opposition rejected the ideas. It is believed that politicians mainly participated in these discourses, because they expected some political benefits from them. For an individual company these issues are hard to manage, company managers can claim that these problems are off their beat and they should accept these circumstances as exogenous conditions.

[5]The protestation against a dam on the River Danube became a symbol of the civil initiatives in Hungary.

5 STAKEHOLDER MANAGEMENT AND NON-MARKET STRATEGIES

In his often cited work, Baron (1995) promotes the usage of integrated strategy which is a fine blend of some market and nonmarket components. Nonmarket strategies—says Baron—are equally important as market ones, since they shape the firm's market environment. Freeman in his landmark book (1984) suggests a stakeholder management approach to avoid these conflicts. In this section some recommendation for companies are given in the base of the two approaches mentioned above.

The investor companies first should draw a stakeholder-map, exploring who the real stakeholders of the company are, and what interests they represent. They have to create a fair compensation system providing appropriate compensation for stakeholders. They also have to consider the different situation of the local interest groups and municipalities. Vari & Caddy (1999) argue that in those cases where the public opinion was investigated and the public participated in seeking alternatives and selection were much more successful. They present the example of the belt highway surrounding the Hungarian capital Budapest. Two different sections of the ring-road were analyzed: one with the negligence of public participation and one where the local interest groups were involved in the decision-making. Where public participation occurred alternative selection was less costly and happened quicker.

In these cases, trust building must be one of the most significant factors of the non-market strategy. As an enormous distrust appeared in Hungary against hazardous facilities, since prior central decisions were made with the exclusion of the public. The individual companies should build the trust almost from the beginning with constant effort. Trust-building activities can conclude fast responsiveness to the local authorities' needs, contribution to local charities, meeting with the public on a regular basis (not only when unexpected events happen), managing other company operations on an environmentally responsible way or with no evidence of withholding information. Although Firm A tried to build up an environmental-friendly image—they created a corporate environmental policy and organized study tours to similar facilities abroad—the efforts were not sufficient. Slovic states that negative events are seen as far more likely to have a powerful effect on trust as positive ones (Slovic, 2000).

6 CONCLUSION

The siting decisions of hazardous waste management facilities face a lot of difficulties almost in every country of the world. This paper explores the decision traps where the investor companies trying to set up a used battery processing plant in Hungary fell into during the last two decades. These decision traps relates to alternative seeking and selection, the different risk perception of the companies and their stakeholders, and the very complex problem environment of the cases. Although there are some existing case studies depicting the individual cases, the author argues that the problem cannot be understood without a comprehensive analysis of the almost 20-year-long process.

ACKNOWLEDGEMENTS

I wish to thank the following people for their comments on earlier drafts of this paper: Robin Keller and Gyorgy Pataki.

REFERENCES

Baron, D.P. 1995. Integrated strategy: Market and non-market components. *California Management Review*, Berkeley, Winter, XXXVII, 2:47–67.

Covello, 1989. *Risk analysis: a guide to principles and methods for analyzing health and environmental risks*. Executive Office of the President of the U.S., Council on Environmental Quality. VA: Springfield.

Freeman, R. 1984. *Strategic Management. A Stakeholder Approach*. Boston: Pitman.

Kirkwood, C.W. 1997. *Strategic Decision Making. Multiobjective Decision Analysis with Spreadsheet*. Belmont: Duxbury Press.

Krimsky, S. & Golding, D. 1991. Factoring Risk into Environmental Decision Making. In R.A. Chechile and S. Carlisle (eds), *Environmental Decision Making. A Multidisciplinary perspective*. New York: Van Nostrand Reinhold.

OECD 1999 *Environment in the transition to a market economy*.

Persanyi, M. 1992. Red Pollution, Green Evolution, Revolution in Hungary. In Environmental Action in Eastern-Europe.

Slovic, P. 2001. The risk game. *Journal of Hazardous Materials* LXXXVI 1: 17–24.

Slovic, P. 2000. Perceived Risk, Trust and Democracy. In T. Connolly—H.R. Arkes—K.R. Hammond (eds.), *Judgment and decision making*. Cambridge University Press.

Szirmai, V. 1993. The structural mechanisms of the organization of ecological-social movements in Hungary. In Environmental and democratic transition: policy and politics in Central and Eastern-Europe.

Tversky, A. & Kahneman, D. 2000. Choices, Values, and Frames. In. T. Connolly—H.R. Arkes—K.R. Hammond (eds.), *Judgment and decision making*. Cambridge University Press.

Vari, A. & Caddy, J. (eds.) 1999. *Public participation in environmental decisions: recent developments in Hungary*. Budapest: Akademiai Kiado.

Yin, R. 1994. *Case Study Research – Design and Methods*. Thousand Oaks, CA: Sage.

Risks Challenging Publics, Scientists and Governments – Menoni (ed)
© 2010 Taylor & Francis Group, London, ISBN 978-0-415-58072-4

Knowledge based tools for stability verification during pressure equipment life cycle

P.A. Bragatto & P. Pittiglio
ISPESL – DIPIA Centro Ricerche Monteporzio Catone, Rome, Italy

S. Ansaldi
Consultant, Rome, Italy

ABSTRACT: In process plants major items design, construction and operation are based on rules driven by standards and engineering codes, aimed to assure reliability and safety. These rules are aimed to assure vessel stability under required pressure and temperature condition, considering construction material. As all pressure components are nowadays designed using CAD systems, we tried to exploit the most advanced capabilities of such systems. In this paper we present a software application for the rules based design of pressure vessels. ISPESL VSR code has been selected, as the most popular in Italian industries. The application has been developed as a customization of the well-known KB-CAD system CATIA V5®. The application is aimed to help both pressure equipment manufactures and inspectors. Manufacturers may design vessels, automatically complying with VSR rules. Inspectors may use equipment digital representation to check the compliance with safety rules along equipment lifetime.

1 DESIGN CODES FOR PROCESS PLANT EQUIPMENT

1.1 *From safety criteria to the technical rules*

The design of a process plant is a complex activity that usually involves many different disciplines over a long time. The design activity goes through many steps from the original research and development phases, through conceptual design, detailed process design and onto detailed equipment mechanical design. Pressure vessels, including tanks, reactors, boilers and heat exchangers, have to meet safety requirements and criteria. General safety criteria come from matter and energy containment principles and are quite difficult to assess in design phase. In past decades, regulators, professional institutions and trade associations developed engineering codes and standards, which cover the design, the construction and the inspection of a wide range of equipment. Engineering codes on Boiler And Pressure Vessels, Low-Pressure Storage Tanks, Heat Exchangers, Steam Generators, Compressors, Pumps, Steam Turbines and Gas Turbines are widely applied in process industry to have reliable and safe components and to meet plant safety criteria. In design phase, the use of sound technical codes is highly recommended, in order to achieve safety objectives by less effort and less uncertainty. For this reason, the adoption of appropriate standards and codes of practice for equipment installed in plants operating inside a major hazard establishment is considered a major issue for the assessment of the Safety Reports, as required by "Seveso II Directive" (COMAH 2004).

2 PED DIRECTIVE AND PRESSURE VESSELS CODES

2.1 *PED directive*

For pressure vessels design and verification, many codes are available. In past decades, in main industrialized countries, national regulator bodies developed their own codes, such as BS5500 in UK, ASME VIII-div.1 in USA and AD-MERKBLATT in Germany. In Italy a set of rules for pressure vessels stability verification, namely VSR code, had been developed by constructors and public bodies and enforced by law in early seventies. For twenty years ISPESL has been in charge to manage and update this code (ISPESL 1995). In 1997 European Commission issued Directive 97/23/CE/ (PED 1997) in order to allow the free trade of pressure equipment across the Union. Directive doesn't enforce any national code, but define only general criteria, named *essential safety requirements* ESR, which have to be verified, to approve pressure equipment. Anyway Directive requires the design to incorporate appropriate safety coefficients using comprehensive methods, such as VSR, which are known to incorporate adequate safety margins against all relevant failure modes in a consistent manner. For that reason, after Directive 97/23 implementation, VSR code had already been harmonized with PED directive and is widely used in Italy for pressure equipment design and approval (Misiti, 2004). In the future EN13445, the new European standard, which unifies major national codes, will probably spread in Italian industry.

2.2 *ISPESL-VSR code*

In order to assure the containment of internal and external pressure under design condition, VSR rules define minimal thickness of cylindrical or spherical shells, of domed or conical heads and covers, of flat ends and plates and state conditions on all vessel details, such as stiffening rings, openings, flanges, jackets, nozzles, valves and tubesheets. In VSR code, conditions are expressed as functions of:

- geometrical parameters, such as diameter, length, width, distance, angle, areas and volume;
- operation parameters such as peak pressure, test pressure, design pressure, operation temperature;
- material feature such as elasticity modulus, Poisson coefficient and corrosion allowance.

3 ENGINEERING RULES IN LAST GENERATION CAD SYSTEMS

3.1 *Knowledge based CAD systems*

Computer Aided Design (CAD) systems are worldwide used in every phases of process plants design. In seventies and eighties, CAD systems were mainly 2D and 3D geometrical modellers. In nineties, CAD systems evolved into more flexible parametric modellers. Last generation CAD systems are now evolving towards models that hold inside geometric parameters and rules. CAD systems, initially suitable only for geometry, have become more sophisticated tools, able to manage all the aspects of an engineering design: shapes, geometric features, functions, constraints, relationship and standards. In last generation CAD systems, geometric rules are not written inside the software code but they are external to the programs and may be managed directly by the user. Every designer, or better designers group, adopt his own rules, without modifying software code. Such rules may be defined by both normalisation bodies and industrial internal standards, resulting from company knowledge and experience. For this reason, those types of CAD systems are emphatically defined "knowledge based CAD" (KB-CAD). Parts designed by a KB-CAD automatically comply with *a priori* defined rules. In other words parts designed by a KB-CAD are automatically compliant with the selected code.

Furthermore a KB-CAD may handle parts previously designed and may verify after compliance with defined rules and show incoherencies. Anyway a KB-CAD is able to manage only detailed and well-defined geometrical rules. Of course generic, vague or indirect criteria cannot be handled (Sprumont et al. 2002).

4 A PROTOTYPE TOOL FOR PRESSURE EQUIPMENT DESISGN AND CHECK

4.1 *Users needs*

Pressure vessels designers need to verify code (VSR or other) compliance, already at very first design phases, in order to avoid errors and time wasting in further design and approval phases. Technical bodies need to verify automatically the code compliance, as soon as they receive the pressure vessel drawings, in order to speed up approval process, as required by National and European regulations. In the marketplace you can find a few product for pressure vessels design but most of them are just 2D automated drawing and there are no product for rule checking.

4.2 *General objectives*

As the needed tool for pressure vessels design and control is not available at the moment, we investigated the feasibility for develop a new one. Objective of the experiment has been to verify the feasibility of an automaton, based on KB-CAD capabilities, for the application of VSR rules, already in the first stages of design process. CATIA V5®, a software product developed by Dassault-Systèmes and marketed worldwide by IBM, has been chosen as a sound KB-CAD. Using customisation tools, available for normal CATIA users, we have developed a software, called *RP*. It may be used by both engineers, to design vessels automatically complying with ISPESL-VSR rules, and inspectors, to verify, also for part designed by less sophisticated CAD systems, the compliance with ISPESL-VSR code. The "knowledge ware" module of CATIA has been used for defining and verifying design rules.

4.3 *"RP" for pressure vessel design*

RP allows engineers to model new vessels, automatically complying with VSR rules. For design purposes, the user indicates, from a guided catalogue, the general type of apparatus with dimensions and operating conditions. User may add ends, shells and any details and define parameters according with VSR notation. As pressure vessel designers are already used to fill in detailed forms, in order to obtain the required VSR compliance certificate, we developed the user interface of *RP* having in our mind those forms. In such a way designers do not need to learn CATIA, which is a complex tool indeed. They have just to fill in the electronic forms and the application automatically generates a true CATIA assembly. Using a very little subset of CATIA commands it is possible to complete the generated assembly with all details, as required by VSR code.

As shown in Figure 1 user inputs general project parameters, such as pressure, temperature, materials and fluids main features. Then user defines vessel items, ends and shells. As shown in Figure 2a and 2b a VSR-like interface facilitates parameters input. A complete assembly automatically generated by *RP* is shown in Figure 3. Of course, details, such as branches openings or nips, may be added on the main items (shell or ends). Detail parameters are defined by a VSR-like interface, then details are positioned using a simple CATIA command, as shown in Figure 4. To do the entire described job, the typical Italian pressure vessel designer uses a technical language that sounds very familiar, namely the language of ISPESL VSR code. Anyway, following this easy path, the designer creates a true solid model of the part. Such a model is the essential input for the containment verification as well as for the stress calculations.

4.4 *"RP" for pressure vessel verification*

RP allows the inspectors to verify the compliance of pressure vessels CAD drawings with VSR rules. In the first release of *RP* we have decided to bypass the problem related to the automatic conversion of different digital representations, as supplied by manufacturers, into a coherent representation schema. We preferred to rebuild a synthetic digital representation of the equipment, basically using just the data supplied by manufacturer in the certification request form. In such way boring problems of format conversion have been avoided. Furthermore details useless for stability purpose are omitted. Using *RP*, inspector builds a synthetic equipment model and transfer it into CATIA, in order to have a visual representation on the computer screen. In this way VSR rules

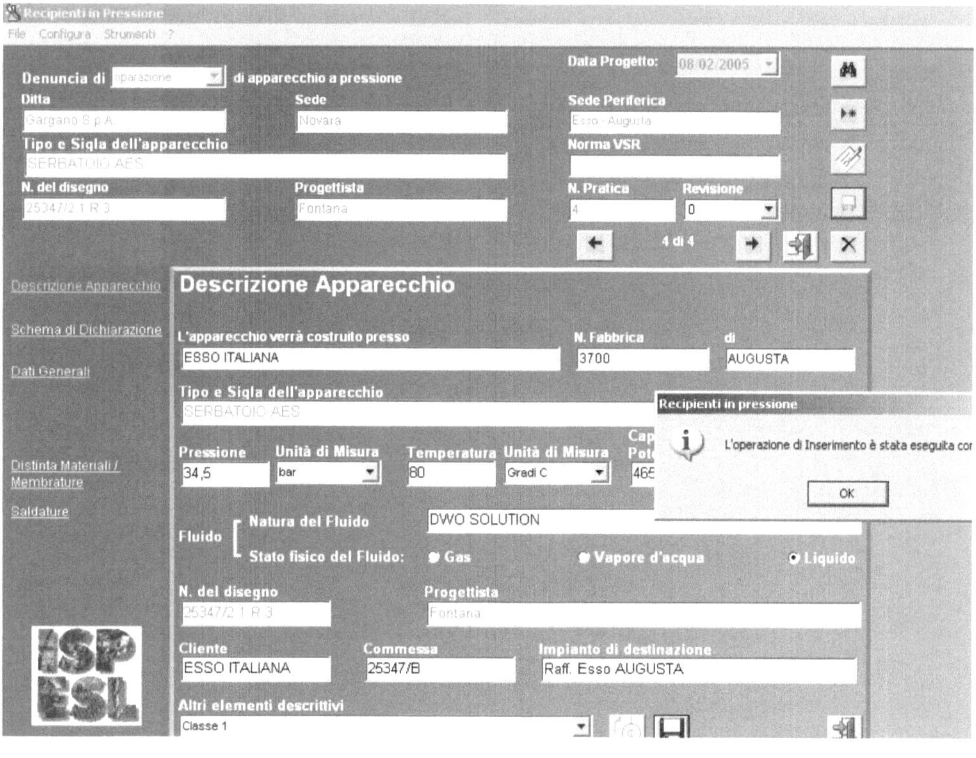

Figure 1. *RP* Input equipment data.

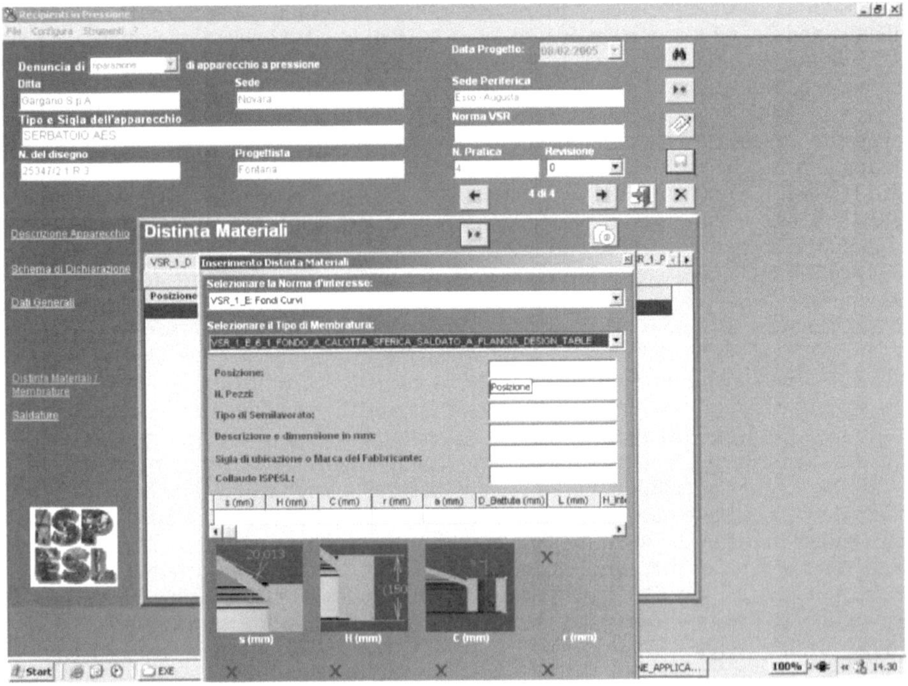

Figure 2a. *RP* input member parameters through specific VSR rules (flanged ends).

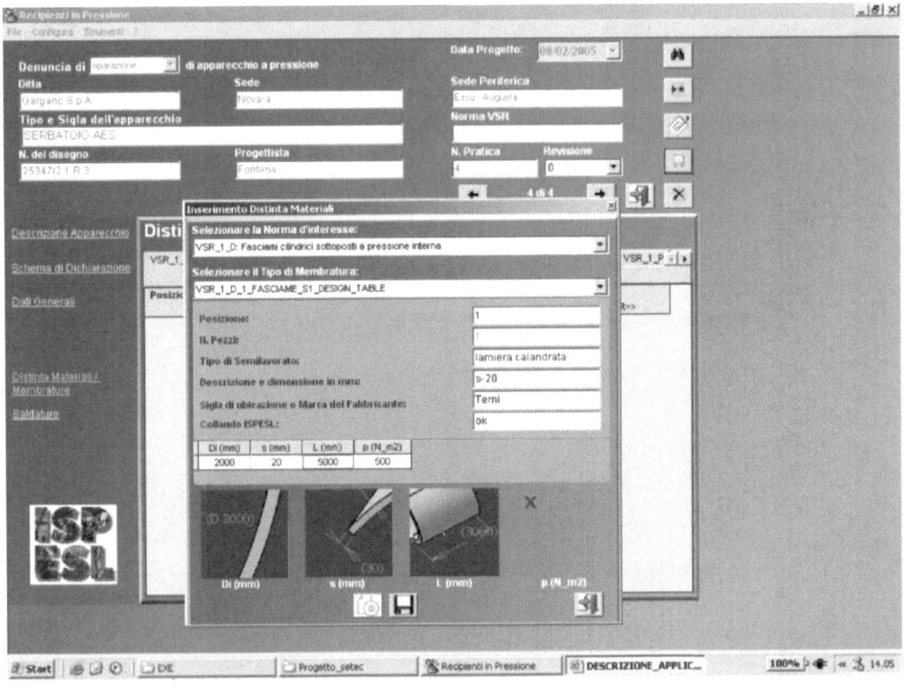

Figure 2b. *RP* input member parameters through specific VSR rules (Shell).

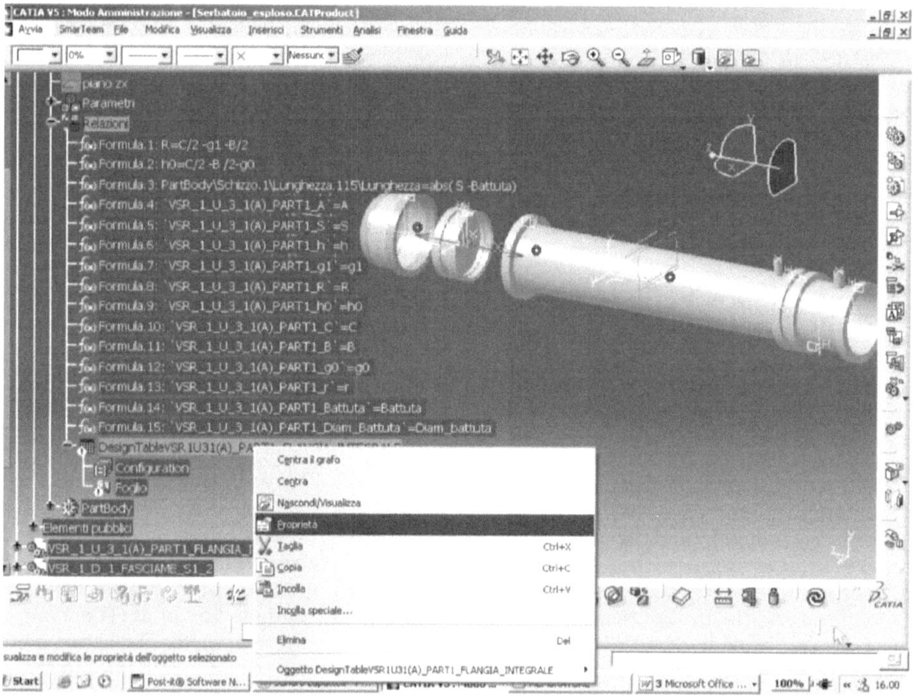

Figure 3. Assembly created by *RP* in CATIA V5®.

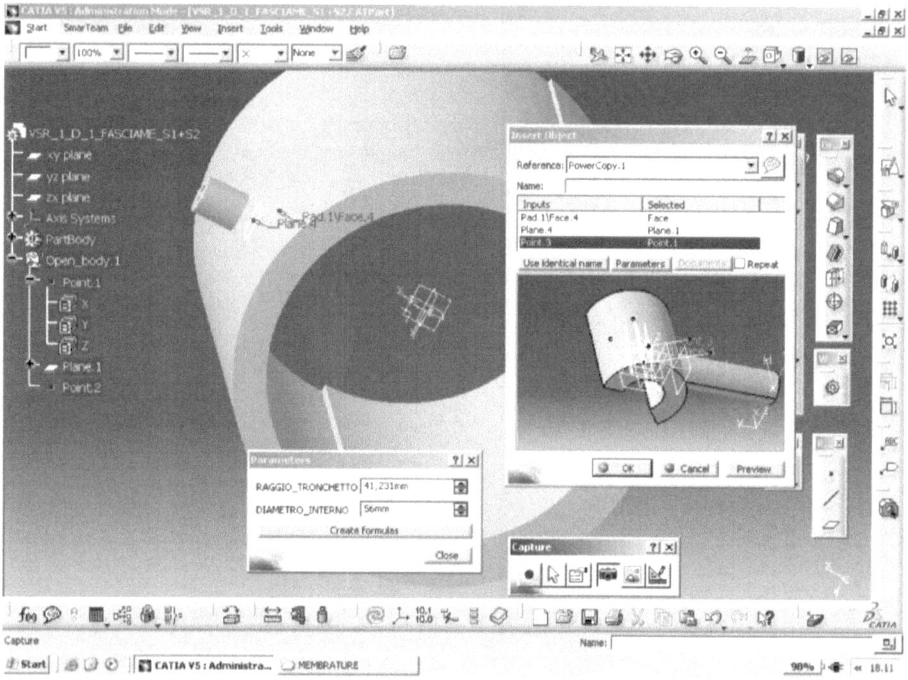

Figure 4. Detail added on a vessel item. Parameters are defined as required by VSR.

are automatically checked. Results are displayed, through the "traffic light" icon. A traffic light is associated to each rule present in the data structure tree, as displayed on the screen by CATIA. Obviously green light means compliance and red one means discrepancy. Figure 5 shows and example of automatic VSR checking on a vessel end. The red light indicates the user a deviation occurs. In order to understand the deviation a full report may be required.

4.5 *Advantages respect to the standard engineering software.*

At now, by means of the best commercial software products, engineering companies are able to verify a single stability rule for a single part of the component. In each individual rule, formulas are related to a few construction details and the verification is considered completed if every details have been verified. Trough *RP*, instead, no detail may be represented if its parameters do not comply with pertinent VSR rules. In this way, VSR verification is inherent in component digital representation. If a non compliance is reported by *RP*, the model have to be updated. For instance the thickness may be increased or the distance between two outlets may be modified, in order to have the "green light", which assures the overall component complies the safety rules.

4.6 *Applicability to complex plants.*

If the components are put together in a complex plant, the stability of each assembled component is assured by *RP*, whilst geometrical congruence of connections it is suppose to be verified trough the usual CATIA capabilities. For instance, in the case of connected components, the same diameter must be entered for the pipe and the nipples in CATIA and, of course, the same pressure parameter must be entered in *RP*.

5 PRESSURE EQUIPMENT LIFECYCLE

VSR safety rules follow all the pressure equipment lifecycle. In the previous chapter we treated just the rules for equipment design, but all stages of equipment life are regulated by rules. There are rules for construction, for testing, for inspection along operation time, as well as for repair and modification. In

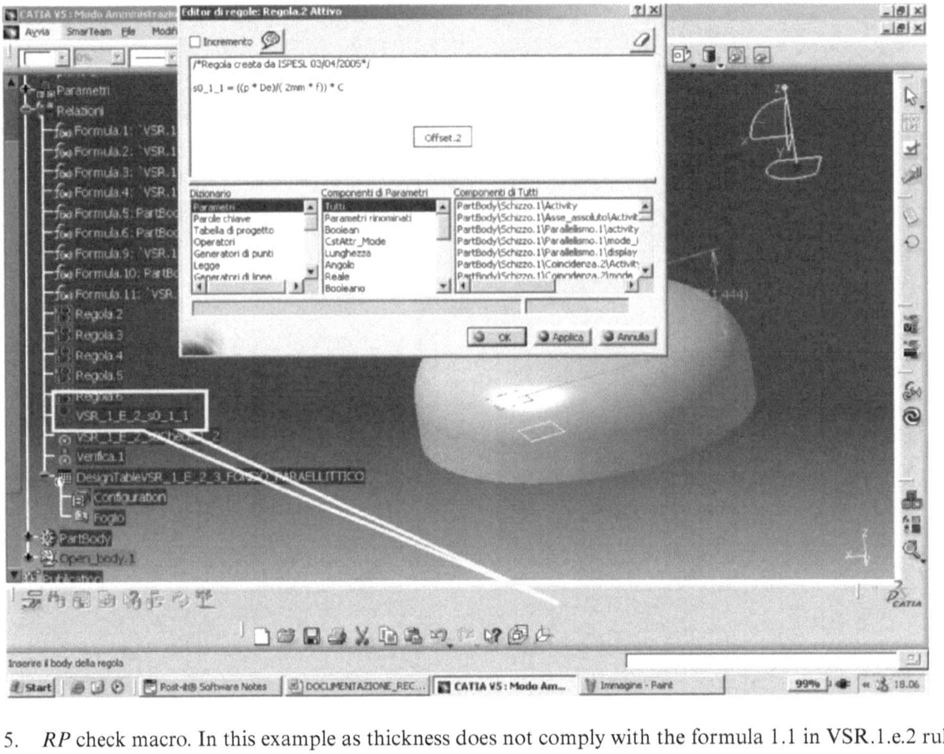

Figure 5. *RP* check macro. In this example as thickness does not comply with the formula 1.1 in VSR.1.e.2 rule, the red light indicates that a deviation occurs.

Table 1. Summary of VSR.1 rules for pressure vessels, made by unalloyed steels and weakly alloyed steels (austenitic steels not included).

VSR.1.D	Cylindrical shells under internal pressure
VSR.1.E	Domed ends
VSR.1.E.2	Domed ends under pressure on inner surface
VSR.1.E.3	Intermediate domed ends welded to cylindrical shells
VSR.1.E.4	Domed ends under pressure on extrados
VSR.1.E.5	Spherical ends welded to conical shells
VSR.1.E.6	Spherical ends welded to flanges
VSR.1.F	Conical ends under internal pressure
VSR.1.G	Spherical shells under internal pressure
VSR.1.H	Members under external pressure
VSR.1.H.2	Cylindrical shells under external pressure
VSR.1.H.3	Stiffening rings for cylindrical under internal pressure
VSR.1.H.4	Spherical shells under external pressure
VSR.1.H.5	Domed ends under pressure on convex surface
VSR.1.H.6	Conical ends under external pressure
VSR.1.H.7	Openings on shell plates under external pressure
VSR.1.K	Openings and nipples on shell plates under internal pressure
VSR.1.K.3	Cylindrical, conical and spherical shells, domed ends with insulated openings
VSR.1.K.4	Cylindrical, conical and spherical shells, domed ends with multiple openings
VSR.1.K.5	Minimal distance of an openings from main plate
VSR.1.L	Flat plates and flat ends
VSR.1.L.2	Not stiffened circular flat plates and flat ends without openings
VSR.1.L.3	Not stiffened circular flat plates and flat ends with openings
VSR.1.L.4	Stiffened circular flat plates and flat ends
VSR.1.L.5	Rectangular, elliptical not stiffened flat plates and flat ends
VSR.1.M	Tubes
VSR.1.N	Tube sheets for heat exchangers
VSR.1.P	Bellows expansion joint
VSR.1.Q	Tubes welded on pressure vessels extrados
VSR.1.R	Bolts, tie rods, screws
VSR.1.S	Square collectors
VSR.1.T	Expansion vases
VSR.1.U	Bolted flanged junctions
VSR.1.V	Chuck screws

order to have a performing and consistent management of these activities, a database of the approved equipment is an essential tool. At now just a few technical parameters may be managed in this database. *RP*, instead, generates a true 3D digital representation of the equipment. This representation, as shown in previous chapter, is generated during design check stage and may be exploited in next stages. For each equipment approved equipment, 3D models may be stored in the database and then may be retrieved for future activities. For instance, they may be very useful to organize tests and inspections, as well to present, in an impressive way, the results of these activities. Furthermore for any modification or repair, the original 3D model could be very useful to understand how changes effects on vessel safety.

6 RESULTS TRANSFER

6.1 *Transfer to other codes*

RP Software architecture is very flexible as rules are not embedded in the main code. For this reason, changes in rules definition are quite easy and do not affect software main code. The developed *RP* prototype works just with VSR, the Italian standard for pressure vessels, but the extension to the new European code, EN 13445, may be considered quite easy, as they have the same logical internal structure. Also steam generators, low-pressure tanks, piping, and pumps could enter in application range of *RP*. At the end, any body of technical rules, leading to formulas, could be considered for a potential extension of *RP*.

The main difficulty is to manage a wide range of different CAD parametric models and convert them into a unique schema, coherent with mathematical formulas defined in standard rules. As demonstrated by the first *RP* prototype version, such a difficulty could be partially turned around using a customized interface to rebuild the part, following geometric notation adopted in standard code.

6.2 *Porting of RP from CATIA V5® to other CAD*

In CATIA V5® rules management is deeply integrated with parametric CAD modeller and that has been well exploited to develop the software prototype. Anyway this fact should not be an obstacle to spread experiment results. Rules management integration is present in a few high-end CAD products; prototype porting into these systems is a quite easy exercise. Furthermore rules managers could be find out in public or free software. A rules manager could be coupled with the preferred

CAD. Integration between CAD systems and rules management systems may require a heavy programming job. Anyway this integration had been demonstrated feasible already in 80's (Judica & Ansaldi, 1989; Falcidieno & Ansaldi, 1990).

7 CONCLUSIONS

It is ingenuous to think that progresses of computer science, just because applied, carry immediate benefits for environment and safety or for any other fields. Anyway some recent computer science advances offer potential chances for safety increasing, that it is opportune to pick up. Among these the possibility to make safety rules automatically embedded the parts designed by CAD systems is interesting indeed. So it would be impossible at all to have the parts not obeying the safety rules. For such a goal, to have sound codes of safety rules is essential (Palmer & Chung, 2001).

Advantages may be expected from using KB-CAD systems for the approval job of the Competent Authorities. Using an automated rules checker, they could make a first coarse evaluation for plant equipment approval like pressure vessels, pumps, tanks and so on. This first step would constitute a good base for a fast approval process. Furthermore, this automated "rules checker" enables regulatory bodies to build a sound database of approved equipment, which could be used also to follow the equipment lifecycle, including operation, inspection, maintenance and repair.

REFERENCES

Ansaldi, S. & Falcidieno, B. 1990. *Extracting and completing feature information in process, planning application in geometric modelling for product engineering.* Amsterdam: North Holland.

COMAH Safety Report assessment manual – Level III Technical Aspects in *www.hse.gov.uk/comah/sragtech.*

ISPESL VSG, 1995. *"Raccolta VSG – Specificazioni tecniche applicative del DM del 21 novembre 1972 per l'impiego dei materiali nella costruzione degli apparecchi e sistemi in pressione"* (in Italian).

ISPESL VSR, 1995. "*Raccolta VSR – Specificazioni tecniche applicative del DM del 21 novembre 1972 per la verifica della stabilità dei recipienti in pressione*" (in Italian).

Judica, N. & Ansaldi, S. 1989. Generatives Arbeit. Plannungssystem mit CAD-Kopplung in *ZWF-CIM Zeitschrift fur Wirtschaftliche Fertigung und Automatisierung* vol 84 Munchen (in German).

Misiti, A. 2004. "Raccomandazioni del Comitato Termotecnica Italiano per l'uso delle raccolte ISPESL rev.95, nell'ambito dell'applicazione della direttiva CE 97/23" *Comitato Termotecnico Italiano* CTI-R0302.

Palmer, C. & Chung, P.W.H. 2001. Verification of process plant models—*Proceedings of the 2001 AAAI Symposium on Model-Based Validation of Intelligence.* Stanford: AAAI Press.

PED; 1997 Directive 97/23/EC of the European Parliament and of the Council of 29 May 1997 on the approximation of the laws of the Member States concerning pressure equipment *EU Official Journal* L 181, P. 0001–0055.

Sprumont, F. & Xirouchakis, P. 2002 Towards a Knowledge-Based Model for the Computer Aided Design Process Concurrent Engineering. *Res & Appl.* 10: 2.

 ISBN 978-0-415-58072-4

Author index

Ale, B. 233
Ansaldi, S. 267
Aromolo, R. 77
Asche, F. 49
Attolico, A. 27, 33
Aven, T. 49

Bagni, G. 69
Balfanz, H.P. 247
Balmforth, H. 119
Bellagamba, S. 211
Beni, C. 77
Berg, H.P. 247
Bixio, A. 27, 33
Boccia, P. 69, 77
Bozek, F. 85
Bragatto, P.A. 267

Carlé, B. 17
Casorri, L. 77
Ceudech, A. 179
Christou, M. 9
Colli, A. 233
Conti, C. 77
Curbach, M. 123

Delaval, M. 9
Delvosalle, C. 41
Dugnoille, X. 41
Dupuis, D. 41

Ernst, B. 247
Evangelista, G. 63

Felici, B. 77
Ferrazza, P. 77
Ficociello, B. 77
Fiévez, C. 41
Fowler, A. 119
Francioni, G. 63
Frischknecht, C. 105

Galderisi, A. 179
Gatteschi, A. 77

Hardeman, F. 17
Hervas, J. 9
Hlusek, J. 85
Hyblerova, K. 85

Kirchsteiger, C. 233
Köhler, U. 123
Krausmann, E. 9
Kundak, S. 155

Lindoe, P. 49
Losak, T. 85

Mara, S. 9
Marconi, S. 69, 77
Markey, A. 57
Martelli, M. 215
Mascini, M. 69
McCleery, D. 57
McManus, H. 119

Nelson, R. 57
Njå, O. 223

Olcina Cantos, J. 191

Pacifico, S. 27
Pesaro, G. 145
Phelan, J. 57
Pigeon, P. 205
Pittiglio, P. 267
Proske, D. 123
Pugliano, A. 3

Rohde-Liebenau, B. 137
Romerio, F. 105
Rusanen, M. 101
Rusanen, T. 101

Scheer, S. 9
Simeoni, C. 211
Sitek, E. 255
Spagnesi, P. 215
Sturchio, E. 69, 77
Sweetnam, R. 57
Szanto, R. 259

Taylan, A. 167
Toft, A. 49
Tonelli, E. 63
Turcanu, C.O. 17

Van Zanten, A. 131
Vatn, J. 223
Vetere Arellano, A.L. 233

Wagner, J.-J. 105
Wiencke, H.S. 49
Windrum, N. 57

Zaruk, D. 93